lence Modelling Approaches - Current State, Development Prospects, Applications
dx.doi.org/10.5772/65527
by Konstantin Volkov

butors

Walters, Galina Ilieva, Lei-Yong Jiang, Gervásio Degrazia, Charles R. P. Szinvelski, Lidiane Buligon, Michel tella Stefanello, Silvana Maldaner, Débora Regina Roberti, Yuriy Nuzhnov, Jesús Manuel Fernández Oro, Andrés -Fernández, Bruno Pereiras García, Hector E. Nistazakis, Lydia Gkoura, George D. Roumelas, Andreas oulos, H.G. Sandalidis, Alexander Vavoulas, George S. Tombras, Tao Wang, Jingsong Bai, Nikolai A. drovich Magnitskii, N. M. Evstigneev

hed by InTech
Trdine 9, 51000 Rijeka, Croatia

shing Process Manager Nina Kalinic
ical Editor SPi Global
InTech Design team

ublished July, 2017
d in Croatia
deposit, Croatia: National and University Library in Zagreb

onal hard copies can be obtained from orders@intechopen.com

lence Modelling Approaches - Current State, Development Prospects, Applications, Edited by Konstantin Volkov

SBN 978-953-51-3349-0
e ISBN 978-953-51-3350-6

AF896798

TURBU
MODE
APPROACHES - CU
STATE, DEVELOP
PROS
APPLICA

Edited by **Konstan**

Turb
http:
Edite

Con
Keith
Bapt
Mea
Tsig
Alex

Pub
Jane

No
Sta
of t
cha
use

Pu
Te
Co

Fir
Pri
Le

Ac

Tu
p.
Pr
O

For EU product safety concerns:
IN TECH d.o.o., Prolaz Marije Krucifikse Kozulić 3, 51000 Rijeka, Croatia,
info@intechopen.com or visit our website at intechopen.com.

Meet the editor

Dr Volkov is a senior lecturer in thermofluids at the Kingston University (London, UK). He holds a PhD degree in fluid mechanics. After completion of his PhD degree, Dr Volkov worked at the Baltic State Technical University (Russia), University of Central Lancashire (UK), and University of Surrey (UK). His areas of expertise cover multidisciplinary areas: from design and optimization of energy systems to fundamental problems focused on modelling and simulation of turbulent multi-phase flows. He is the Chartered Engineer and a member of Institute of Physics, Institution of Mechanical Engineers, and Combustion Institute. He is the author of more than 120 scientific papers and a member of the editorial board and scientific committee of a number of journals and conferences.

Contents

Preface

Despite intense development of computational technologies, achievements in the area of construction of numerical methods and development of commercial and open-source software, improvements of experimental methods, and high-performance computing facilities, the problem of modelling and simulation of turbulence remains one of the most complex and important problems of fluid dynamics. In contrast to laminar flows whose computation has already become a routine procedure, reliable prediction of turbulent flows is more art than rigorous science for numerous reasons including three-dimensional nature of the flow, stochastic character, and a wide spectrum of spatial and temporal scales.

A detailed knowledge of the flow regimes is of crucial interest for improving performance of many engineering devices including mixing chambers, turbine blade passages, and engines. Accurate prediction of turbulent flows remains a challenging task despite considerable work in this area and the acceptance of computational fluid dynamics (CFD) as a design tool. The application of computational techniques to the design and optimization of engineering devices is difficult because of the complexity of flows subjected to, among others, phenomena, confinement, boundary layers, rotation effects, Ekman layers on rotating surfaces, stagnation flow heat transfer, heat transfer in the presence of steep pressure gradients both favourable and adverse, free stream turbulence, and three-dimensional effects such as tip leakage flow and secondary flows.

The quality of CFD calculations of the turbulent flows and heat transfer in boundary layers, free mixing layers, and free and impinging jets strongly depends on the proper prediction of turbulence phenomena. Investigations of heat transfer, skin friction, flow separation, and reattachment effects demand a reliable simulation of the turbulence, reliable numerical methods, accurate programming, and robust working practices. A necessary step consists in the verification and validation of numerical algorithms and turbulence models for idealized geometries, which can be performed only if experimental data are available for comparisons.

CFD provides three main options to turbulence simulation, including direct numerical simulation (DNS), large-eddy simulation (LES), and solution of Reynolds-averaged Navier-Stokes (RANS) equations.

DNS implies solving the full (unsteady and three-dimensional) Navier-Stokes equations, which allows obtaining instantaneous characteristics and resolving all scales of a turbulent flow, if numerical and other types of errors can be avoided. The resultant statistics is used to validate turbulence models, to develop methods of flow control, and to study the laminar-turbulent transition. As the capabilities of measurement equipment are limited, DNS is considered as an additional source of experimental data, e.g., pressure fluctuations, vorticity, and dissipation rate of the turbulent kinetic energy. Limitations in the use of DNS are high requirements to finite difference schemes, satisfaction of initial and boundary conditions, and limited computational resources. Time and mesh steps are of the order of Kolmogorov's scales of time and length and decrease with increasing Reynolds number. Obtaining a statistically steady flow pattern requires tens and hundreds of hours of processor time. The use of unstructured meshes also contributes to consumption of computer memory and processor time. It is difficult to implement computations that involve DNS except for low Reynolds numbers and simple flow geometry.

Solution of RANS equations requires much lower computational resources and is successfully used in engineering practice. The issues of closure are solved at different levels of complexity. Turbulence models are classified in terms of the number of equations introduced in addition to the RANS equations. An increase in the number of equations requires additional semi-empirical information to be involved, which spoils model universality. Available turbulence models do not possess acceptable universality and, therefore, cannot be used to solve a wide range of applied engineering problems.

The lack of a universal turbulence model suitable for computing all or at least most turbulent flows shifted the focus in turbulence research. Improved capabilities of CFD tools and high-performance resources stimulated the search for and application of approaches that are more rigorous and universal than RANS.

LES is a compromise between DNS and solution of RANS equations. LES implies solution of filtered Navier-Stokes equations. Large eddies, being under a direct action of boundary conditions and carrying the maximum Reynolds stresses, are computed. Small eddies have a more universal structure and are modelled by sub-grid scale (SGS) models based on the eddy viscosity or other rational approximations of transport processes. SGS models are normally characterized by significant diffusion and dissipation, which allows one to overcome computational problems caused by presentation of small eddies on a chosen mesh and to stabilize numerical computations. As LES excludes direct computations of small eddies, the time and mesh steps are much greater (approximately by an order of magnitude) than Kolmogorov's scales of length and time. Higher Reynolds number than that in the DNS can be achieved with a fixed computational memory.

A large number of SGS models, filters, boundary conditions, and finite-difference schemes have been tested in numerous computations. Nevertheless, neither the optimal choice of the SGS model is clear nor the choice, if made, is justified. There are no universal near-wall functions providing a decrease in the number of nodes in the near-wall region; therefore it is difficult to use LES for computing flows with small separation regions and transition points. Yet, LES is a promising direction in the development of methods for computing turbulent flows and seems to be a serious alternative to DNS and RANS.

There are also hybrid approaches that combine some features of DNS, RANS, and LES, in particular, Detached Eddy Simulation (DES) and hybrid RANS-LES approaches. The main motivation for hybridizing the two methods is to decrease the cost of the traditional LES method, which is large because of the requirement to directly capture all the scales of motion responsible for turbulence production and the observed inability of most SGS models to correctly account for anisotropy and non-equilibrium nature of the flows.

DES is a modification of a RANS model in which the model switches to SGS formulation in regions fine enough forLEScalculations. Regions near solidboundaries and where the turbulent length scale is less than the maximum mesh dimension are assigned the RANSmode of solution. As the turbulent length scale exceeds the mesh dimension, the regions are solved using the LES mode. Therefore, the mesh resolution is not as demanding as pure LES, thereby considerably cutting down the cost of thecomputation.

The book gives an overview of various approaches to turbulence simulation and highlights possible weaknesses in the CFD codes and suggests some possible improvement paths. The current scientific status of simulation of turbulent flows as well as some advances in computational techniques and practical applications of turbulence research is reviewed and considered in the book. The book also covers issues related to development, verification, and validation of turbulence models and focuses on development of the best practice for engineering calculations.

Critical assessment of hybrid RANS-LES modelling for attached and separated flows

The dynamic hybrid RANS and LES modelling framework is assessed in the chapter. Computations of two benchmark test problems, turbulent channel flow and backward-facing step flow, are performed to assess the model for attached and separated turbulent flows. This investigation attempts to evaluate the ability of the hybrid method to reproduce the detailed physics of attached and separated turbulent flows as well as to resolve the delayed break down of separated shear layers. The computed results are compared with experimental and computational data based on RANS and DNS. The comparison demonstrates that the model addresses many of the weaknesses inherent in common models.

Numerical analysis of laminar-turbulent bifurcation scenarios in Kelvin-Helmholtz and Rayleigh-Taylor instabilities for compressible flow

The laminar-turbulent transition in compressible flows triggered by Kelvin-Helmholtz and Rayleigh-Taylor instabilities is considered in the chapter. Floquet theory is applied to the linearized problem using matrix-free implicitly restarted Arnoldi method. The numerical calculations are performed for some benchmark test problems, and laminar-turbulent development of compressible Kelvin-Helmholtz and Rayleigh-Taylor instabilities as the bifurcation scenarios is shown.

Interface instability and turbulent mixing

Numerical investigation of the Richtmyer-Meshkov instability and turbulent mixing is presented in the chapter. The verification and validation of numerical method and computer code, the growth laws and mechanics of turbulent mixing, the effects of initial conditions, and the dynamic behaviour and some new phenomenon for Richtmyer-Meshkov instability and turbulent mixing are discussed.

Statistical modelling for the energy-containing structure of turbulent flows

The development of statistical theory for the energy-containing structure of turbulent flows, taking into account the phenomenon of internal intermittency, is proposed in the chapter. New differential equations for conditional means of turbulent and non-turbulent flow are established. A new principle of constructing mathematical models as the method of autonomous statistical modelling of turbulent flows is presented. Testing of the method is accomplished on the example of constructing a mathematical model for the conditional means of turbulent flow in a mixing layer.

Turbulence and its effects on aerodynamics of flow through turbine stages

The turbulence features and their impact on fluid dynamics, streaming of blades, and efficiency performance are carried out in the chapter. The turbulence effects and transition onset in turbine stages, approaches to their modelling, and how they affect efficiency and flow parameter distribution and lead to an innovative design are discussed.

Turbulence transport in rotor-stator and stator-rotor stages for axial flow fans

Recent developments concerning numerical simulation of rotor-stator and stator-rotor interactions in low-speed axial fans using LES techniques are presented in the chapter. A post-processing framework is introduced to segregate the deterministic and turbulent components of the unsteady flow, allowing an accurate description of both phenomena. The ability of LES computations to disclose flow turbulence in rotor-stator environments at off-design conditions is demonstrated.

RANS modelling of turbulence in combustors

In this chapter, some widely used RAMS turbulence models are discussed and validated against a comprehensive experimental database from a model combustor. The results obtained show that the flow features are captured by all turbulence models. However, in terms of quantitatively predicting the velocity, temperature, and species fields, various degrees of agreement with the experimental data are observed. It is found that the turbulent Prandtl and Schmidt numbers have a significant effect on the predicted temperature fields in the combustor and the temperature profile.

Testing physical and mathematical criteria in a new meandering autocorrelation function

An alternative formulation for the low-wind speed meandering autocorrelation function is presented in the chapter. This expression for the meandering autocorrelation function reproduces well-observed wind meandering data measured in a micrometeorological site located in a Pampa ecosystem area. The comparison shows that the alternative relation for the meandering autocorrelation function is suitable to provide meandering characteristic parameters.

Underwater optical wireless communication systems: a concise review

Underwater optical wireless communication (UOWC) has gained a considerable interest as an alternative means for broadband inexpensive submarine communications. It was demonstrated that UOWC networks are feasible to operate at high data rates for medium distances up to a hundred meters. However, it is not currently available as an industrial product, and mainly test-bed measurements in water test tanks have been reported so far. The chapter summarizes the recent advances in channel modelling and system analysis and design in the area of UOWC.

The book covers the current state, development prospects, and engineering applications of turbulence science representing the latest research of various groups of internationally recognized experts. This book is intended for engineers and technical workers whose work is related to predictions of turbulent flow properties in science and engineering. It will be of interest to academics working in environmental engineering and to industrial practitioners in companies concerned with design and optimization of engineering systems.

The open exchange of scientific data, results, and ideas will hopefully lead to improved predictions of turbulence impact on industrial and technological processes. The book presents necessary data and helpful suggestions to advance understanding of the turbulent phenomena.

Dr Konstantin Volkov, MEng, MSc, PhD, DSc, CEng, MIMechE, MInstP
Department of Mechanical and Automotive Engineering
School of Mechanical and Aerospace Engineering
Faculty of Science, Engineering and Computing
Kingston University
London, United Kingdom

Chapter 1

Critical Assessment of Hybrid RANS-LES Modeling for Attached and Separated Flows

Mohammad Faridul Alam, David Thompson and Dibbon Keith Walters

Additional information is available at the end of the chapter

http://dx.doi.org/10.5772/intechopen.68387

Abstract

The objective of this study is to evaluate a recently proposed dynamic hybrid Reynolds-averaged Navier-Stokes (RANS)-Large Eddy Simulation (LES) modeling framework that seeks to effectively address issues regarding RANS-to-LES transition and explicit grid dependence inherent in most current hybrid RANS-LES (HRL) models. RANS-to-LES transition in the investigated dynamic HRL (DHRL) model is based on the physical concept of maintaining continuity of total turbulence production using two rigorously separated turbulent stress parameters, where one is obtained from the RANS model and the other from the LES model. Computations of two canonical test cases —two-dimensional turbulent channel flow and backward facing step flow—were performed to assess the potential of the DHRL model for predicting both attached and separated turbulent flows. This investigation attempts to evaluate the ability of the DHRL method to reproduce the detailed physics of attached and separated turbulent flows, as well as to resolve the issues concerning log-layer mismatch and delayed break down of separated shear layers. The DHRL model simulation results are compared with experimental and DNS data, along with the computational results for other HRL and RANS models. In summary, these comparisons demonstrate that the DHRL framework does address many of the weaknesses inherent in most current HRL models.

Keywords: computational fluid dynamics, turbulence modeling, Reynolds-averaged Navier-Stokes, large eddy simulation

1. Introduction

Turbulent shear flows, including attached boundary layers and separated shear layers, are important application test cases for computational fluid dynamics prediction since they are observed for a wide array of engineering processes and systems. In particular, turbulence modeling for these systems is an important aspect of the simulation that is often responsible for predictive error. Reynolds-averaged Navier-Stokes (RANS) models are known to perform relatively well for turbulent boundary layer flow [1] due to the somewhat universal nature of wall-bounded turbulence. RANS models typically do not perform as well in regions of separated flow due to the presence of adverse pressure gradients, strong three-dimensionality, shear layer reattachment, and large-scale unsteadiness [2–8]. Large eddy simulation (LES) models theoretically provide greater accuracy than RANS approaches in these regions, but their application to attached wall-bounded flows suffers from significantly increased computational expense relative to RANS. As a consequence, the LES approach is still not widely used for industrial analysis and design for high Reynolds number flows, especially those with attached boundary layers [9]. Hybrid RANS/LES (HRL) [1] approaches offer a potentially attractive alternative to RANS or LES, since they attempt to combine the advantages of both RANS and LES modeling in an optimized manner that resolves both attached and separated flows effectively. Specifically, HRL models have the potential for greater accuracy than RANS and less expense than LES. Interest in HRL methods has, therefore, grown substantially over the past several years.

Hybrid RANS-LES models are categorized as either zonal or nonzonal. For zonal models, the RANS and LES models are separately employed in selected regions of the computational domain which are determined a priori. Development of effective methods for coupling the two model types at their interface remains a challenge and is an ongoing area of research [10, 11]. Nonzonal methods are generally simpler to implement and do not require the user to decide where LES or RANS is used in a particular simulation. In a nonzonal model, the eddy viscosity adopts a value representative of a RANS model in the near-wall region and a value representative of an LES subgrid stress (SGS) model in separated regions. Detached-eddy simulation (DES) [12] is probably the most commonly used nonzonal modeling methodology. Blending between RANS and LES model types in the original DES formulation is a function of the local grid size and has been shown to suffer from inaccuracy in attached boundary layers [13]. While several ad hoc modifications have been implemented to address these limitations, the modifications often only mitigate weaknesses while not completely resolving them. For example, to address the problem of reduced levels of eddy viscosity in attached boundary layers due to premature switching to LES mode, Spalart et al. [13] introduced a modified version of the baseline DES model denoted as Delayed DES (DDES). DDES adopts a definition for length scale different from that in the original baseline DES model, as a function of local mean flow and turbulence model variables. Shur et al. [14] introduced a DES version with an additional modification—improved delayed DES (IDDES)—in an attempt to eliminate the well known "log layer mismatch" issue that arises in classical DES and wall-modeled LES (WM-LES). The IDDES model behaves similar to DDES except that it replicates a WM-LES type model in boundary layer regions when resolved turbulent fluctuations are present.

The key challenge for nonzonal HRL modeling is that of effectively specifying the transition between RANS and LES modes in the simulation. Typically, such a transition is defined based on the value of the eddy viscosity that varies between the Reynolds stress model and the SGS value. Significantly, the Reynolds stress and subgrid stress are mathematically distinct and physically different; therefore, any method that switches between the two using a single parameter (i.e., eddy viscosity) is not straightforward. This is because the Reynolds stress is based on an ensemble averaging of all turbulent scales present in the flow field, while the subgrid stress is based on (typically spatial) filtering of all turbulence scales that are not directly resolved in the simulation for a given mesh size. The difficulty in defining zonal transition using a single variable for eddy viscosity has been identified as a major weakness for commonly used HRL models [15–17]. Additionally, many of the currently used HRL models incorporate the local grid size directly as a model variable. This necessitates that great care is taken when constructing grids for HRL modeling. For the best performance, the grid should be built with foreknowledge of the expected model behavior and design of the specific grid used as the method of forcing RANS-to-LES transition in the desired locations of the domain [15].

The transition from a purely modeled RANS stress to a resolved dominating LES stress has been recognized as a major concern [15]. The problem is exacerbated if separation occurs at a clearly defined location such as a sharp point (e.g., a backward-facing step) and the separating boundary layer lacks any initial level of fluctuating turbulence content. Paterson and Peltier [16] studied the issues for RANS-to-LES transition for cases where no geometrically imposed separation point exists. The authors note that a delay in the development of resolved fluctuations stress terms is present in the RANS-to-LES transition region upstream of the separation location; therefore, the (SGS) turbulent viscosity values obtain dominance over the RANS eddy viscosity values prematurely. This effect is the well known "modeled-stress depletion" described by Spalart et al. [13]. Nitikin et al. [17] also demonstrated the difficulties inherent in calculating the appropriate level of grid resolution for the "gray region" that lies between the RANS and LES modes, particularly for RANS-to-LES transition that occurs in the wall normal direction in boundary layer flows.

Previous researchers have attempted to resolve the RANS-to-LES transition issue [13, 14, 18–21] including efforts discussed above [13, 14]. Menter et al. [18, 19] introduced the concept of scale-adaptive simulation (SAS), which is significant as it provides the potential to develop turbulence models that are applicable in both RANS and LES modes without including any explicit grid-dependence. Hamba [20] proposed that rapid variation of the filter width near the interface of the RANS and LES zones is the primary reason for the log-layer mismatch in channel flow simulations and that the problem can be resolved by using an additional filtering operation. Piomelli et al. [21] have proposed the use of a pseudo-random forcing function as a backscatter model in the interface region as a means of resolving the underlying issues of the a transition layer between RANS and LES. It is important to note that most of these prior attempts should be viewed as ad hoc modifications rather than fundamental changes to the basic modeling approach. Celik [22], in a review paper, proposes that entirely new criteria are needed in order to effectively address the RANS-to-LES transition issue in HRL models in a more fundamental way.

The dynamic hybrid RANS-LES (DHRL) modeling methodology presented in this paper was specifically developed as an attempt to resolve the aforementioned weaknesses, including explicit grid dependence, that have been documented for most previous HRL models. Furthermore, it is assumed that these issues are fundamental in nature and unlikely to be effectively addressed using ad hoc modifications. The novel features of the DHRL modeling framework are that (1) it is a general framework which allows coupling of any particular RANS model with any particular LES model; (2) it does not include mesh size as a variable in the model formulation; (3) the blending between RANS and LES modes is enforced through the assumption of continuity in total turbulence energy production; and (4) it exactly reduces to the underlying RANS model in flow regions that exhibit numerically steady-state results.

In the present paper, a detailed investigation of the DHRL model is presented using the finite-volume based commercial CFD software Ansys FLUENT®. The DHRL model was implemented into FLUENT using the user-defined function (UDF) capability available with that solver. For the present cases, the DHRL framework was used to integrate Menter's SST k-ω model [23] as the RANS component, with monotonically integrated LES (MILES) [24] as the LES model component. To evaluate the performance of the DHRL model in both attached and separated flow regions, the test cases considered were turbulent channel flow matching the DNS case of Hoyas and Jimenez [25] and backward-facing step flow matching the experimental case of Driver and Seegmiller [26]. Simulation results using the DHRL model are compared with the corresponding DNS and experimental data, and with the results of companion simulations using other RANS and HRL turbulence models available in FLUENT.

2. Model formulation

The DHRL model formulation is summarized in this section. Readers are referred to Bhushan and Walters [27] for a more detailed description. Other models, including DES, DDES, and SST k-ω, are also summarized since they are used to perform companion simulations, and the results are compared with DHRL. In addition, the SST k-ω model is used as the RANS component of the DHRL model, while MILES is used as the LES component. The models used here were chosen for comparison purposes because they are widely used examples of HRL and RANS, respectively. Since they have been previously thoroughly documented, only a brief description and appropriate references are provided in Subsections 2.2–2.5.

2.1. Dynamic hybrid RANS-LES (DHRL) methodology

The description of the DHRL model presented here focuses on single-phase, incompressible, Newtonian flow with no body forces. Extension to compressible flow or flows with gravitational effects is straightforward. First, applying an (undefined) filtering operation to the momentum equation results in:

$$\frac{\partial \widehat{u}_i}{\partial t} + \widehat{u}_j \frac{\partial \widehat{u}_i}{\partial x_j} = -\frac{1}{\rho}\frac{\partial \widehat{P}}{\partial x_i} + \frac{\partial}{\partial x_j}\left(2\nu \widehat{S}_{ij}\right) - \frac{\partial}{\partial x_j}\left(\tau_{ij}\right) \tag{1}$$

where u_i and $\widehat{u}_i$ are the instantaneous and filtered velocity, respectively. The last term on the right-hand side represents the turbulent stress, which corresponds in general to any residual stress tensor obtained from either Reynolds averaging or filtering. This can be expressed as follows:

$$\tau_{ij} = \widehat{u_i u_j} - \widehat{u}_i \widehat{u}_j \tag{2}$$

The turbulent stress term must be modeled in order to close the filtered momentum equation.

Most hybrid RANS-LES models, including the commonly used DES model, use a single scalar variable in the filtered momentum equation to model the turbulent stress using the Boussinesq hypothesis. This term is denoted as the eddy viscosity, and in theory obtains a value appropriate for a modeled Reynolds stress in the RANS region (typically near the wall) and a value appropriate for a modeled subgrid stress in the LES region (typically far from the wall).

As discussed above, blending the effects of ensemble-averaged velocity fields (Reynolds stress) and spatially filtered velocity fields (subgrid stress) using a single scalar variable introduces ambiguity into the model. The DHRL modeling methodology seeks to avoid this ambiguity. The model development begins with decomposition of the velocity field in such a way that the effects of ensemble-averaged velocity fields and spatially filtered velocity fields retain a distinct separation in the transitional or "gray" zones.

First is introduced a simulation-specific decomposition for the instantaneous velocity (u_i):

$$u_i = \underbrace{\overline{u}_i + u''_i}_{\widehat{u}_i} + u'_i \tag{3}$$

where $\widehat{u}_i$ is the velocity resolved in the simulation, $\bar{u}_i$ is the mean (Reynolds-averaged) velocity, u''_i is the resolved fluctuating velocity, and u'_i is the unresolved or modeled fluctuating velocity. The Reynolds-averaged velocity and resolved fluctuating velocity can be obtained directly from the simulation. The modeled fluctuating velocity is defined using the turbulent stress term. Substitution of the decomposed instantaneous velocity (u_i) in Eq. (3) into Eq. (2), along with the assumption that resolved and unresolved velocity fluctuations are uncorrelated, yields the following expression for the residual stress:

$$\tau_{ij} = \left(\widehat{\widehat{u}_i \widehat{u}_j} - \widehat{u}_i \widehat{u}_j\right) + \widehat{u'_i u'_j} \tag{4}$$

The scale similarity concept can be used to model both of the terms on the right-hand side of Eq. (4). The result is an expression for the subfilter stress term as:

$$\tau_{ij} = \alpha\left(\widehat{u_i u_j} - \widehat{u}_i \widehat{u}_j\right) + \beta \overline{u'_i u'_j} \tag{5}$$

The first (both parts inside parenthesis) and the second terms on the right-hand side of Eq. (5) can be modeled as linear functions of the subgrid stress (SGS) and Reynolds stress, respectively. The SGS and Reynolds stresses can be obtained from any suitable SGS and RANS model. The proportionality constants α and β can be both spatially and temporally varying, but are assumed to be complementary, such that the residual stress term can be expressed as a weighted average of both the SGS and RANS stress as follows:

$$\tau_{ij} = \alpha \tau_{ij}^{SGS} + (1 - \alpha)\tau_{ij}^{RANS} \tag{6}$$

To determine the value of the blending coefficient α, a secondary filtering operation is applied, conceptually similar to the method of Lilly [28] for dynamic model coefficient evaluation for subgrid stress modeling. Based on the following assumptions:

$$\tau_{ij}^{RANS} = \overline{u_i u_j} - \overline{u_i}\,\overline{u_j} \tag{7}$$

$$\overline{\widehat{u_i u_j}} = \overline{u_i u_j} \tag{8}$$

the Reynolds-averaging operation can be applied to Eq. (2) as a secondary filter and combined with Eq. (7) to yield:

$$\tau_{ij}^{RANS} - \overline{\tau_{ij}} = \left(\overline{u_i u_j} - \overline{u_i u_j}\right) - \left(\overline{\widehat{u_i u_j}} - \overline{\widehat{u}_i \widehat{u}_i}\right) = \overline{\widehat{u}_i \widehat{u}_i} - \overline{u_i}\,\overline{u_j} = \overline{u_i'' u_j''} \tag{9}$$

The Reynolds-averaged form of Eq. (6) is combined with Eq. (9) to eliminate $\overline{\tau_{ij}}$. The scalar product of the result with the mean (Reynolds-averaged) strain rate yields the expression for α:

$$\alpha = \left(\underbrace{\overline{u_i'' u_j''}\,\overline{S_{ij}}}_{\substack{\text{Resolved turbulent} \\ \text{Production}}} \right) \Big/ \left(\underbrace{\tau_{ij}^{RANS}\overline{S_{ij}}}_{\substack{\text{RANS} \\ \text{Production}}} - \underbrace{\overline{\tau_{ij}^{SGS}}\,\overline{S_{ij}}}_{\substack{\text{Inhomogeneous} \\ \text{SGS Production}}} \right) \tag{10}$$

The local value of the coefficient α is determined based on the relative contribution to turbulence production by the resolved scales, the mean (Reynolds-averaged) component of the subgrid model stress, and the RANS model stress. For stability, the value of α is limited in practice to lie between 0 and 1. Eq. (10) shows that the value of α becomes zero in regions with no resolved fluctuations, i.e., numerically steady-state flow. In those regions, therefore, a pure RANS model is recovered. In regions for which turbulence production by resolved velocity fluctuations is significant, the RANS stress contribution decreases, and an LES subgrid stress contribution appears in the momentum equation. This ensures a smooth variation of turbulent production through the transition between RANS and LES. If the resolved turbulent production in any region is sufficiently large, α increases to a value of 1, and a pure LES model is recovered.

One final key aspect of the DHRL methodology is the manner in which the RANS model component is computed. In contrast to most other hybrid methods, the DHRL approach computes the RANS terms based only on the Reynolds-averaged flowfield. For stationary flows, the velocity field used to compute all RANS terms can be obtained from a time-averaging operation that runs concurrent with the simulations. Other appropriate averaging methods can be adopted for nonstationary flows. For the current study, stationary flows are considered, and the RANS model is computed using the time-averaged flowfield in all of the turbulence model terms.

2.2. Detached-eddy simulation (DES)

The DES model is based on the one-equation eddy-viscosity RANS model developed by Spalart and Allmaras [29] (SA), which has been used extensively in applied CFD analysis. The SA model includes one additional transport equation for an eddy-viscosity variable, $\widetilde{\nu}$. The transport equation includes terms for diffusive transport, production, and destruction, where the latter is given in simplified form as follows:

$$Dest = -C_d\left(\frac{\widetilde{\nu}}{d}\right)^2 \tag{11}$$

where C_d is a variable model coefficient and d is the distance to the nearest wall. The wall distance also appears in other model coefficients.

The DES model replaces the wall distance with an effective length scale, $\widetilde{d}$, that is a function of both wall distance and local mesh size:

$$\widetilde{d} \equiv \min(d, C_{DES}\Delta) \tag{12}$$

where Δ is the local characteristic mesh spacing and C_{DES} is a model constant. The effect of the modification is to increase destruction of $\widetilde{\nu}$ in regions with Fine mesh, allowing the development of resolved turbulent fluctuations, while recovering the original SA RANS formulation in regions with large mesh spacing. The reader is referred to [12] for further details.

2.3. Delayed detached-eddy simulation (DDES)

The DDES model was proposed as a modification to the baseline DES model in order to address the issue of grid-induced premature switching to the LES mode in attached boundary layers [13]. This anomalous activation of LES mode occurs in the DES model when a highly refined grid is used. In contrast to the baseline DES model, DDES redefines the length scale such that it depends on both the local grid size as well as the eddy viscosity:

$$\widetilde{d} \equiv d - f_d \max(0, d - C_{DES}\Delta) \tag{13}$$

The function f_d recovers a value of zero inside the boundary layer, ensuring that the RANS mode is activated; however, the value of f_d limits to one outside the boundary layer, thus recovering the underlying DES mode. The functional form of f_d is

$$f_d \equiv 1 - \tanh\left([8r_d]^3\right) \tag{14}$$

where the variable r_d is obtained as:

$$r_d \equiv \frac{\nu_t + \nu}{\sqrt{U_{i,j}U_{i,j}}\,\kappa^2 d^2} \tag{15}$$

2.4. Shear-stress transport (SST) k-ω

The SST k-ω model proposed by Menter [23] is conceptually based on the transport of the principal shear stress and was developed to improve prediction of flows with adverse pressure gradients. It has been used for a large number of practical RANS CFD simulations of complex turbulent flows [30]. In the SST model, the eddy viscosity from the baseline k-ω model is modified within the framework of the SST model as follows:

$$\nu_t = \frac{a_1 k}{max(a_1 \omega,\ SF_2)} \tag{16}$$

where F_2 is a blending function, a_1 is a constant, and S represents an invariant measure of the strain rate magnitude. F_2 recovers a value of one in attached boundary layers, and a value of zero in free shear layers. The model improves prediction in adverse pressure gradient boundary layers by ensuring that production of turbulent kinetic energy is larger than dissipation.

Two transport equations are adopted in the SST model, one for the turbulent kinetic energy (k) and one for the specific turbulence dissipation rate (ω):

$$\frac{D\rho k}{Dt} = \tau_{ij}\frac{\partial u_i}{\partial x_j} - \beta^* \rho \omega k + \frac{\partial}{\partial x_j}\left[\left(\mu + \sigma_k \mu_t\right)\frac{\partial k}{\partial x_j}\right] \tag{17}$$

$$\frac{D\rho\omega}{Dt} = \frac{\gamma}{\nu_t}\tau_{ij}\frac{\partial u_i}{\partial x_j} - \beta\rho\omega^2 + \frac{\partial}{\partial x_j}\left[\left(\mu + \sigma_\omega \mu_t\right)\frac{\partial \omega}{\partial x_j}\right] + 2(1-F_1)\rho\sigma_{\omega 2}\frac{1}{\omega}\frac{\partial k}{\partial x_j}\frac{\partial \omega}{\partial x_j} \tag{18}$$

The function F_1 plays a similar role as the blending function F_2, acting as an indicator variable for near-wall and farfield regions of the flow. Near the wall, $F_1 = 1$, and the k-ω model form is obtained. In farfield regions, $F_1 = 0$ and the model behaves similar to a k-ε model. Refer to Ref. [23] for further model details. For the DHRL model used in the present study, the SST model was used as the RANS component.

2.5. Monotonically integrated large eddy simulation (MILES)

MILES refers to an approach for implicit LES first documented by Boris et al. [31] for which the subgrid stress is assumed to be accurately approximated by the dissipation inherent in the discretization error for the convective terms. For upwind-biased schemes in particular, the numerical error can be shown mathematically to be similar in form to an explicit eddy-viscosity model for such an approach [24], and several studies have been documented in the literature in which practical LES solutions have been successfully obtained. For the present study, the MILES method is used for the reference LES simulations and also as the LES component in the DHRL implementation. Two different convective discretization schemes are used (these are discussed below). Both schemes are designed to preserve monotonicity through upwinding and flux limiting, which is appropriate for a MILES approach.

3. Numerical method

Simulations in the present study were all run using the commercial CFD code Ansys FLUENT. The DHRL methodology was implemented using the User-Defined Function subroutines available with that solver. Other turbulence models were all available as built-in modeling options in FLUENT. All of the test cases considered are incompressible, so the segregated pressure-based solver was used with the SIMPLE pressure-velocity coupling method [32]. Second-order implicit (three-point backward difference) discretization was used for the unsteady term in all time-dependent simulations including HRL and LES. A second-order upwind, linear reconstruction scheme with a Least-Squares gradient computation and conventional slope limiting [33] was used as the baseline discretization scheme for the convective terms in the momentum equations. Mass fluxes were computed using the momentum-weighted interpolation method of Rhie and Chow [34]. For some of the test cases, a less dissipative bounded central differencing spatial discretization scheme based on the Normalized Variable Diagram [35] was used to evaluate the influence of discretization scheme on the HRL results. A second-order centered discretization was used for the pressure gradient terms, and second-order central differencing was used for all diffusion terms.

Steady RANS simulations were run until a converged solution was obtained, with convergence determined based on stabilization of all flow variables to constant values as iterations increased, and a decrease in the L2 norm of all residuals by at least six orders of magnitude less from the initial value. For the hybrid RANS-LES (HRL) models, the converged steady-state RANS results were used as the initial conditions. The HRL model simulations were time-dependent and were run until a statistically steady-state flow field was obtained, as judged by time-averaged variables obtaining constant values with increasing time steps. The time-step size for unsteady simulations was selected to maintain maximum convective CFL number of approximately one, based on the smallest streamwise mesh dimension and the freestream velocity. For the time-dependent cases, the L2 norm of all residuals was verified to be reduced by at least three orders of magnitude during each time step. For both attached and separated flow test cases, an additional HRL simulation was run with the time-step size reduced by half, and results were compared to the default time step cases. No significant difference was observed for the resulting statistical quantities, so it was concluded that the use of a convective CFL number of approximately one is appropriate.

4. Attached turbulent flow test case

The selected attached flow test case is a fully developed turbulent channel flow, corresponding to the experimental study of Hoyas and Jimenez [25]. Results for this case using the DHRL model were previously obtained using a pseudo-spectral solver and a combination of the Spalart-Allmaras one-equation model [29] for the RANS component and the dynamic Smagorinsky model [28] for the LES component [27]. This paper investigates in more detail DHRL simulation using a general-purpose finite-volume flow solver.

The simulations were performed for the case of Reynolds number equal to 2003, where

$$Re = \frac{u_\tau \delta}{\nu}, \tag{19}$$

u_τ is the wall friction velocity, δ is the channel half-height, and ν is the kinematic viscosity. The domain extent was $2\pi\delta$ in the streamwise and $\pi\delta$ in the spanwise directions. Three different structured meshes were investigated, with streamwise × spanwise × wall-normal cell counts of 32 × 24 × 36 (Coarse), 64 × 48 × 72 (Medium), and 128 × 96 × 144 (Fine). The grid spacing was uniform in the streamwise and spanwise directions. Stretching was employed in the wall-normal direction so that the first near-wall cell centroid was located with y^+ of 2, 1, and 0.5 for the Coarse, Medium, and Fine grids, respectively, and cells located at the channel centerline had aspect ratios of approximately unity.

Hybrid RANS-LES simulations were performed with three different models—the DHRL method described above, the original detached-eddy simulation (DES) model based on the Spalart-Allmaras RANS model [12], and the more recently proposed delayed detached-eddy simulation (DDES) model [13]. The DHRL model combined the k-ω SST model [23] as the RANS component and monotonically integrated LES (MILES) [24] as the LES component. For comparison purposes, steady RANS simulations were obtained using k-ω SST model alone, and LES simulations were obtained using the MILES method. Results for all models were evaluated based on comparison with the available DNS data [25]. Quantitative comparisons were made primarily for mean velocity and turbulent kinetic energy. Of particular interest was the effect of grid resolution level and discretization scheme for the different modeling approaches investigated. Unless otherwise noted, all results shown are plane-averaged, though all statistical quantities showed little variation in the streamwise and spanwise directions once the simulation had run sufficiently long for a stationary state to be reached.

Figure 1 shows the mean velocity profile predicted by each of the models on the Coarse, Medium, and Fine grids. Note that the results with the SST model are only shown for the Coarse grid, since no noticeable differences were observed between that and the two finer grids. The MILES results are only shown for the Medium and Fine grids, since the results on the Coarse grids were comparable to laminar flow, and dramatically overpredicted the velocity in the middle portion of the channel. In fact, it is apparent from **Figure 1 (b)** that even results on the Medium grid are qualitatively incorrect, and only on the Fine grid does the characteristic log-layer behavior begin to be observed, though mean velocity of it is still significantly overpredicted.

For the Coarse grid, the SST, DDES, and DHRL models all show qualitative agreement with the DNS data, with the DDES model showing the best agreement, and both the SST and DHRL models showing a similar, slight overprediction in the log-layer region. The DES model shows the characteristic overshoot that has previously been noted in the literature, which is due to the performance of the model in the transition region between the (near-wall) RANS zone and the far-field (LES) region in the middle of the channel. As the grid is refined (**Figure 1 (b, c)**),

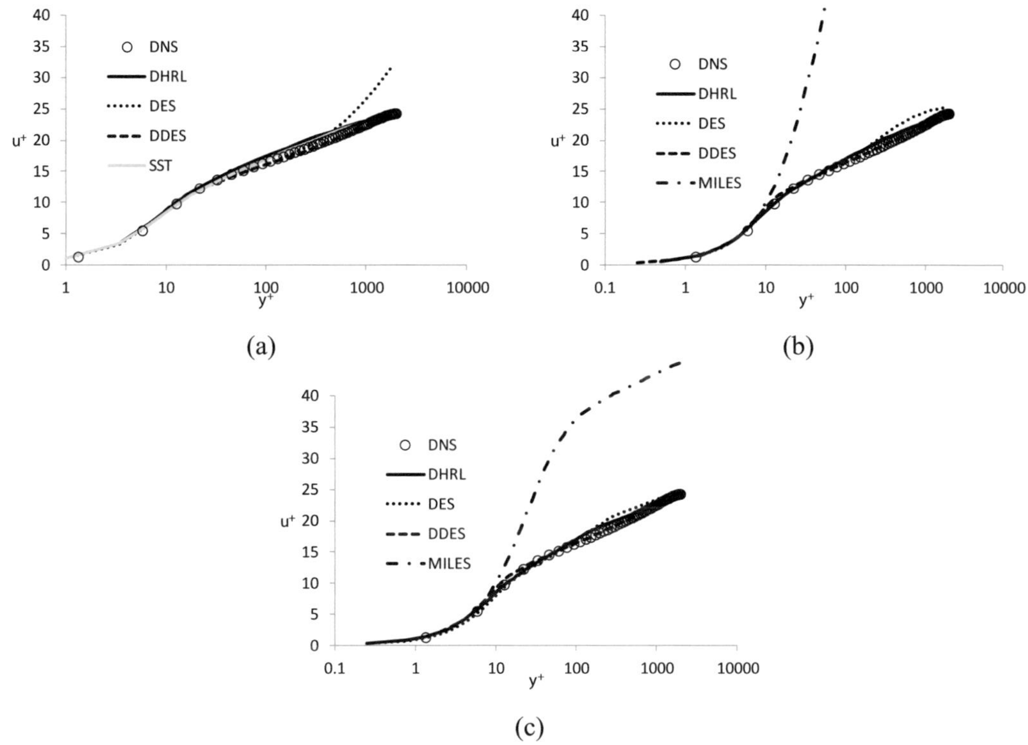

Figure 1. Mean velocity profiles for turbulent channel flow test case using different turbulence models: (a) Coarse mesh; (b) Medium mesh; (c) Fine mesh.

the DES results agree more closely with the DNS data, and for the Fine grid, the log-layer mismatch is much smaller, though still apparent.

For all three grid sizes, the DDES model obtained a numerically steady-state result, effectively operating as a pure RANS model. This is not surprising, since the goal of the Delayed DES formulation is to inhibit the development of LES behavior in attached flow regions, while allowing its development in separated flow regions. As a consequence, the DDES mean velocity results were limited to a Spalart-Allmaras model equivalent, which is in good agreement with the reference data.

The predicted turbulent kinetic energy (k) profiles for the hybrid RANS-LES models are shown in **Figure 2**. Since the DDES model returned a steady RANS solution, no result is plotted. For the DHRL model, both resolved turbulence and total (resolved + modeled) turbulence is shown, where the total k is computed as:

$$k_{\text{total}} = \alpha k_{\text{resolved}} + (1 - \alpha) k_{\text{modeled}}. \tag{20}$$

For both the DES and DHRL models, the overall trends are correctly predicted, with a peak near the wall and decay to a minimum value at the channel centerline. Both models tend to

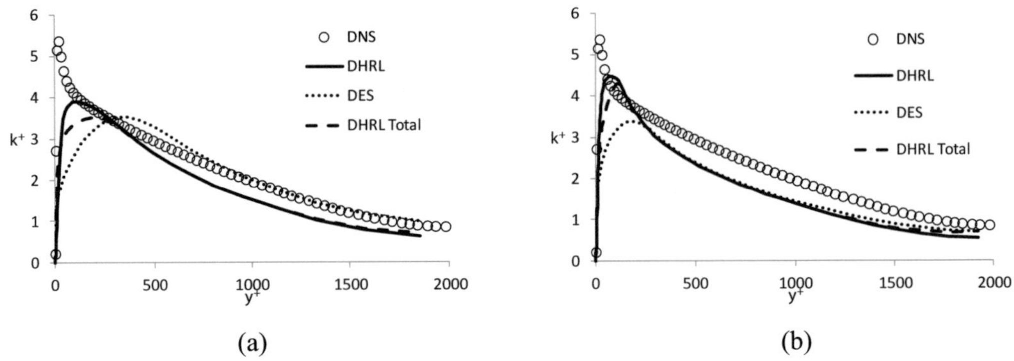

Figure 2. Turbulent kinetic energy (k) profiles for turbulent channel flow test case comparing results with DES and DHRL models: (a) Medium mesh; (b) Fine mesh. For DHRL, both resolved k and total k (resolved + modeled contributions) are shown.

underpredict the peak value of k near the wall as well as predicting the location of the peak to be too far from the wall, and for the Fine grid case show an underprediction throughout the channel. The DES result shows nontrivial sensitivity to the mesh refinement level, and surprisingly shows better agreement throughout most of the channel on the Coarse grid than on the Fine grid. The DHRL model shows relatively little dependence on the mesh resolution level, except for an improvement in the near-wall prediction as the grid is refined (as does the DES model).

The mesh sensitivity is further highlighted in **Figure 3**, which shows the change in the mean velocity profiles for each model as the mesh is refined from 27,648 total cells (Coarse) to 1,769,472 cells (Fine). For the DES model (**Figure 3 (a)**), the improvement with increasing mesh resolution is apparent, as the characteristic overshoot becomes smaller. The DDES model is not shown since it yielded a steady-state result that was insensitive to mesh refinement level for the grids considered here. The DHRL model (**Figure 3 (b)**) shows very little sensitivity to grid refinement level and shows better agreement with the DNS data than the DES model for all refinement levels. Also shown in **Figure 3 (b)** are the steady RANS results from the SST model, which is the RANS component for the DHRL hybridization. The results from DHRL agree

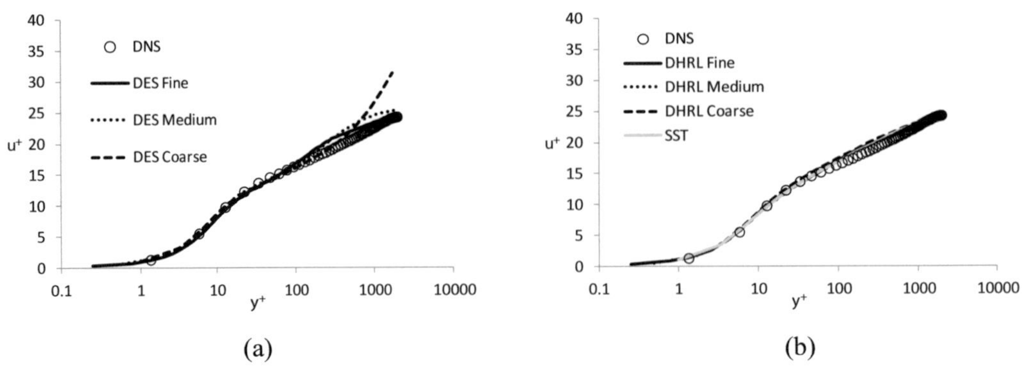

Figure 3. Effect of grid resolution on predicted mean velocity profiles for DES (a) and DHRL (b) models.

quite closely with the SST for all mesh refinement levels, despite the fact that a significant portion of the turbulent kinetic energy is resolved as unsteady velocity fluctuations (**Figure 2**) in contrast to the steady SST result.

The effect of the choice of discretization scheme (second-order upwind versus NVD-based Bounded Central Difference) for the convection term is shown in **Figure 4**, for both the DES and DHRL models. While the BCD scheme has been shown to significantly reduce numerical dissipation on structured meshes [35, 36], the effect on the mean flowfield using hybrid RANS-LES models was found to be insignificant. For the DES model, as the numerical dissipation increases, the local shear stresses decrease in magnitude, resulting in lower overall magnitudes of the eddy viscosity. This decrease in model eddy viscosity helps to compensate for the higher numerical viscosity of the second-order upwind scheme. For the DHRL model, a similar effect occurs. As seen in Eq. (10), the blending coefficient α responds to changes in the resolved flowfield to adjust the RANS contribution accordingly.

The underlying method by which the DHRL model operates is highlighted in **Figure 5**, showing the distribution of the blending parameter α through the channel for all three grids investigated. Recalling that $\alpha = 0$ corresponds to a pure RANS mode and $\alpha = 1$ corresponds to a pure LES mode, it is apparent that both the Medium and Fine grid cases have a significant portion of the channel, up to about $y+ = 1000$, for which the model operates as pure MILES. The MILES region extends further toward the wall for the Fine grid case ($y+ \sim 100$) than for the Medium grid case ($y+ \sim 300$). In the near-wall region, $\alpha \rightarrow 0$ as $y \rightarrow 0$ and the RANS model is recovered. Likewise, near the channel centerline ($y+ > 1000$), $\alpha < 1$ and some portion of the RANS model stress is introduced into the momentum equations. The behavior of the Coarse grid case is quite different. Since the mesh is too coarse to resolve a significant level of turbulent production, the value of α is quite low throughout the channel, and a significant portion of the RANS model stress is introduced. As seen in **Figure 1 (a)**, the effect is to

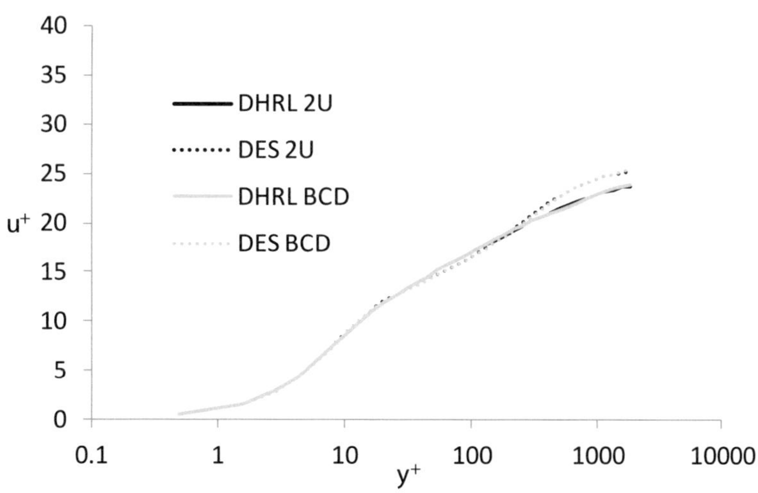

Figure 4. Effect of discretization scheme (second-order upwind vs. bounded central difference) on predicted mean velocity profiles for DES and DHRL models.

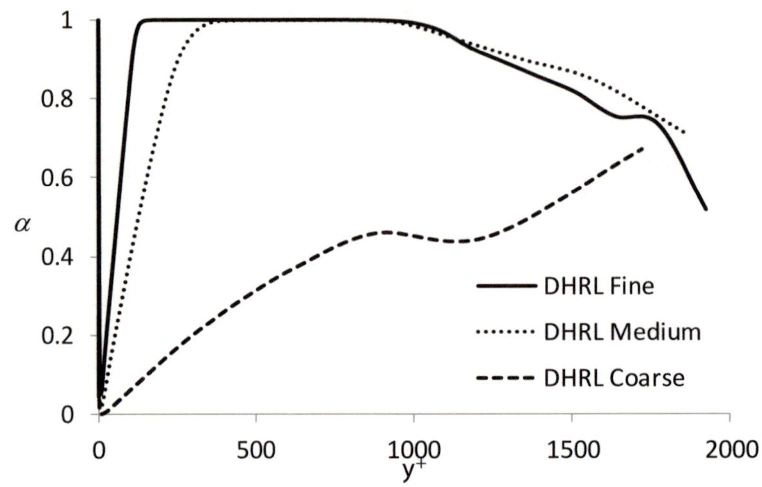

Figure 5. Variation of DHRL blending parameter (α) across the channel for three different mesh sizes.

minimize overshoot of the mean velocity profile in the middle portion of the channel and recover a result close to the steady RANS SST result.

Figures 6 and **7** illustrate some of the features of the resolved velocity field in the hybrid RANS-LES simulations. **Figure 6** shows isosurfaces of Q-criterion on the Fine grid, for both the DES and DHRL models, to highlight the presence of vortical turbulent structures. Such structures are visible for both models; however, for an equivalent value of Q-criterion, the structures predicted by the DHRL model are more prevalent and of a smaller scale than those predicted by the DES model. This was found to be consistently true for all cases. The reason is that the DHRL model operates in a MILES mode with respect to the fluctuating velocity field and as such the dissipation of small-scale turbulent features is less than with the DES model. **Figure 7** shows qualitatively the effect of mesh refinement on the resolution of the fluctuating

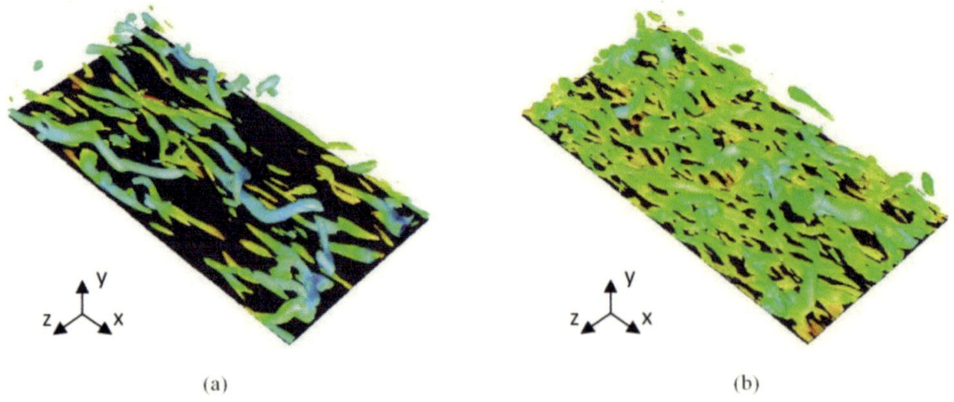

Figure 6. Isosurfaces of Q-criterion showing resolved turbulent structures on the Fine grid: (a) DES and (b) DHRL.

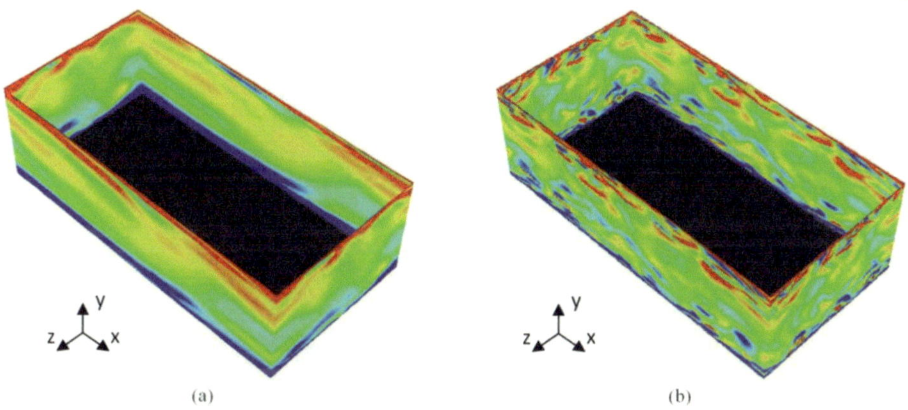

Figure 7. Contours of spanwise vorticity predicted using DHRL model: (a) Coarse mesh; (b) Fine mesh.

velocity field using the DHRL model by showing contours of the spanwise vorticity on the periodic boundary planes. It is apparent that for the Fine grid (**Figure 7 (b)**), a significant portion of the turbulent spectrum is resolved, similar to the Q-criterion contours in **Figure 6 (b)**. However, for the Coarse grid (**Figure 7 (a)**), only large-scale, low-frequency fluctuations are present in the flow. It is this lack of resolution of the turbulence field that results in the decrease of the blending parameter α (**Figure 5**) in the Coarse grid case, and the introduction of a relatively large RANS stress contribution into the momentum equations.

5. Separated turbulent flow test case

The separated flow test case is the backward facing step flow matching the experimental study of Driver and Seegmiller [26], which is a commonly used benchmark case for turbulence model verification and validation. Two different three-dimensional meshes were created using the commercial meshing tool Ansys GAMBIT® in order to evaluate grid sensitivity issues common to most current HRL models [15], for example, the delay of instability and breakdown of separated shear layers. For both grids, the computational domain size extended from 4 step heights (H) upstream of the step to 32H downstream of the step, 16H vertically from the wall downstream of the step, and 6H total in the spanwise direction. A structured multiblock mesh was generated, with a baseline Coarse mesh containing 744,960 total cells, and a Fine mesh containing 7,946,400 total cells. An average $y+$ value less than one was enforced to satisfy the recommended $y+$ requirements for the RANS turbulence model and for accurate HRL modeling. Nearly isotropic quadrilateral cells were used in the LES region (from x/H = 0.0 to 10.0) for maximum accuracy. A 2D planar slice of the Fine mesh is shown for illustrative purposes in **Figure 8**.

Simulations were performed using the DHRL, DDES, and SST k-ω models, with all simulation parameters and boundary conditions matching the experimental study. Profiles of inlet flow

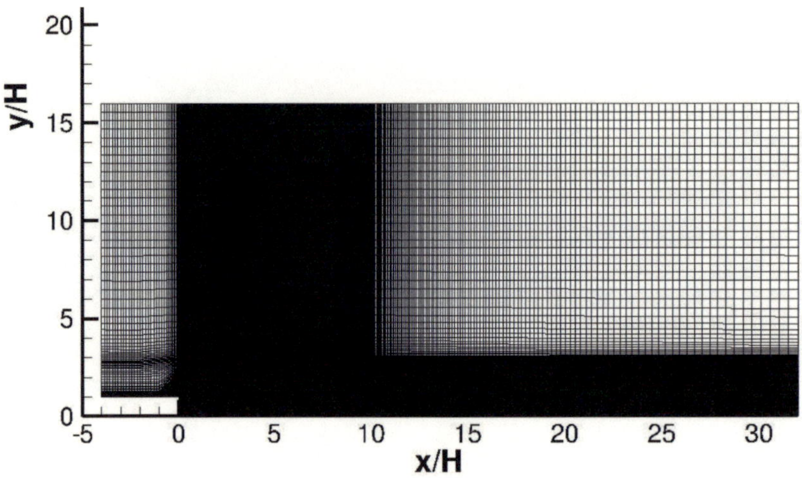

Figure 8. Fine mesh for backstep case on spanwise periodic plane.

variables including mean inflow velocity, turbulent kinetic energy, and specific dissipation rate for the DHRL and RANS model simulations, and modified eddy viscosity for the DDES model simulation were created to match the distributions in the experimental data. For the spanwise boundaries, a periodic boundary condition was applied in the transient HRL simulations, and a symmetry boundary condition was used for the steady RANS simulation.

Figure 9 shows the spanwise-averaged mean wall static pressure distribution for both the Coarse mesh and the Fine mesh, respectively, using three different turbulence models. In the

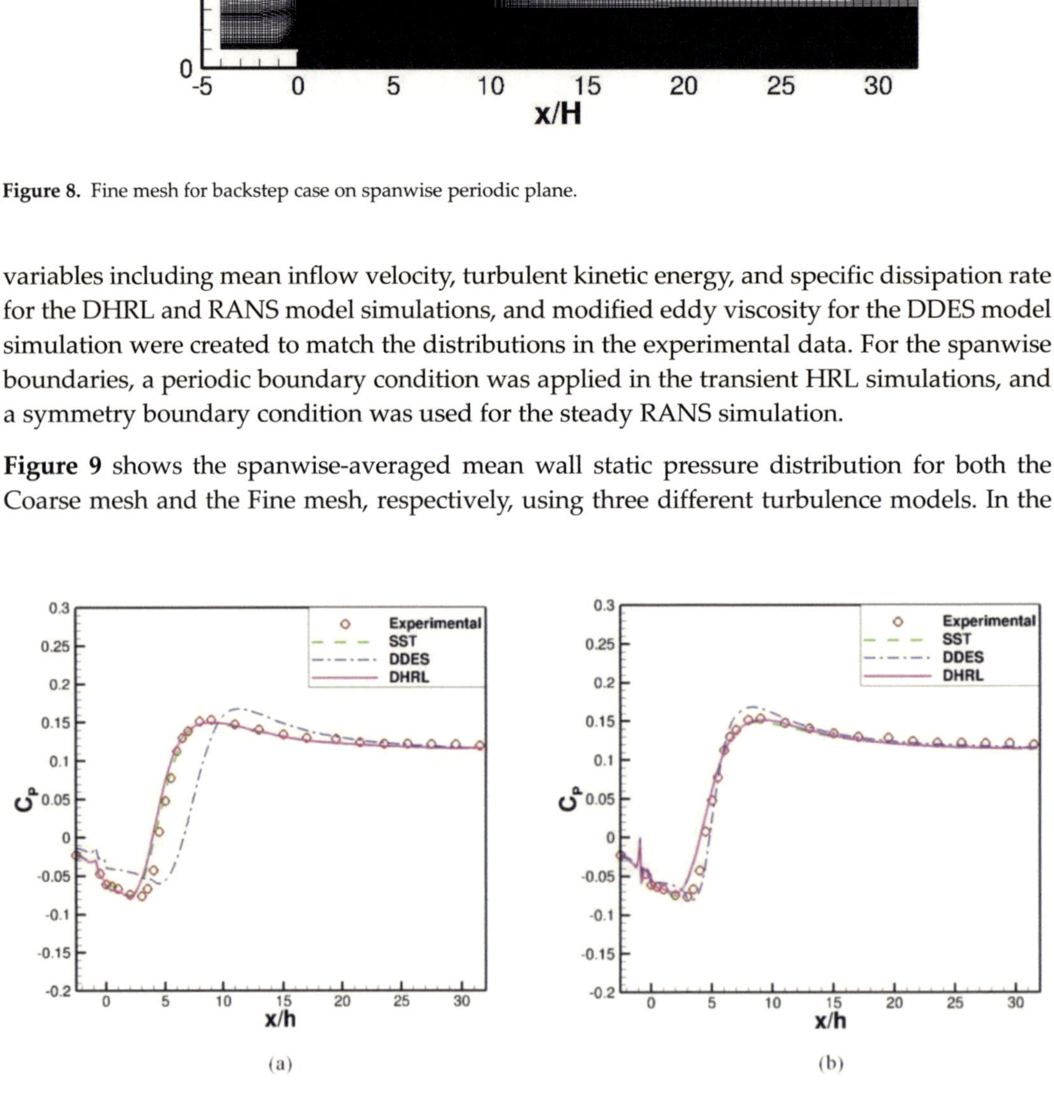

Figure 9. Static pressure distribution on wall downstream of step: (a) Coarse mesh; (b) Fine mesh.

recirculation region (x/H > 0), the Coarse mesh results shown in **Figure 9 (a)** indicate that both the DHRL and RANS computations predict a smooth pressure decrease and resolve the negative peak pressure with reasonable accuracy compared with the experimental data. The DDES results in contrast show overprediction of the pressure decrease and an offset in the location of negative peak pressure. The pressure rise predicted by both the DHRL and RANS models in the separated flow region shows similar behavior to the experimental results, but the wall pressure in the DDES computation shows a delayed pressure recovery. The pressure distribution after flow reattachment is similar to the experimental results for all simulations, except for a small overprediction by the DDES model in the region 10 < x/H < 16. **Figure 9 (b)** shows that the Fine mesh computations compare reasonably well with the experimental data for all models used. Simulations using the Fine mesh show a substantial improvement relative to the Coarse mesh for the DDES model. The differences underscore the mesh sensitivity inherent in DDES. In contrast, the Coarse and Fine mesh results using the DHRL model are similar, indicating that the DHRL model is relatively insensitive to mesh resolution, which agrees with the results from the attached flow test case above.

Figure 10 shows the spanwise-averaged mean wall skin friction distribution on both the Coarse mesh and Fine mesh. **Figure 10 (a)** indicates that the wall skin friction result on the Coarse grid in the separated flow region agrees closely with the experimental data for both the DHRL model and the RANS model. In particular, the negative peak C_f values are quite close to the experimental result. In contrast, the DDES model shows an overprediction of the size of the flow separation region along with an overprediction of skin friction. The reattachment point was determined in the experiments using a linear interpolation of the oil-flow laser skin-friction measurements and was found to be x/H = 6.38 [26]. The DHRL model results show an early prediction of flow reattachment and a reattachment point of approximately x/H = 5.60, an underprediction of 9%. The RANS model predicts the flow reattachment

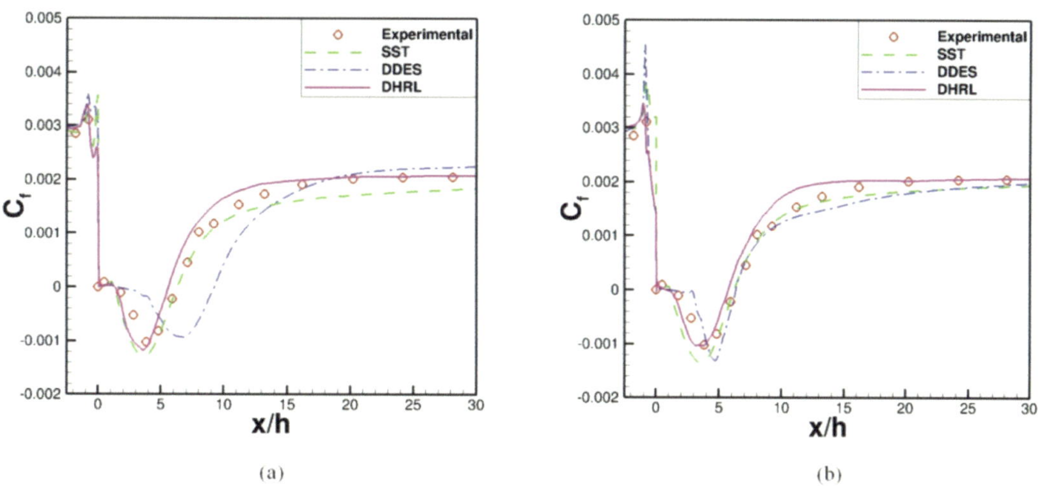

Figure 10. Skin friction distribution on wall downstream of step: (a) Coarse mesh; (b) Fine mesh.

location at x/H = 6.30, which is closer to the experimental measurements. The DDES results show a delayed flow reattachment at x/H = 9.23. The grid is too coarse for the DDES model to accurately resolve the Reynolds stress contribution in either LES or RANS mode, leading to the delay in flow reattachment. The skin friction is predicted reasonably well by the DHRL model downstream of the reattachment point. The RANS model accurately predicts skin friction immediately downstream of the reattachment point but shows an underprediction in the region farther downstream. The differences in the DHRL model prediction, as compared to the SST model, can be attributed to the presence of resolved unsteady turbulent fluctuations in the flow that lead to more rapid mixing of the separated shear layer and transport of momentum in the wall-normal direction. This is also clearly seen in the mean velocity and turbulent kinetic energy profiles shown below.

The predicted skin-friction distribution for the Fine mesh, shown in **Figure 10 (b)**, indicates that results with all three models agree reasonably well with the experimental data. However, with regard to mesh sensitivity, some models show quantitative or even qualitative differences between the Coarse and Fine meshes. The DHRL model predictions are closest to being mesh insensitive, as the C_f calculations for both meshes are quite similar. The reattachment location (x/H = 5.70) calculated using the Fine mesh is less than 2% different than the result on the Coarse mesh. The RANS model predictions for both the Coarse and Fine meshes show very close agreement with one another in all regions, including the separation point. This is similar to the attached flow results above, for which the steady RANS results were effectively mesh independent. In contrast, a significant level of mesh sensitivity is shown by the DDES model predictions. Skin friction over most of the downstream wall shows significant differences between the Coarse and Fine grid results. Mesh refinement significantly improves the prediction of reattachment point (x/H = 6.31) for the DDES model.

Previous studies have shown that DDES model predictions have reported delayed shear layer breakup and limited resolution of the Reynolds stresses in separated flow regions, particularly on coarse grids [2, 37]. One key advantage of the DHRL model is that it inherently addresses these potential problems. **Figure 11** shows contours of instantaneous spanwise vorticity (z-vorticity) obtained from the DHRL and DDES model simulations, respectively, on the Coarse mesh. The DHRL model results indicate a better ability to resolve turbulent eddies in the separated shear layer and a more obvious breakup and transition to LES mode than the DDES model. The relatively poor predictions of wall static pressure and skin friction predictions by the DDES model shown above are partly explained by these results. **Figure 12** shows the contours of instantaneous spanwise vorticity (z-vorticity) on the Fine mesh for both models. DHRL and DDES are both able to resolve turbulent scales in the separation region and as a result are able to provide reasonably accurate prediction of C_p and C_f. With regard to mesh sensitivity, the DHRL model contours are similar on both the Coarse and Fine mesh computations except for the increased resolution of small scales on the Fine mesh. The DDES model contours, in contrast, show even qualitative differences between results on the Coarse and Fine meshes. **Figure 13** shows the spanwise vorticity (z-vorticity) contours predicted by the RANS simulations on the Coarse and Fine meshes, respectively. It is apparent that no turbulent scales are directly resolved, as expected. Rather, the effect of all turbulent scales on the mean flow is included via the eddy-viscosity term.

http://dx.doi.org/10.5772/intechopen.68387

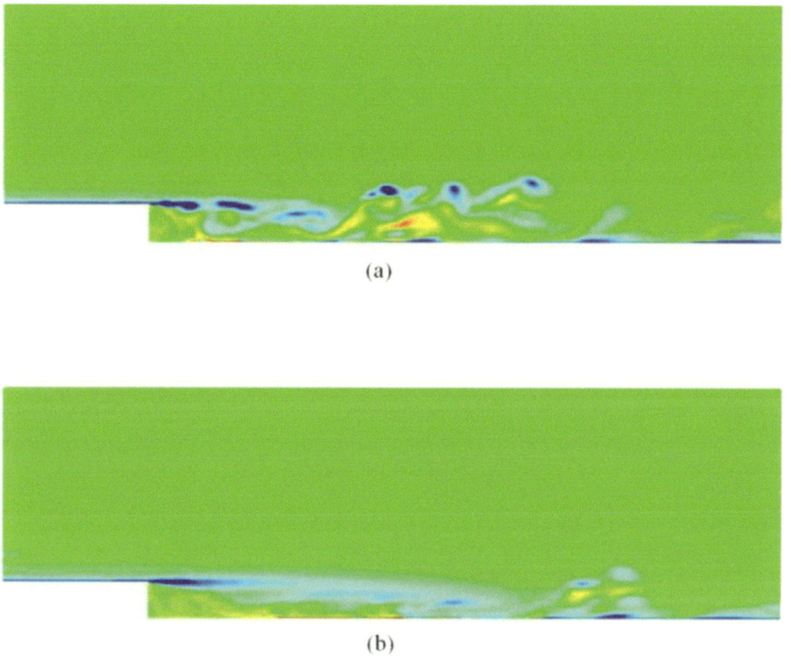

Figure 11. Contours of instantaneous resolved z-vorticity for the Coarse mesh simulations: (a) DHRL; (b) DDES.

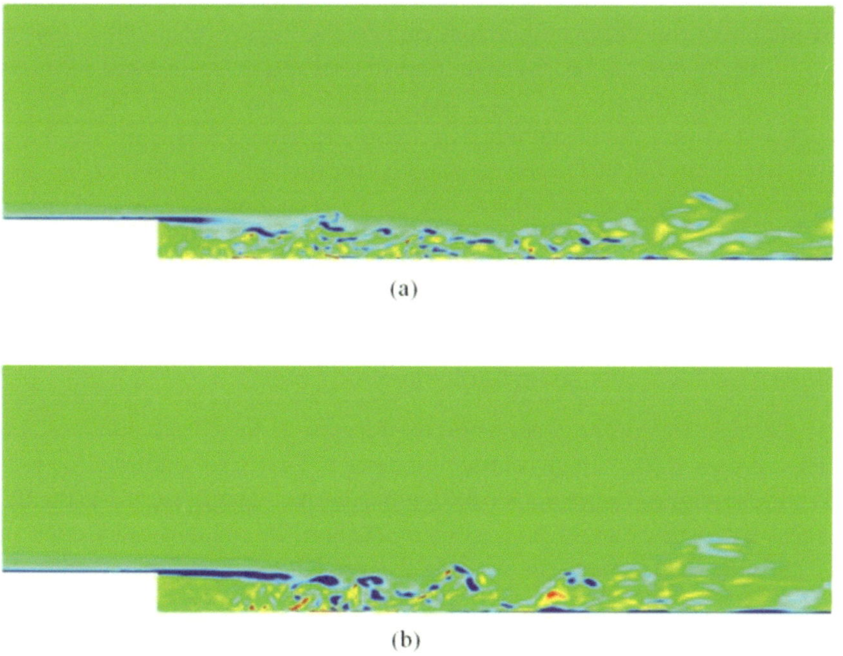

Figure 12. Contours of instantaneous resolved z-vorticity for the Fine mesh simulations: (a) DHRL; (b) DDES.

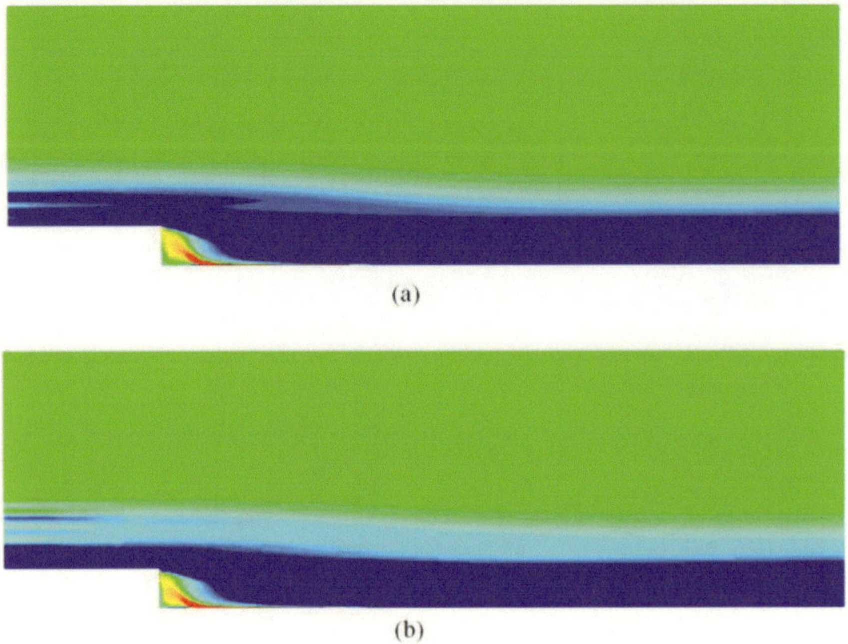

Figure 13. Contours of instantaneous resolved z-vorticity using RANS model: (a) Coarse mesh; (b) Fine mesh.

In order to minimize the dissipation errors due to the numerical method, low dissipation schemes are generally used for convective term discretization in LES and HRL simulations. The bounded central differencing (BCD) [35] scheme is an example of a low-dissipation scheme developed to provide stable results on both structured and unstructured grids. Fine mesh simulations using the BCD scheme were performed along with baseline second-order upwind simulations to investigate the effect of discretization on the HRL model results. **Figure 14** shows the distribution of pressure in the streamwise direction, obtained using the Fine mesh. Very little difference due to discretization scheme is observed for the DHRL results. The DDES simulations, in contrast, show more sensitivity to the scheme, with the BCD discretization method producing a slightly overpredicted negative C_p peak and a delayed pressure recovery relative to the second-order upwind results.

Figure 15 shows skin friction predictions on the Fine mesh for DHRL and DDES using both discretization schemes. Again, there is little difference in DHRL simulations due to the numerical scheme, though slightly better agreement with experiments is observed in the BCD scheme predictions just downstream of the reattachment location. The reattachment point (x/H = 5.75) for the DHRL with BCD scheme is close to the predicted reattachment with the upwind scheme (x/H = 5.70). For the DDES model, the BCD scheme results show a streamwise offset for the Cf prediction in the separated flow region compared to the upwind scheme. The BCD scheme results also show a significantly delayed reattachment point (x/H = 6.80) relative to the upwind scheme (x/H = 6.31). To summarize, the effect of discretization scheme in the DHRL

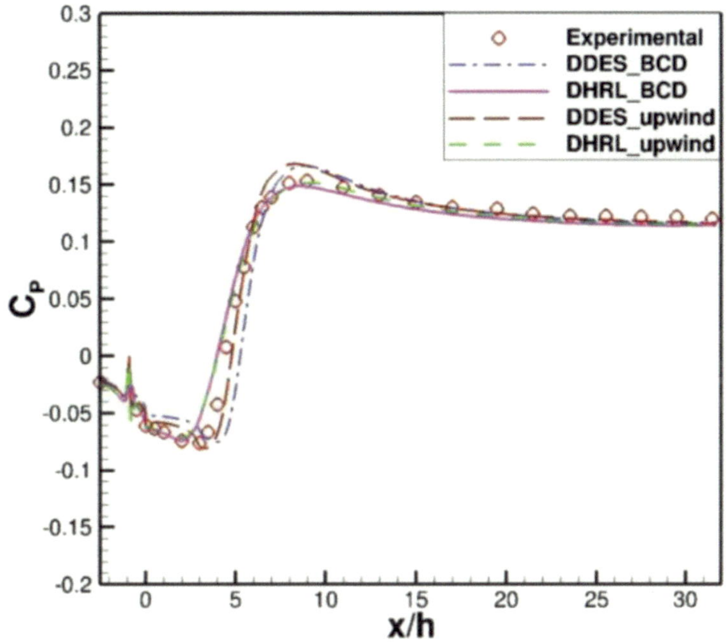

Figure 14. Static pressure distribution on downstream bottom wall comparing results with BCD and second-order upwind discretization scheme on the Fine mesh.

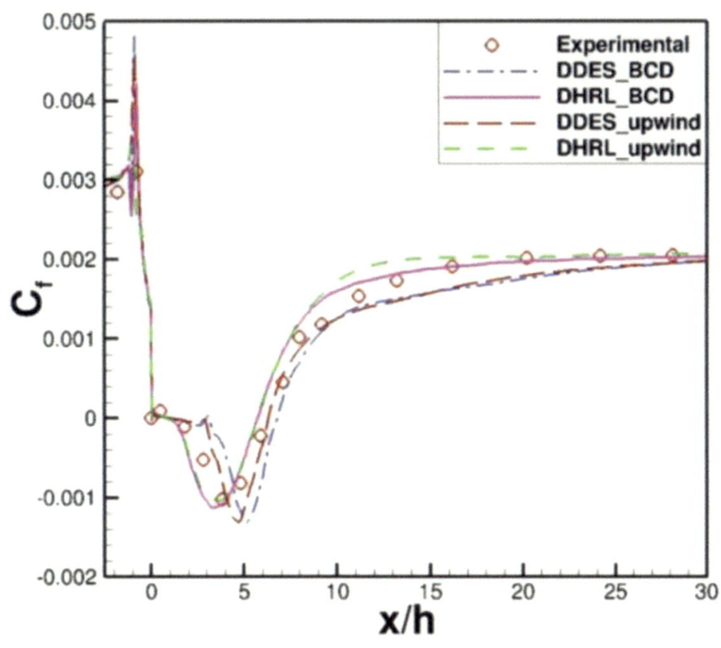

Figure 15. Skin friction distribution on downstream bottom wall comparing results with BCD and second-order upwind discretization scheme on the Fine mesh.

simulations appears to be minimal, with a slight improvement in the results when using the BCD scheme. The use of the BCD scheme for the DDES simulations appears to result in slightly less agreement with experimental data, which is surprising. The apparent reason for this behavior may be the violation of the convection boundedness criterion of the BCD method due to the very low sub-grid scale turbulent diffusivity produced by the DDES model simulations.

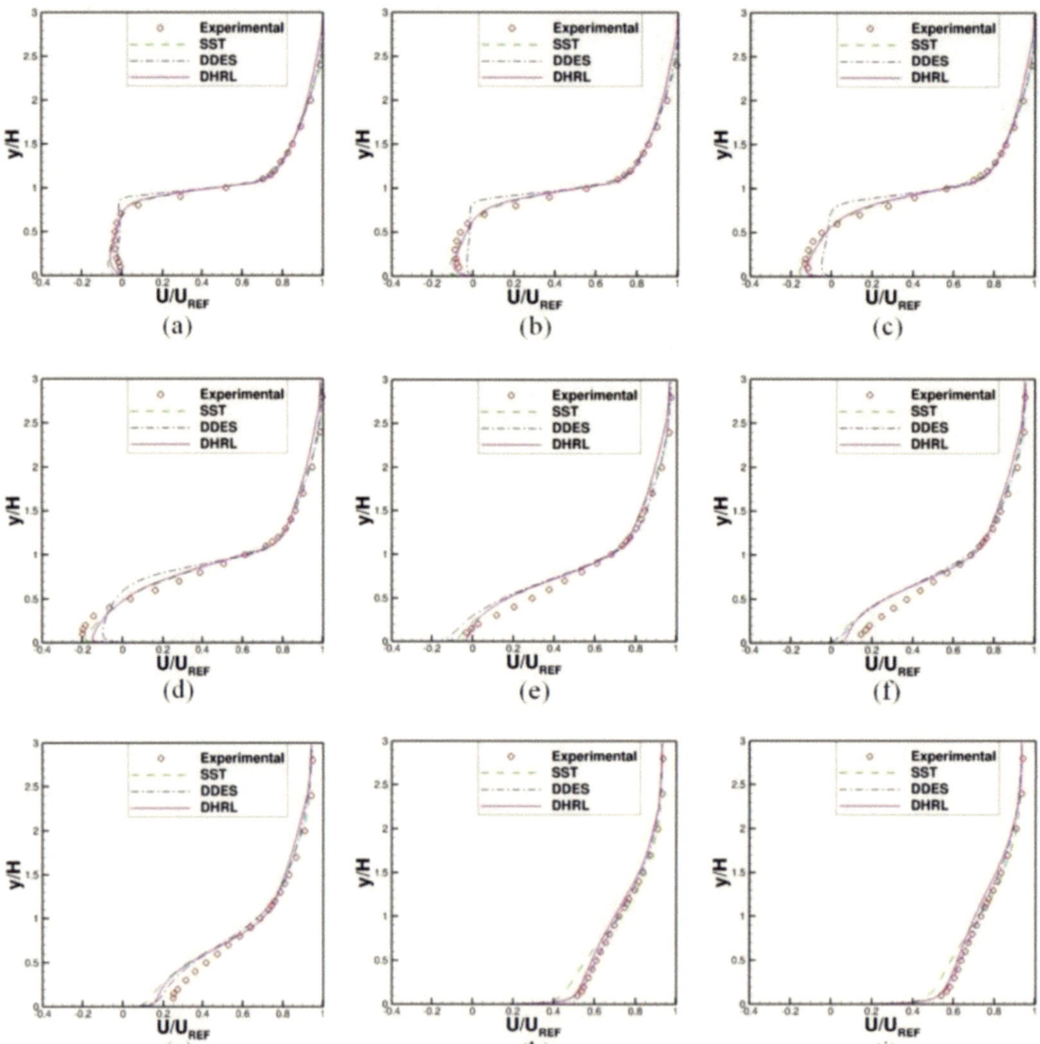

Figure 16. Predicted (Fine mesh) and measured mean velocity profiles at measurement stations downstream of the step. (a) x/H = 1.0. (b) x/H = 1.5. (c) x/H = 2.0. (d) x/H = 3.0. (e) x/H = 5.0. (f) x/H = 6.0. (g) x/H = 7.0. (h) x/H = 16.0. (i) x/H = 20.0.

Figure 16 shows mean streamwise velocity profiles at several streamwise measurement stations for simulations using the Fine mesh. In the separated flow region (stations x/H = 1.0, 1.5, 2.0, and 3.0), the profiles clearly show that characteristics including boundary layer growth and size of the separation bubble in the wall normal direction are well resolved by both the DHRL and RANS simulations in comparison to the experimental measurements. The DDES simulation shows a small underprediction of the negative velocity peak in the near wall region and an overprediction of the size of the separation bubble. Simulations using each of the models show good agreement with the experimental data far from the wall. Immediately upstream of the reattachment point (station x/H = 5.0), DHRL and RANS results show reasonably good agreement with the data in terms of the negative velocity peak and separation-bubble size, but the DDES model somewhat overpredicts the flow behavior. It is apparent that in the separated flow region near the wall, the DHRL simulation shows nearly RANS-like behavior, which is expected since the DHRL model tends to limit to the RANS result close to the wall unless the grid is highly refined. It is interesting to note that the experimental velocity profile at station x/H = 6.0 clearly shows that the flow has reattached, which is early relative to the experimentally determined reattachment location of x/H = 6.38 based on the oil-flow laser skin-friction measurements [26]. The reason for this apparent discrepancy in the experimental data is not clear. The DHRL results also predict reattached flow at station x/H =6.0, while the RANS and DDES simulations show separated flow. Farther downstream (x/H = 7.0), all models clearly show not only reattached flow but also an underprediction of the velocity very close to the wall. In the wake recovery region (x/H = 16.0 and 20.0), the two HRL models show good agreement with the experimental data while the RANS model results show an underprediction in the rate of wake recovery, presumably due to lower predicted levels of turbulent mixing.

Figure 17 shows computed turbulent kinetic energy profiles (normalized by square of mean inlet velocity) obtained for the Fine mesh simulations, compared with experimental data, at the measurement stations identical to those used to compare the mean velocity profiles. In the separated shear layer just downstream of the step at x/H = 1.0, DHRL results show the sudden rise in turbulent kinetic energy seen in the experimental data, indicating relatively rapid breakup of the shear layer characteristic of DHRL simulations. In contrast, the DDES model results show a relatively small peak in turbulent kinetic energy, which is characteristic of the more delayed shear layer break up characteristic of that model. This is also reflected in the delayed mixing of mean momentum shown at the same location in **Figure 16**. RANS model results show a relatively rapid increase in modeled turbulent kinetic energy downstream of flow separation. Both the DHRL and RANS results show a steady increase in turbulent kinetic energy similar to the experimental data up to station x/H = 5.0. The DDES simulation shows a relatively slow increase at stations x/H = 1.5 and 2.0, but a more rapid increase at x/H = 5.0, characteristic of a delayed but sudden breakup of the shear layer. Near the reattachment location at stations x/H = 6.0 and 7.0, the computed turbulent kinetic energy profiles decay similar to the experimental results, but the peak value of turbulent kinetic energy is underpredicted by the DHRL and RANS models results and significantly overpredicted by the DDES results. Farther downstream (stations x/H = 16.0 and 20.0), results obtained with all three models show qualitative agreement with the experimental data.

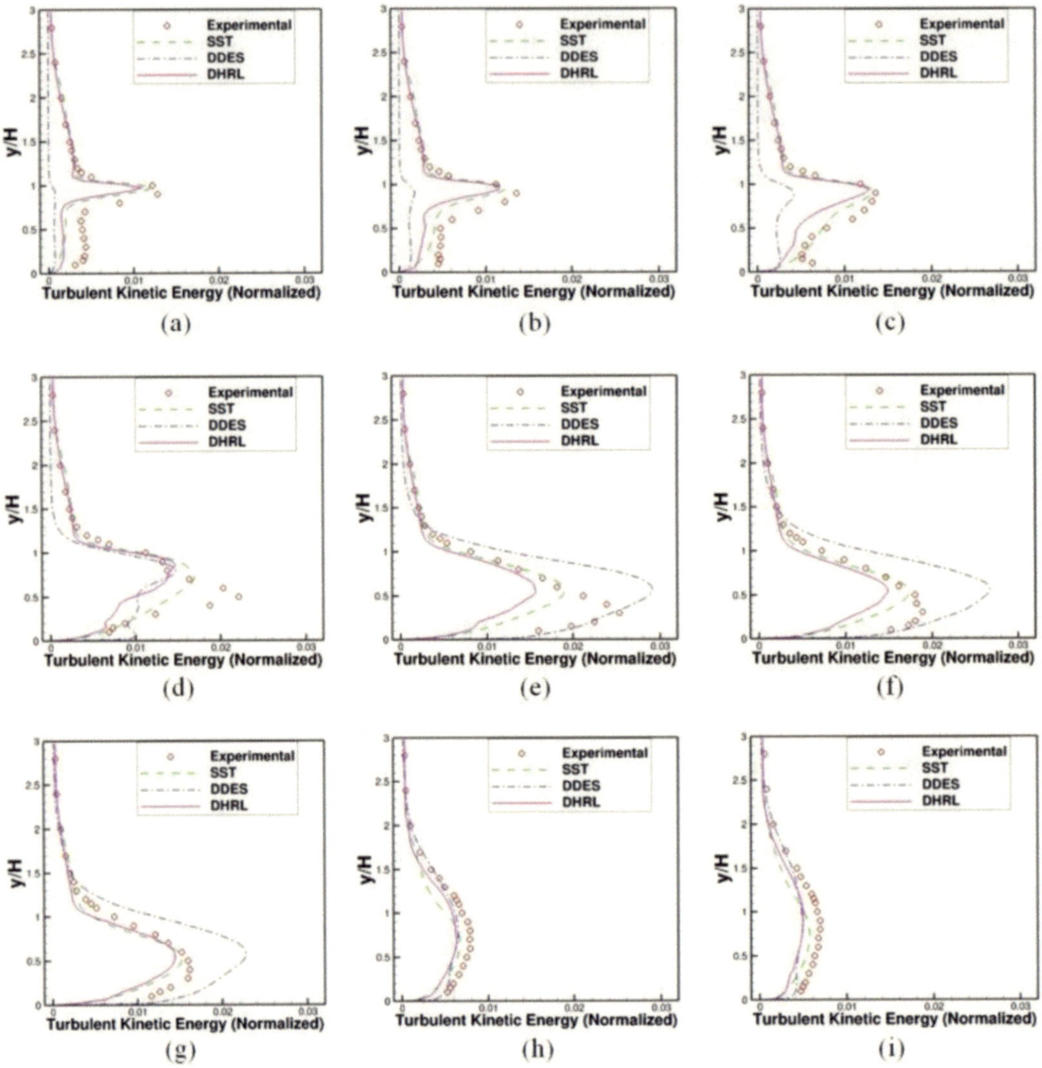

Figure 17. Predicted (Fine mesh) and measured turbulent kinetic energy profiles at measurement stations downstream of the step. (a) x/H = 1.0. (b) x/H = 1.5. (c) x/H = 2.0. (d) x/H = 3.0. (e) x/H = 5.0. (f) x/H = 6.0. (g) x/H = 7.0. (h) x/H = 16.0. (i) x/H = 20.0.

6. Summary and conclusions

The recently proposed dynamic hybrid RANS/LES modeling framework (DHRL) has been investigated for two test cases, turbulent channel flow and flow over a backward facing step, in order to investigate the capability of the model to accurately predict both attached and separated turbulent flows. The test cases elucidated the behavior of the DHRL model in comparison with a more traditional hybrid RANS-LES model, including model response to mesh refinement level and changes in discretization scheme.

For the channel flow test case, the DES and DHRL models showed significant resolved velocity fluctuations for all mesh refinement levels and for both discretization schemes, on the order of the turbulence levels in the reference DNS data. In contrast, the DDES model yielded a steady RANS-type solution, which is consistent with its intended performance. The DES model exhibited the well-known log-layer mismatch in the mean velocity, which was significantly reduced for the DHRL model. Furthermore, the DES model showed significant mesh dependence, with results improving as the mesh was refined, as expected. The DHRL model showed almost no mesh sensitivity with regard to the mean flow results, although the resolved turbulent kinetic energy levels, as well as the range of resolved scales of fluctuating motion, increased with increasing mesh refinement. The DDES model and SST models showed no mesh dependence due to the fact that each yielded a steady-state result. Both the DES and DHRL models showed almost no sensitivity to discretization scheme for the attached flow case.

For the backward facing step test case, the DHRL results showed rapid breakdown of the shear layer from RANS to LES type flow even for the Coarse mesh simulations, and relatively small differences in results on both the Coarse and Fine meshes, as well as for two different discretization schemes. The DDES model results showed a delayed breakdown of the shear layer and substantial mesh dependence, as well as nontrivial dependence of the results on the discretization scheme. The mean flow results predicted using the DHRL model were similar to the SST model results in the separated flow region, but showed improved prediction downstream of the reattachment point. The DHRL results were also comparable to SST results and superior to DDES results in terms of predicted turbulent kinetic energy downstream of the step, although it should be noted that for the SST model turbulence was completely modeled while for the DHRL model turbulent kinetic energy was resolved as unsteady velocity fluctuations. A key benefit to the DHRL model is the potential to effectively address grid dependence issues present for most currently used HRL models in predictions of the mean flow, while providing for increasing resolution of unsteady turbulence fluctuations while operating in LES mode as the grid size is reduced.

Overall, the results indicate that the DHRL model performs as designed. The attached flow case demonstrates a smooth variation between the RANS and LES zones, perpendicular to the flow direction, controlled by the model blending parameter α. Likewise, the separated flow case showed a similar behavior, with transition between RANS and LES regions occurring in the streamwise and wall-normal directions. The lack of explicit grid dependence in the DHRL model terms resulted in minimal grid dependence of the mean flow results, in contrast with the DES and DDES models. Furthermore, the DHRL model was shown to agree closely with the RANS component (SST) results for the attached flow case, while resolving significant levels of turbulent fluctuations in the boundary layer, at least on well-refined grids. Results indicate that the DHRL model may provide a more flexible and universal method for hybrid RANS-LES than existing standalone models.

Author details

Mohammad Faridul Alam[1], David Thompson[1] and Dibbon Keith Walters[2*]

*Address all correspondence to: d.keith.walters@ou.edu

1 Mississippi State University, Starkville, MS, USA

2 University of Oklahoma, Norman, OK, USA

References

[1] Spalart PR. Strategies for turbulence modelling and simulations. International Journal of Heat and Fluid Flow. 2000;**21**:pp. 252–263

[2] Probst A, Radespiel R, Wolf C, Knopp T, Schwamborn D. A Comparison of Detached-Eddy Simulation and Reynolds-Stress Modelling Applied to the Flow over a Backward-Facing Step and an Airfoil at Stall. AIAA Paper No. AIAA-2010-920. 2010

[3] Wilcox DC. Turbulence Modeling for CFD. 2nd ed. La Canada, California, 5354 palm Drive: DCW Industries Inc; 1998

[4] Dunn T, Loth E. Effects of Simulated-Spanwise-Ice Shapes on Airfoils: Computational Investigation. AIAA Paper No. AIAA-1999-0093. 1999

[5] Dunn T, Loth E, Bragg M. Computational investigation of simulated large-droplet ice shapes on airfoil aerodynamics. Journal of Aircraft. 1999;**36**:pp. 836–843

[6] Kumar S, Loth E. Aerodynamic simulations of airfoils with upper-surface ice-shapes. Journal of Aircraft. 2001;**38**:pp. 285–295

[7] Thompson S, Mogili P, Chalasani S, Addy H, Choo Y. A Computational Icing Effects Study for a Three-Dimensional Wing: Comparison with Experimental Data and Investigation of Spanwise Variation. AIAA Paper No. AIAA-2004-0561. 2004

[8] Hunt JCR. The structure of velocity and pressure fields in turbulent flow over bluff bodies, hills and waves. Journal of Wind Engineering and Industrial Aerodynamics. 1990;**36**:pp. 245–253

[9] Hirsch C, Tartinville B. Reynolds-Averaged Navier-Stokes modeling for industrial applications and some challenging issues. International Journal of Computational Fluid Dynamics. 2009;**23**:pp. 295–303

[10] Spille-Kohoff A, Kaltenbach HJ. Generation of turbulent inflow data with a prescribed shear stress profile. In: Liu, C., Sakell, L., Beutner, T. (eds.), DNS/LES Progress and Challenges. Columbus: Greyden Press; 2001. pp. 319–326

[11] Schluter JU, Pitsch H, Moin P. LES inflow conditions for coupling with Reynolds-averaged flow solvers. AIAA Journal. 2004;**42**:pp. 478–484

[12] Spalart PR, Jou WH, Stretlets M, and Allmaras SR. 1997, Comments on the feasibility of LES for wings and on the hybrid RANS/LES approach, In: Liu, C. and Liu, Z. (eds.), Advances in DNS/LES, Proceedings of the First AFOSR International Conference on DNS/LES, pp. 137–147, Greyden Press, Columbus

[13] Spalart PR, Deck S, Shur ML, Squires KD, Strelets M, Travin A. A new version of detached-eddy simulation, resistant to ambiguous grid densities. Theoretical and Computational Fluid Dynamics. 2006;**20**:pp. 191–195

[14] Shur ML, Spalart PR, Strelets M, Travin A. A hybrid RANS-LES model with delayed DES and wall-modeled LES capabilities. International Journal of Heat and Fluid Flow. 2008;**29**: pp. 1638–1649

[15] Spalart P. Young-Person's Guide to Detached Eddy Simulation Grids, NASA Contractor Report CR-2001-211032, U.S. National Aeronautics and Space Administration

[16] Paterson EG, Peltier LJ. Detached-eddy simulation of high-Reynolds-number beveled-trailing-edge boundary layers and wakes. ASME Journal of Fluids Engineering. 2005;**127**:pp. 897–906

[17] Nitikin NV, Nicoud F, Wasistho B, Squires KD, Spalart PR. An approach to wall modeling in large eddy simulations. Physics of Fluids. 2000;**12**:pp. 1629–1632

[18] Menter FR, Kuntz M, Bender R. A Scale-Adaptive Simulation Model for Turbulent Flow Predictions. AIAA Paper No. AIAA-2003-0767. 2003

[19] Menter FR, Egorov Y. The scale-adaptive simulation method for unsteady turbulent flow predictions. Part 1: Theory and model description. Flow Turbulence Combust. 2010;**85**: pp. 113–138

[20] Hamba F. A hybrid RANS/LES simulation of high-Reynolds-number channel flow using additional filtering at the interface. Theoretical and Computational Fluid Dynamics. 2006;**20**:pp. 89–101

[21] Piomelli U, Balaras E, Pasinato H, Squires KD, Spalart PR. The inner-outer layer interface in large-eddy simulations with wall-layer models. International Journal of Heat and Fluid Flow. 2003;**24**:pp. 538–550

[22] Celik I. RANS/LES/DES/DNS: The future prospects of turbulence modeling. ASME Journal of Fluids Engineering. 2005;**127**

[23] Menter FR. Two-equation eddy-viscosity turbulence models for engineering applications. AIAA Journal. 1994;**32**:pp. 1598–1605

[24] Fureby C, Grinstein FF. Monotonically integrated large eddy simulation of free shear flows. AIAA Journal. 1999;**37**:pp. 544–556

[25] Hoyas S, Jimenez J. Scaling of the velocity fluctuations in turbulent channels up to $Re\tau$ = 2003. Physics of Fluids. 2006;**18**:No. 011702

[26] Driver DM, Seegmiller HL. Features of a reattaching turbulent shear layer in divergent channel flow. AIAA Journal. 1985;**23**:pp. 163–171

[27] Bhushan S, Walters DK. A dynamic hybrid RANS/LES modeling framework. Physics of Fluids. 2012;**24**:No. 015103

[28] Lilly DK. A proposed modification of the Germano subgrid-scale closure method. Physics of Fluids. 1992;**4**:pp. 633–635

[29] Spalart P, Allmaras S. A One-Equation Turbulence Model for Aerodynamic Flows. AIAA Paper No. AIAA-92-0439. 1992

[30] Menter F. Review of the shear-stress transport turbulence model experience from an industrial perspective. International Journal of Computational Fluid Dynamics. 2009;**23** (4):pp. 305–316

[31] Boris JP, Grinstein FF, Oran ES, Kolbe RL. New insights into large eddy simulation. Fluid Dynamics Research. 1992;**10**(4–6):pp. 199–228

[32] Patankar SV, Spalding DB. A calculation procedure for heat, mass, and momentum transfer in three-dimensional parabolic flows. International Journal of Heat Mass Transfer. 1972;**15**:pp. 1787–1806

[33] Barth TJ, Jespersen DC. The Design and Application of Upwind Schemes on Unstructured Seshes. AIAA Paper No. AIAA-89-0366. 1989

[34] Rhie CM. and Chow WL. Numerical study of the turbulent flow past an airfoil with trailing edge separation. AIAA Journal. 1983;**21**:pp. 1525–1532

[35] Jasak H, Weller HG, Gosman AD. High resolution NVD differencing scheme for arbitrarily unstructured meshes. International Journal for Numerical Methods in Fluids. 1999;**31**:pp. 431–449

[36] Adedoyin AA, Walters DK, Bhushan S. Assessment of modeling and discretization error in finite-volume large eddy simulations. ASME Paper No. IMECE2006-14918. 2006

[37] Breuer M, Jovicic N, Mazaev K. Comparison of DES, RANS and LES for the separated flow around a flat plate at high incidence. International Journal for Numerical Methods in Fluids. 2003;**41**:pp. 357–388

Chapter 2

Numerical Analysis of Laminar-Turbulent Bifurcation Scenarios in Kelvin-Helmholtz and Rayleigh-Taylor Instabilities for Compressible Flow

Nikolay Mihaylovitch Evstigneev and
Nikolai Alexandrovitch Magnitskii

Additional information is available at the end of the chapter

http://dx.doi.org/10.5772/67918

Abstract

In the chapter, we are focused on laminar-turbulent transition in compressible flows triggered by Kelvin-Helmholtz (KH) and Rayleigh-Taylor (RT) instabilities. Compressible flow equations in conservation form are considered. We bring forth the characteristic feature of supersonic flow from the dynamical system point of view. Namely, we show analytically and confirm numerically that the phase space is separated into independent subspaces by the systems of stationary shock waves. Floquet theory analysis is applied to the linearized problem using matrix-free implicitly restarted Arnoldi method. All numerical methods are designed for CPU and multiGPU architecture using MPI across GPUs. Some benchmark data and features of development are presented. We show that KH for symmetric 2D perturbations undergoes cycle bifurcation scenarios with many chaotic cycle threads, each thread being a Feigenbaum-Sharkovskiy-Magnitskii (FShM) cascade. With the break of the symmetry, a 3D instability develops rapidly, and the bifurcations includes Landau-Hopf scenario with computationally stable 4D torus. For each torus, there exist threads of cycles that can develop chaotic regimes, so the flow is more complicated and difficult to study. Thus, we present laminar-turbulent development of compressible RT and KH instabilities as the bifurcations scenarios.

Keywords: laminar-turbulent transition, bifurcations in fluid dynamics, compressible flow bifurcations, Kelvin-Helmholtz instability, Rayleigh-Taylor instability, dynamical systems, direct numerical simulation, eigenvalue solver

1. Introduction and definitions

1.1. Introduction

In the chapter, we continue the work started in 2006 with the numerical analysis of laminar-turbulent transition using dynamical system approach. The results of the previous work are partially collected in [1]. One must point out that the bifurcation analysis of laminar-turbulent transition becomes a very interesting subject of research that opens a new way of looking at the problem of turbulence. This can be traced by the growing amount of publications in this field [2]. From mechanical point of view, one must study not only secondary flow (that bifurcates from the primary flow) but all other secondary flows and ways of their formation until they lead to the "turbulent" flow. At this point, the definition of "turbulence" must be introduced. It usually changes from paper to paper since there is no strict definition. For example, statistical definition is usually given in engineering papers and attractor-related definition is given in mathematical papers. We stick to the latter definition and introduce definitions after a short review of the progress in this field.

1.2. Problem overview

One of first ideas of bifurcation formation of complex flow structures originates from the 2D periodic forced flow problem by A.N. Kolmogorov [3], formulated in 1958 in the "Selected topics in analysis" seminar at Lomonosov Moscow State University. In 1960, the problem of the stability for the primary flow was solved by Meshalkin and Sinai [4] using chain fractions. It was stated that if the aspect ratio of the domain $\alpha = Y/X < 1$, then the flow is unstable. It was later confirmed [5, 6] that the flow is indeed unstable and forms oscillatory solutions. Analysis of laminar-turbulent transition was conducted later in Refs. [7, 8]. It was found that the system undergoes supercritical pitchfork bifurcation with further formation of limited cycle and torus through Hopf bifurcation. The flow complication continuous on invariant cycles (or tori if the symmetry is not exploited) cascades with Feigenbaum-Sharkovkskiy sequence up to the torus period three with sequential emergence of the chaos. Another classical problem is the Rayleigh-Benard convection problem. The linear stability was analyzed in 1916 by Lord Rayleigh himself [9]. The emergence of secondary stationary and oscillatory flows was later confirmed by many authors, e.g., [10–12]. Last two references contain bifurcation scenarios for 2D and 3D rectangular domains for laminar-turbulent transition ("frozen turbulence" in 2D case since stretching of vortex tubes is locked). See detailed monograph [13] for more information. Bifurcations, as laminar-turbulent transition mechanism, are also considered for ABC flow [14]. The appearance of Landau-Hopf bifurcations up to the stable invariant torus of co-dimension three with Feigenbaum bifurcations on two-dimensional invariant tori was observed. It is known that the formation of the turbulence occurs through bifurcations in Cuette-Taylor flow [15–18] and experimental bifurcation mode analysis in Ref. [19]. It is shown that the flow undergoes either subcritical or supercritical pitchfork bifurcation with further flow development through series of limited cycles or tori up to co-dimension two. Analytical research of the Cuette-Taylor problem was conducted in Ref. [18]. Nonlinear theory of secondary flow is constructed and its stability is analyzed. Bifurcation analysis of the flow over a backward-facing step problem is presented in Refs. [20–22]. One shows the appearance of stability loss due to the Taylor-Gortler instability (corresponds to the Hopf bifurcation) with further formation of the stable invariant three-dimensional torus. We were able to trace the formation of 3D torus of period two before chaos in [22]. It was also found that the bifurcation scenario of limited circles and limited two-dimensional and three-dimensional tori is an initial stage of transition to chaos in 2D Poiseuille flow [23, 24].

In this chapter, we concentrate on compressible flow features and instabilities triggered by Kelvin-Helmholtz (KH) and Rayleigh-Taylor (RT) mechanisms. Rayleigh-Taylor instability is the instability of an interface when a lighter fluid (gas) supports a heavier one in a gravitational field or some external potential. Direction of the potential gradient vector is from the heavier fluid to the lighter one. It can also occur when a lighter fluid pushes a heavier one. It was first studied experimentally by Lord Rayleigh [25] and later, theoretically, by Taylor [26]. Kelvin-Helmholtz instability is the instability of a shear layer (a tangential discontinuity for inviscid fluid) that can occur when there is velocity shear in a single continuous fluid or if there is a velocity difference across the interface between two fluids. It was initially studied by Helmholtz [27] and Lord Kelvin [28]. Those two instabilities are often considered together. Indeed, RT instability causes movement of adjusted fluids in different directions with the appearance of the shear layer that is subject to KH instability.

The problem of RT and KH instabilities was first attacked by linear stability analysis. We give brief information about it, following [29–31] and a highly recommended book [32]. The first stability conditions for the RT and KH instability were derived from the problem of interface stability of inviscid potential flow of two fluids. The setup is as follows: interface is at $y = 0$, flow properties are for $y > 0 : u_1, \rho_1$ and for $y < 0 : u_2, \rho_2$. Gravity vector is $\mathbf{g} = (0, -g, 0)^T$, directed downward. The analysis gives the following stability condition:

$$k^2\rho_1\rho_2(u_1 - u_2)^2 < kg(\rho_2^2 - \rho_1^2), \tag{1}$$

where k is a wavenumber. It is clear, that if $u_1 \neq u_2$ then there exists such $k = k_{crit}$, that the flow is unstable and KH instability always occurs (including cases with $\rho_1 = \rho_2$ and $g = 0$). If $u_1 = u_2$, then RT instability occurs if $\rho_1 > \rho_2$, e.g., the lighter fluid supports the heavier one. This result is obtained from very simple problem formulation that is merely adequate. In most cases of linear analysis, Squire's theorem [32–34] is applied in order to drive the problem of linear analysis of a flat-parallel 3D flow to a 2D flow analysis. The first improvement of the result (Eq. (1)) for KH is the consideration of boundary condition problem instead of initial-condition one (e.g., flow of two co-directional fluids with the shear). In this case, one can derive the critical wavelength ($\lambda = 2\pi/k$, $\lambda_{crit} = 2\pi/k_{crit}$) for instability [35] to be:

$$\lambda_{crit} = \frac{\pi\rho_0(u_1 - u_2)^2}{g(\rho_2 - \rho_1)}\left[1 + \frac{u_1 + u_2}{\sqrt{2(u_1^2 + u_2^2)}}\right]^{-1} \tag{2}$$

It is clear that in this case, the critical wavelength will be longer than for the initial value problem case. The necessary instability condition for the stratified shear flow instability is derived from the energy balance equation and consideration of continuous shear velocity and density profiles [32]. Then, from the stream-function formulation, one derives Taylor-Goldstein equation, which governs the vertical structure of the wave perturbation as a function of density gradient. The condition for instability is balanced with the velocity vertical gradient (both of them varying linearly in the mixing height H), resulting in the following necessary condition:

$$Ri = \frac{gH(\rho_2 - \rho_1)}{\rho_0(u_1 - u_2)^2} < \frac{1}{4}, \tag{3}$$

where Ri is called Richardson number, [36]. Extension of the analysis is dedicated to the viscous KH or RT instabilities. In this case the mode analysis shows [37–39], that the neutral curve for KH instability may have shorter regions of stability, depending on the values of viscosity and wavenumber. It is interesting to note that viscosity may play a destabilizing role[38], where air-water stratified system is considered. This analysis usually done, again, by introducing Squire's reduction, showing that the most dangerous three-dimensional disturbance is flat (two-dimensional). Inclusion of the surface tension additionally stabilizes the problem [38] in the region of small wavenumbers; however, this subject is beyond the scope of the chapter. In the later stages of instability, viscosity stabilizes the flow. The RT instability remains unstable under inverse condition (Eq. (1)) despite the action of viscosity.

The RT and KH instabilities in compressible fluid are more complex to study. One of the first results of KH instability is presented in [33] for ideal gas with basic thermodynamic state characterized by the constant speed of sound $c = \gamma P/\rho$, where P is pressure, ρ is density, and γ is the adiabatic index (ratio of specific heats). It is noted that there are three types of disturbances associated with the eigenvalues of the linearized operator: subsonic ($M < 1$), sonic ($M = 1$), and supersonic ($M > 1$), where Mach number $M = u/c$, u is the velocity of the basic parallel flow, i.e., $u = u_1 - u_2$. Supersonic and sonic disturbances are responsible for another type of instability called Richtmyer-Meshkov instability that is not considered here. So, considering subsonic disturbances, it was noticed that for a wavenumber k, the increase of M, complex kM increases, and it leads to the decrease of such wavemodes that are linearly unstable. It is shown that the Fjortoft's theorem [40] (let there exists a point y_s, that $u''(y_s) = 0$. If the flow is unstable, then $u''(u - u(y_s)) < 0$ for some point $y \in [y_{min}, y_{max}]$) also presents sufficient condition for stability to subsonic disturbances. Results of eigenmodes and growth rates for KH instabilities for compressible flow are presented in [33, 41, 42].

The inviscid incompressible RT instability gives exponential growth of the amplitude ($\theta(k)$) of infinitesimal perturbations [29]:

$$\theta(k) \sim e^{\alpha(\rho,k,g)t}, \alpha(\rho, k, g) = \sqrt{A(\rho)gk}, \tag{4}$$

with

$$A = \frac{\rho_2 - \rho_1}{\rho_2 + \rho_1}, \tag{5}$$

where α is the temporal growth rate, k is the spatial wavenumber, and A is called the Atwood number. The growth rate is an important property of such instabilities. It can be estimated as the magnitude of the least stable eigenvalue. The growth rate of a linearized system can be detected experimentally or numerically while the secondary solution remains stable. With further complication of the flow, i.e., development of solution bifurcating branches, the value of growth rate changes. In accordance with Eq. (4), the growth of modes is more intensive for high k. Inclusion of viscosity μ in the model results in the formation of maximum growth rate for specific $k(\mu)$ with $\theta(k) \to 0$ with $k \to \infty$. So the growth rate and the mixing layer length become functions of time that can be compared with experimental or numerical results. The mixing length is a function of initial wavenumber disturbance, Atwood number, gravity constant, and diffusion coefficient.

Compressibility effects of the RT instability include both stabilizing and destabilizing effects compared to the incompressible flow [31, 43–46]. For infinite domains, the growth rate $\theta(k)$ obtained for the compressible case is bounded by the growth rates obtained for the corresponding incompressible flows. This result is not affected by the presence of surface tension or viscosity. As the speed of sound is increased, the rate of growth increases toward the value obtained for the corresponding constant density incompressible flow, while as γ increases, $\theta(k)$ decreases toward the value obtained for the corresponding incompressible flow with exponentially varying density. Another effect of compressibility is the decrease of the average local Atwood number for onset of instability. It is interesting to note [47] that the growth rate for compressible RT instability of two immiscible inviscid barotropic fluids was found to be $e^{t\sqrt{|k|}}$. In addition, solutions that grow arbitrarily quickly in the Sobolev space W_2^k, which leads to an ill-posedness result for the linearized problem were presented in [47]. Here, it is shown that the nonlinear problem does not admit reasonable estimates of solutions for small time in terms of the initial data. The compressible inviscid RT instability must be regularized, e.g., with viscosity.

There are also some papers on the bifurcation analysis of RT and KH instabilities. Very good experimental results for RT instability are presented in [48], where stabilization is performed using zimuthally rotating magnetic field for ferrofluids. A subcritical pitchfork bifurcation was found during the stability loss, where this bifurcation depended on the magnetic field strength as a parameter. Furthermore, it is shown on phase space portrait (**Figure 7** in [48]) that further increase of complexity goes through Andronov-Hopf bifurcation. Another paper containing some information on bifurcation for KH instability is [49], where authors obtained reliable results (DNS and eigenvalue analysis) confirming the emergence of self-oscillatory solution. Hence, Andronov-Hopf bifurcation for KH instability is confirmed. Another paper on KH bifurcation analysis in MHD flow is presented in Ref. [50], where Andronov-Hopf bifurcation was found.

Finally, we can conclude the following. Initial stability loss is well studied for RT and KH instabilities. One can use such data as stability criteria, growth rates, and eigenmodes to verify the linear analysis in numerical methods. There is evidence of bifurcation route to chaos in RT and KH instabilities, but this analysis is very complex and only initial stages (pitchfork and Andronov-Hopf bifurcations) are observed. The subject of this chapter is the analysis of compressible flow from the dynamical systems point of view and the attempt to bring together data on the bifurcation analysis of laminar-turbulent transition of KH and RT instabilities.

1.3. Definitions

Let us introduce some definitions and notations that are required for the analysis of laminar-turbulent transition from the nonlinear dynamic system point of view. For some of these definitions, we are using [51, 52]. We are considering a system (infinite-dimensional, in general):

$$\frac{\partial \mathbf{u}(\mathbf{x}, t)}{\partial t} = \mathbf{F}(\mathbf{u}, \mathbf{x}, \mathbf{R}, t). \tag{6}$$

Let $\mathbf{x} \in \mathbb{M} \in X$, t is time, $\mathbb{M}$ is a phase space, and $\varphi^t(\mathbf{x}) : \mathbb{M} \times \mathbb{R} \to \mathbb{M}$ is a phase flow on $\mathbb{M}$ and $\mathbf{R}$ is a vector of parameters (e.g., similarity complexes in case of fluid dynamics) of size Rp. Let $\gamma(t) \in \mathbb{M}$ be a phase trajectory of a solution of Eq. (6).

Definition 1. *Compact set $B \in \mathbb{M}$, invariant with respect to φ^t is an attracting set if there exists a neighborhood U of the set B such that for almost all $\mathbf{x} \in U$, $\varphi^t(\mathbf{x}) \to B$, $t \to \infty$.*

Definition 2. *Attractor of the system* (6) *is an indecomposable attracting set, [51, 52].*

Definition 3. *Regular attractors of the system* (6) *are stable singular points, stable invariant limited cycles, and stable invariant tori of any dimension.*

Definition 4. *Singular attractor of Eq.* (6) *is an attractor that is not regular.*

Definition 5. *System* (6) *has a chaotic solution if and only if it has a singular attractor.*

Definition 6. *Bifurcation points of the system* (6) *are those and only those points in which there is no continuous dependence of the phase portrait of the system by its parameters.*

Linearize the system (4) on its regular attractor $\mathbf{u}_0(\mathbf{x}, t)$ or the parameter value $\mathbf{R}_0$:

$$\frac{\partial \mathbf{y}(\mathbf{x}, t)}{\partial t} = \mathrm{L}(\mathbf{u}_0, \mathbf{x}, t, \mathbf{R}_0)\mathbf{y}(\mathbf{x}, t),$$

where $\mathrm{L} : \mathbb{M} \to \mathbb{M}$ is a linear operator and

$$\mathbf{y}(\mathbf{x}, t) = \mathbf{u}_0(\mathbf{x}, t) - \mathbf{u}(\mathbf{x}, t).$$

If the spectrum of operator L lies in the left complex half-plane when $\forall j \in Rp$, $R_j \leq R_{0,j}$, and if $\exists j$, $R_j > R_{0,j}$ with one or few its eigenvalues that are moving in the right half-plane, then the system (6) has a bifurcation at $\mathbf{R}_0$.

Definition 7. *The infinite sequence of bifurcations: $\forall j \in Rp$, $R_{j,n} \to R_{j,*}$, $n \to \infty$ is known as the cascade of bifurcations.*

Any singular attractor is a limit of a cascade of bifurcations of regular attractors. For example, the simplest singular Feigenbaum attractor is a nonperiodic trajectory, which is the limit of the cascade of period doubling bifurcations of stable cycles. So, any cascade of bifurcations is known as a scenario of transition to chaos.

Definition 8. *Let the trajectories of the Navier-Stokes dynamical system in the phase space be located at least on one singular attractor (there can be more attracting states since the system exhibits multistability phenomenon), then such solution of the system is called turbulent.*

2. Governing equations and numerical methods

2.1. Governing equations

We are considering conservation laws for viscous gas dynamics. This system can be written in the following differential form [53]:

$$\begin{gathered}\frac{\partial \rho}{\partial t}+\frac{\partial}{\partial x_j}\left[\rho u_j\right]=0\\ \frac{\partial}{\partial t}(\rho u_i)+\frac{\partial}{\partial x_j}\left[\rho u_i u_j+p\delta_{ij}-\tau_{ji}\right]=\rho g_i, \quad \{i,j,k\}=1,2,3;\\ \frac{\partial}{\partial t}(\rho E)+\frac{\partial}{\partial x_j}\left[\rho u_j E+u_j p-u_i\tau_{ij}\right]=\rho u_k g_k,\\ \rho E=\frac{1}{2}\rho\mathbf{u}^2+\rho e,\\ p=(\gamma-1)(E-1/2\rho\mathbf{u^2}).\end{gathered} \tag{7}$$

Here, we are considering some bounded domain $\Omega\subset\mathbb{R}^3$ with boundary $\partial\Omega$; scalar functions f are defined as $f:\Omega\times[0,\mathcal{T}]\to\mathbb{R}$; vector-functions $\mathbf{f}$ are defined as $\mathbf{f}:\Omega\times[0,\mathcal{T}]\to\mathbb{R}^3$, where $\mathcal{T}$ is some defined finite time. Then, ρE is a scalar function of the full gas energy; e is a scalar function of the internal gas energy; γ is the adiabatic index (value 1.4 is used for calculations); p is a scalar pressure function; $\mathbf{u}$ is a gas velocity vector function; ρ is a scalar function of gas density; $\mathbf{g}$ is an external forcing. We assume the Einstein summation rule. We also assume that the gas is Newtonian with the following viscous tensor:

$$\tau_{ij}=2\mu S_{ij}, \tag{8}$$

where μ is a dynamic gas viscosity and strain rate tensor is defined as:

$$S_{ij}=\frac{1}{2}\left(\frac{\partial u_i}{\partial x_j}+\frac{\partial u_j}{\partial x_i}\right)-\frac{1}{3}\delta_{ij}\frac{\partial u_k}{\partial x_k}. \tag{9}$$

Please note that we ignore secondary viscosity coefficient due to complexity. We use the integral form of Eq. (7) for problems that may have discontinuous solutions. We use the main bifurcation parameter that is considered in the paper—the Reynolds number. It is found as:

$$R=\frac{L_C||\mathbf{u}_C||\rho_C}{\mu}, \tag{10}$$

where L_C is the characteristic length scale of the problem, $||\mathbf{u}_C||$ is the norm of the characteristic velocity vector, ρ_C is a characteristic density. With this in mind, we can rewrite Eq. (8) as follows:

$$\tau_{ij} = \frac{2}{R} S_{ij}. \tag{11}$$

We will use Eq. (8) if we use the term "viscosity coefficient" and we use Eq. (11) if we use the term "Reynolds number".

This method of analysis is based on the construction of phase space and Poincaré sections. The method is analogous to [51, 1]. In order to construct multidimensional Poincaré sections, we take data from different points of the physical space and use slices with the given gap thickness δ. For all results presented here, we use $\delta = 5 \cdot 10^{-7}$.

2.2. Numerical methods

In this section, we give a brief description of numerical methods that are used to perform simulation of problems and bifurcation analysis. We are forced to relay to the numerical methods since one can basically derive only primary or secondary solutions analytically. All other bifurcations of the main solution cannot be derived.

2.2.1. *Main flow solver*

The flow solver is constructed using finite volume-finite element method. The inviscid part is solved using modified WENO schemes of order 7 or 9; for more information, see [54]. The Riemann solver is a Godunov exact solver, see [55]. For viscous part of equations, we use finite element nodal method that is described for Poisson equation in Ref. [12]. Time integration is explicit (WENO)-implicit (FEM) method of the third order. The method is implemented on multiGPU architecture. The efficiency of the method and specific features of implementation are outlined in Ref. [56]. The inflow and outflow boundary conditions are set using characteristic reconstruction with damping zones; for more information see Ref. [57]. All discretization of the problems is done using results of modified linear analysis for WENO schemes and "burglization" numerical experiments, provided in Ref. [58].

2.2.2. *Eigenvalue solver*

In order to study the linear stability of the problem, we are using implicitly restarted Arnoldi (IRA) method designed to be used on multiGPU architecture. We use matrix-free method to call the main solver and analytical linearization of governing equations in order to construct Jacobi matrix-vector operator. For troublesome problems, we apply either exponential or Cayley transformations. See [12, 59] for more details on implementation features and application examples. In all calculations, unless specified otherwise, we use $k = 10$ target eigenvalues with additional Krylov vectors totaling $m = 8$, i.e., Krylov subspace dimension is $\dim(\mathcal{K}) = k \cdot m = 80$.

http://dx.doi.org/10.5772/67918

3. Verification

Verification of the main numerical scheme is conducted in [55, 58], and IRA eigensolver is verified in Ref. [59]. In this section, we present verification results for KH and RT instabilities.

3.1. KH instability verification

The primary problem that can be used as a benchmark is the numerical calculation of the growth rate. It can be done using IRA eigenvalue solver for the periodic KH problem in X-direction and nonreflecting boundary condition in Y-direction with $\Omega = [L \times H]$ and initial mixing layer thickness δ. This problem originated from [33] and was numerically calculated in Refs. [41, 42]. The domain length in X-direction should fit at least one entire wavelength of the desired mode and was chosen, analogously to [42], to be exactly one wavelength of the least stable mode scaled by mixing layer thickness δ. Assuming a wavenumber k of the least stable mode, the length of the domain in X-direction would result as $L = 2\pi\delta/k$. The domain length in Y-direction is chosen as $H = 8L$.

$$\begin{aligned} \rho_0 &= 1.24, \\ u(x,y)_0 = u(y)_0 &= \tanh\left(\frac{y - L/2}{\delta}\right). \end{aligned} \tag{12}$$

Pressure is chosen as to satisfy Mach number that will vary depending on the particular problem.

The solution is performed on 32×256 grid with $k = 8$ and $m = 5$ IRA parameters, that results in the $dim(\mathcal{K}) = 40$. The results are presented in **Table 1** for different Mach numbers and modulus of the least stable eigenvector is presented in **Figure 1**. We can see good convergence properties of the growth rate for all least stable modes. The eigenvector clearly indicates the formation of two symmetric billows.

M	Computed α	Analytical α [33]	α [41]	α [42]
0.1	0.187312	0.187	0.188	0.189
0.5	0.1411	0.141	0.1411	0.1411
0.9	0.054743	0.055	0.054723	0.0547

Table 1. Results of the validation for growth rates α as functions of Mach number.

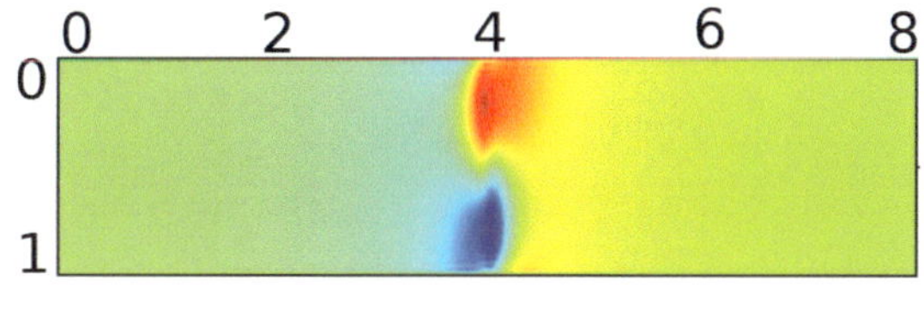

Figure 1. Modulus of the least stable normalized eigenvector of total energy for KH 2D test problem. The problem is rotated 90° CW.

Another test case considered is with regard to [60], where a 2D periodic KH problem is simulated with very high accuracy. For this purpose, benchmark code in [60] was used with 4096×4096 grid points. The problem is set as KH instability of shear flow in rectangular unit domain Ω with periodic boundary conditions in all directions. Initial conditions are set as:

$$\rho(\mathbf{x},0)=\begin{cases}\rho_1-\rho_m e^{\frac{y-1/4}{L}}; 1/4>y\geq 0,\\ \rho_2+\rho_m e^{\frac{-y+1/4}{L}}; 1/2>y\geq 1/4,\\ \rho_2+\rho_m e^{-\frac{3/4-y}{L}}; 3/4>y\geq 1/2,\\ \rho_1-\rho_m e^{-\frac{y-3/4}{L}}; 1>y\geq 3/4,\end{cases}$$
$$u_x(\mathbf{x},0)=\begin{cases}U_1-U_m e^{\frac{y-1/4}{L}}; 1/4>y\geq 0,\\ U_2+U_m e^{\frac{-y+1/4}{L}}; 1/2>y\geq 1/4,\\ U_2+U_m e^{-\frac{3/4-y}{L}}; 3/4>y\geq 1/2,\\ U_1-U_m e^{-\frac{y-3/4}{L}}; 1>y\geq 3/4\end{cases} \tag{13}$$

and

$$u_y(\mathbf{x},0)=0.01\sin(4\pi x), p(x,y,0)=2.5. \tag{14}$$

Here $U_m=(U_1-U_2)/2$, $\rho_m=(\rho_1-\rho_2)/2$, $\rho_1=1$, $\rho_2=2$, $U_1=1/2$, $U_2=-1/2$, $L=0.025$.

For our simulation, we use a 3D version of the problem and set up grid 256×256×256 such that $u_z(\mathbf{x},0)=0$ with periodic boundary conditions in z-direction. The problem is executed until $T=1.5$ is reached. We also compare these results with spectral code that is used for incompressible Navier-Stokes simulation. We monitor the maximum y-direction kinetic energy of perturbations, as $k_y(t)=\max_{\mathbf{x}\in\Omega}\rho u_y^2/2$, and we monitor the growth rate of the amplitude for the unstable u_y mode using an FFT. The results are presented in **Figure 2**. One can see a very reasonable agreement with reference data. Small difference in the evolution of growth rate α_y in the initial stage can be explained by the usage of a 3D code instead of 2D as in reference data.

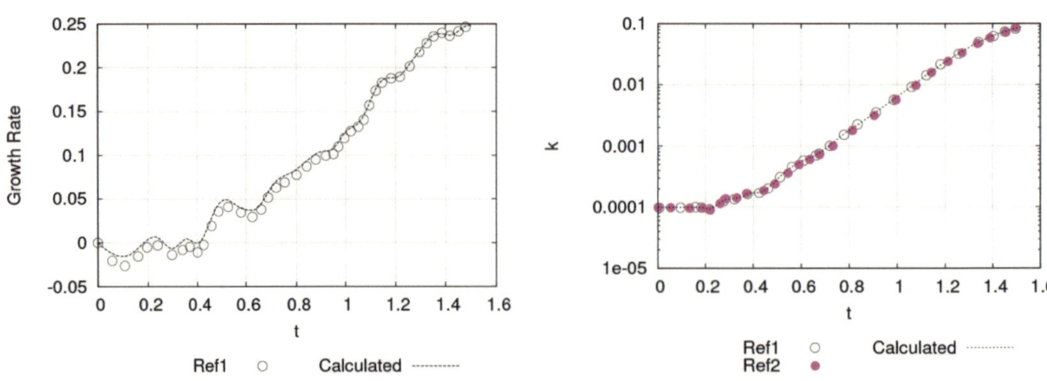

Figure 2. Growth rate α_y and maximum y-direction kinetic energy k_y. Ref1 is [60] and Ref2 is data from spectral code.

3.2. RT instability verification

Verification data on RT instability are freely available in the literature, e.g., [61–67] and many others. One of the benchmarks is the turbulent mixing layer length. It is known that the RT process is basically divided into three stages: (1) linear stage related to the linear stability analysis where initial perturbations $\theta(k)$ are growing exponentially as in Eq. (4); (2) nonlinear stage where the formation of spikes and bubbles occurs. In this stage bubbles are traveling toward heaver fluid and spikes, toward the lighter one. Growth rate of bubbles is proportional to $\theta_B \sim \sqrt{2\pi g/k}t$, and growth rate of spikes is proportional to $\theta_J \sim gt^2$; (3) stage of dissipation of regular structures. In this stage, KH instability is developing on the contacts moving with shear velocity and destabilization of lighter bubbles in heavier fluid due to the secondary RT instability. The onset of turbulence depends on the initial perturbation wavenumber and viscosity coefficient.

We follow the problem setup in [67] and investigate the mixing length L_m as a function of time with $L_m = \alpha g A g t^2$, where α is a constant. We consider the domain $\Omega = [3 \times 1 \times 1]$ with gravity direction in $-x$, and we use $900 \times 300 \times 300$ number of elements. We test two setups with random and sinusoidal initial perturbations. The latter results are presented in **Figure 3**.

In **Figure 4**, we can see the mixing length for random perturbations and for sinusoidal perturbations with comparison to the literature on different length of waves. One can see that the

Figure 3. Density for RT instability test case with $A = 1/3$ and harmonic initial disturbances: $\tau_{Left} = 0.3$ s, $\tau_{Right} = 1.9$ s.

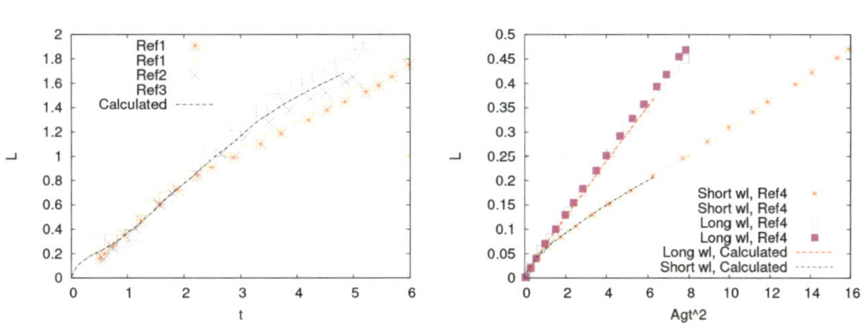

Figure 4. Mixing length of RT instability for random (left) and sinusoidal (right) initial perturbations. Ref 1,2—[67], Ref 3—[61, 63], Ref4—[62].

results are close to the ones in the literature. We identified the constants α for sinusoidal perturbations: $\alpha_{short} = 0.0271$ for short wavelength and $\alpha_{short} = 0.057$ for long wavelength. These results are very close to those provided in the literature.

4. Bifurcation analysis

4.1. Phase space for supersonic flow

4.1.1. Inviscid flow phase space properties

Let us consider the following vector form of inviscid system (7) for 2D case:

$$\begin{gathered}\frac{\partial \mathbf{U}}{\partial t} + \frac{\partial \mathbf{F}}{\partial x} + \frac{\partial \mathbf{G}}{\partial y} = \mathbf{0}, \\ \mathbf{U} = [\rho; \rho u_x; \rho u_y; \rho E]^T, \\ \mathbf{F} = [\rho u_x; \rho u_x^2 + p; \rho u_x u_y; (\rho E + p) u_x]^T, \\ \mathbf{G} = [\rho u_y; \rho u_y u_x; \rho u_y^2 + p; (\rho E + p) u_y]^T.\end{gathered} \tag{15}$$

The system can be linearized:

$$\begin{gathered}\frac{\partial \mathbf{U}}{\partial t} + \mathrm{A}\frac{\partial \mathbf{U}}{\partial x} + \mathrm{B}\frac{\partial \mathbf{U}}{\partial y} = \mathbf{0}, \\ \mathrm{A} = \partial \mathbf{F}/\partial \mathbf{U}; \mathrm{B} = \partial \mathbf{G}/\partial \mathbf{U},\end{gathered} \tag{16}$$

with eigenvalues being:

$$\lambda_1 = u \cdot n - c, \lambda_2 = u \cdot n, \lambda_3 = u \cdot n, \lambda_4 = u \cdot n + c, \tag{36}$$

where $c = \sqrt{\frac{\gamma p}{\rho}}$ is the speed of sound, n is a unit directional vector. All eigenvalues for all x, y are real since system (16) is hyperbolic. We are considering a special case with the base flow being parallel to x axis, i.e., $\mathbf{u} = (u_0, 0)^T$ and $\mathbf{n} = (1, 0)^T$. In this case, the system of eigenvectors becomes one dimensional, and we can reduce Eq. (16) to one-dimensional case. Let us first consider a simple initial value problem for system of advection equations:

$$\begin{gathered}\mathbf{U}_t + \mathbf{A}\mathbf{U}_x = 0, \\ x \in \mathbb{R}; t \in \mathbb{R}^+, \\ \mathbf{U} = (u_1, u_2)^T, \\ u_1(x,0) = 1, u_2(x,0) = \begin{cases} e^{ikx}; x \geq 0 \\ 1; x < 0. \end{cases},\end{gathered} \tag{17}$$

where k is a wavenumber.

Lemma 1: *Let all eigenvalues* $\lambda_j \in \mathbb{R}$ *of* A *in* Eq. (17) *satisfy* $\lambda_j \geq 0$. *Then,* u_1 *and* u_2 *are constant for* $x < 0, \forall t \in \mathbb{R}^+$.

Proof: Let linear operator $\mathbb{A} = \begin{pmatrix} a & b \\ c & d \end{pmatrix}$; $a, b, c, d \in \mathbb{R}$. Eigenvalues and eigenvectors are found as:

$$\lambda_{1,2} = \begin{pmatrix} 1/2d + 1/2a + 1/2\sqrt{(d-a)^2 + 4cb} \\ 1/2d + 1/2a - 1/2\sqrt{(d-a)^2 + 4cb} \end{pmatrix},$$
$$H_{1,2} = \begin{pmatrix} \dfrac{2b}{d - a + \sqrt{(d-a)^2 + 4cb}}; & \dfrac{2b}{d - a - \sqrt{(d-a)^2 + 4cb}} \\ 1 & 1 \end{pmatrix} \tag{18}$$

Let us assume that $\lambda_1 > 0, \lambda_2 \geq 0$. We apply characteristic variables $W_0 = H^{-1}U_0$:

$$\begin{pmatrix} W_{0,1} \\ W_{0,2} \end{pmatrix} = \begin{pmatrix} \begin{cases} \dfrac{1}{4}\dfrac{(d-a+\sqrt{(d-a)^2+4cb})(-d+a+\sqrt{(d-a)^2+4cb}+2b)}{b\sqrt{(d-a)^2+4cb}}; x < 0. \\ \dfrac{1}{4}\dfrac{(d-a+\sqrt{(d-a)^2+4cb})(-d+a+\sqrt{(d-a)^2+4cb}+2b\cdot e^{ikx})}{b\sqrt{(d-a)^2+4cb}}; x \geq 0. \end{cases} \\ \begin{cases} \dfrac{1}{4}\dfrac{(d-a-\sqrt{(d-a)^2+4cb})(-d+a-\sqrt{(d-a)^2+4cb}+2b)}{b\sqrt{(d-a)^2+4cb}}; x < 0. \\ \dfrac{1}{4}\dfrac{(d-a-\sqrt{(d-a)^2+4cb})(-d+a-\sqrt{(d-a)^2+4cb}+2b\cdot e^{ikx})}{b\sqrt{(d-a)^2+4cb}}; x \geq 0. \end{cases} \end{pmatrix} \tag{19}$$

Matrix $\mathbb{A}$ is diagonalized as $\Lambda = H^{-1}AH$. Then, we can rewrite (17) as:

$$H^{-1}\mathbf{U}_t + H^{-1}A\mathbf{U}_x = \mathbf{0};$$
$$\mathbf{\Lambda} = diag(\lambda_i);$$
$$\mathbf{W}_t + \mathbf{\Lambda}\mathbf{W}_x = \mathbf{0},$$

and the last vector equation can be separated:

$$(W_i)_t + \lambda_i (W_i)_x = 0; i = 1, 2.$$

Then characteristic solution is:

$$W_i = W_{0,i}(x - \lambda_i t), i = 1, 2.$$

with the application of initial data (Eq. (19)):

$$\begin{pmatrix} W_1 \\ W_2 \end{pmatrix} = \left(\begin{cases} A+; \left(x - \dfrac{d+a+\sqrt{(d-a)^2+4cb}}{2}t \right) < 0. \\ B+; \left(x - \dfrac{d+a+\sqrt{(d-a)^2+4cb}}{2}t \right) \geq 0. \end{cases} \quad \begin{cases} A-; \left(x - \dfrac{d+a-\sqrt{(d-a)^2+4cb}}{2}t \right) < 0. \\ B-; \left(x - \dfrac{d+a-\sqrt{(d-a)^2+4cb}}{2}t \right) \geq 0. \end{cases} \right)$$

$$A\pm = \frac{1}{4} \frac{\left(d - a \pm + \sqrt{(d-a)^2+4cb}\right)\left(-d + a \pm \sqrt{(d-a)^2+4cb} + 2b\right)}{b\sqrt{(d-a)^2+4cb}},$$

Here:

$$B\pm = \frac{1}{4} \frac{\left(d - a \pm \sqrt{(d-a)^2+4cb}\right)\left(-d + a \pm \sqrt{(d-a)^2+4cb} + 2b \cdot e^{ik\left(x - \frac{d+a\pm\sqrt{(d-a)^2+4cb}}{2}t\right)}\right)}{b\sqrt{(d-a)^2+4cb}}, \tag{20}$$

Now we turn to conservative variables $\mathbf{U} = H\mathbf{W}$:

$$\begin{pmatrix} u_1 \\ u_2 \end{pmatrix} = \begin{pmatrix} \dfrac{2b}{d-a+\sqrt{(d-a)^2+4cb}} W_1 + \dfrac{2b}{d-a-\sqrt{(d-a)^2+4cb}} W_2 \\ W_1 + W_2 \end{pmatrix}. \tag{21}$$

It is clear from the obtained solution that eigenvectors in Eq. (21) are independent of $x - \lambda_i t$. So, the solution can be reformed as:

$$\begin{pmatrix} u_1 \\ u_2 \end{pmatrix} = \begin{pmatrix} \eta_1 \begin{cases} W_1 \\ W_1(e^{ik(x-\lambda_1 t)}) \end{cases} + \eta_2 \begin{cases} W_2 \\ W_2(e^{ik(x-\lambda_2 t)}) \end{cases} \\ \begin{cases} W_1 \\ W_1(e^{ik(x-\lambda_1 t)}) \end{cases} + \begin{cases} W_2 \\ W_2(e^{ik(x-\lambda_2 t)}) \end{cases} \end{pmatrix}, \tag{22}$$

where η_1, η_2 are values in Eq. (21) that are independent of x and t: $\eta_1 = \frac{2b}{d-a+\sqrt{(d-a)^2+4cb}}$, $\eta_2 = \frac{2b}{d-a-\sqrt{(d-a)^2+4cb}}$. Let us consider two possible cases:

1. $\lambda_1 > 0, \lambda_2 = 0$

$$
\begin{aligned}
&1. x \geq \lambda_1. \Rightarrow \begin{pmatrix} u_1 \\ u_2 \end{pmatrix} = \begin{pmatrix} \eta_1 W_1(e^{ik(x-\lambda_1 t)}) + \eta_2 W_2(e^{ikx}) \\ W_1(e^{ik(x-\lambda_1 t)}) + W_2(e^{ikx}) \end{pmatrix}. \\
&2. x > 0, x < \lambda_1. \Rightarrow \begin{pmatrix} u_1 \\ u_2 \end{pmatrix} = \begin{pmatrix} \eta_1 W_1 + \eta_2 W_2(e^{ikx}) \\ W_1 + W_2(e^{ikx}) \end{pmatrix}. \\
&3. x \leq 0. \Rightarrow \begin{pmatrix} u_1 \\ u_2 \end{pmatrix} = \begin{pmatrix} \eta_1 W_1 + \eta_2 W_2 \\ W_1 + W_2 \end{pmatrix}.
\end{aligned}
\tag{23}
$$

2. $\lambda_1 > 0, \lambda_2 > 0, \lambda_1 \geq \lambda_2$

$$
\begin{aligned}
&1. x \geq \lambda_1. \Rightarrow \begin{pmatrix} u_1 \\ u_2 \end{pmatrix} = \begin{pmatrix} \eta_1 W_1(e^{ik(x-\lambda_1 t)}) + \eta_2 W_2(e^{ik(x-\lambda_2 t)}) \\ W_1(e^{ik(x-\lambda_1 t)}) + W_2(e^{ik(x-\lambda_2 t)}) \end{pmatrix}. \\
&2. \lambda_1 \geq x \geq \lambda_2. \Rightarrow \begin{pmatrix} u_1 \\ u_2 \end{pmatrix} = \begin{pmatrix} \eta_1 W_1 + \eta_2 W_2(e^{ik(x-\lambda_2 t)}) \\ W_1 + W_2(e^{ik(x-\lambda_2 t)}) \end{pmatrix}. \\
&3. x \geq 0, x < \lambda_2. \Rightarrow \begin{pmatrix} u_1 \\ u_2 \end{pmatrix} = \begin{pmatrix} \eta_1 W_1 + \eta_2 W_2 \\ W_1 + W_2 \end{pmatrix}. \\
&4. x < 0. \Rightarrow \begin{pmatrix} u_1 \\ u_2 \end{pmatrix} = \begin{pmatrix} \eta_1 W_1 + \eta_2 W_2 \\ W_1 + W_2 \end{pmatrix}.
\end{aligned}
\tag{24}
$$

After some cumbersome but simple manipulations one can show that $\eta_1 W_1 + \eta_2 W_2 = 1$ and $W_1 + W_2 = 1$. $\forall \{a, b, c, d\} \in \mathbb{R}$. □

Lemma 2: *Let there exist at least one eigenvalues λ_0 of $\mathbb{A}$ from* Eq. (17), *that $\lambda_0 < 0$. Then values of u_1 or u_2 for $x < 0$ are not constant for $t \in \mathbb{R}^+$.*

Proof: For the proof, we are considering one example and then use the proof of Lemma 1. Let $\lambda_1 > 0, \lambda_2 < 0$ by the Lemma formulation. Let $\mathbb{A} = \begin{pmatrix} -1 & 1 \\ 0 & 1 \end{pmatrix}$. Then, $\lambda = \begin{pmatrix} 1 \\ -1 \end{pmatrix}$ and $H = \begin{pmatrix} \frac{1}{2} & 1 \\ 1 & 0 \end{pmatrix}$. Using characteristic variables:

$$
\begin{pmatrix} W_{0,1} \\ W_{0,2} \end{pmatrix} = \begin{pmatrix} \begin{cases} 1; x < 0 \\ e^{ikx}; x \geq 0 \end{cases} \\ \begin{cases} \frac{1}{2}; x < 0 \\ 1 - \frac{1}{2} e^{ikx}; x \geq 0 \end{cases} \end{pmatrix}
\tag{25}
$$

Diagonalizing operator $\mathbb{A}$: $\Lambda = H^{-1}\mathbb{A}H = \left(\begin{pmatrix} 0 & 1 \\ 1 & -\frac{1}{2} \end{pmatrix} \cdot \begin{pmatrix} -1 & 1 \\ 0 & 1 \end{pmatrix}\right) \cdot \begin{pmatrix} \frac{1}{2} & 1 \\ 1 & 0 \end{pmatrix} = diag(\lambda_i)$

and arriving at two independent equations. The solution in characteristic variables is:

$$\begin{pmatrix} W_1 \\ W_2 \end{pmatrix} = \begin{pmatrix} \begin{cases} 1; (x-t) < 0 \\ e^{ik(x-t)}; (x-t) \geq 0 \end{cases} \\ \begin{cases} \frac{1}{2}; (x+t) < 0 \\ 1 - \frac{1}{2}e^{ik(x+t)}; (x+t) \geq 0 \end{cases} \end{pmatrix}. \tag{26}$$

Returning to the original variables:

$$\begin{pmatrix} u_1 \\ u_2 \end{pmatrix} = \begin{pmatrix} \begin{cases} \frac{1}{2}; (x-t) < 0 \\ \frac{1}{2}e^{ik(x-t)}; (x-t) \geq 0 \end{cases} + \begin{cases} \frac{1}{2}; (x+t) < 0 \\ 1 - \frac{1}{2}e^{ik(x+t)}; (x+t) \geq 0 \end{cases} \\ \begin{cases} 1; (x-t) < 0 \\ e^{ik(x+t)}; (x-t) \geq 0 \end{cases} \end{pmatrix}. \tag{27}$$

Let us consider the case only for $x < 0$.

$$\begin{aligned} &1. x \geq (x+t); x < 0. \Rightarrow \begin{pmatrix} u_1 \\ u_2 \end{pmatrix} = \begin{pmatrix} \frac{3}{2} - \frac{1}{2}e^{ik(x+t)} \\ 1 \end{pmatrix}. \\ &2. x < (x+t); x < 0. \Rightarrow \begin{pmatrix} u_1 \\ u_2 \end{pmatrix} = \begin{pmatrix} 1 \\ 1 \end{pmatrix}. \end{aligned} \tag{28}$$

We now turn to random real variables in matrix $\mathbb{A}$ that satisfy the condition $\exists \lambda_i < 0$. Solutions (27) and (22) only depend on matrix variables and are identical in the other. So, this particular example is valid for any values in matrix $\mathbb{A}$ and only depends on signs of eigenvalues of $\mathbb{A}$. □

Lemma 3: *Assume that matrix $\mathbb{A}$ is diagonalizable. Then Lemmas 1 and 2 are valid for* $\mathbb{A} \in \mathbb{R}^{3\times 3}$.

Proof: The proof is straightforward, since the operator only depends on eigenvalues. Then follow the proofs for Lemmas 1 and 2. □

Now we turn our attention to Eq. (15). We assume one space dimension of the system and consider a Riemann problem for some line $x = x_0$ and consider variables on the left (L) and on the right (R) from this line. We assume that U_L is an unperturbed constant vector and that $\mathbf{U}_R$ contains perturbations. We also assume that $p_L < p_R$ and that there is a stationary shock wave with shock speed $S = 0$ at line $x = x_0$. In this case, we can conclude the following:

Proposition 1: *Let there be such solution to the system* (15), *that at some line* $x = x_0$, *there is a stationary shock wave with* $p_L < p_R$ *and* $\mathbf{U}(x, 0), \forall x \le x_0$ *is unperturbed. Then,* $\mathbf{U}(x, t), \forall x \le x_0$ *remains unperturbed at* $\forall t \in \mathbb{R}^+$.

Proof: In accordance with the condition of the proposition, the local solution of Eq. (15) at x_0 is a shock wave solution that is located in the first characteristic field, i.e., $\lambda_1 = u - ct$, since $p_L < p_R$. The second characteristic field is a contact discontinuity with $\lambda_1 = u$ with the second contact discontinuity being parallel to the flow, since $\mathbf{n} = (1, 0)^T$. The third characteristic field contains an expansion wave, $\lambda_3 = u + ct$. Then, $\lambda_1 = 0$, since $S = 0$, and so the local Jacobi matrix eigenvalues are ordered as:$\lambda_1 = 0 \le \lambda_2 \le \lambda_3$.

Since the system is hyperbolic and the local Jacobi matrix has linear independent eigenvalues, we can transform, locally, system (15) to the characteristic variables with Jacobi matrix diagonalization. We can then apply Lemmas 3 and 1 to complete the proof.

It is more difficult to prove the same proposition for a general case; however, the same conclusion is valid for 3D problem in case of a flat plane shock wave solution. It concludes that the phase space of the system for hyperbolic equations is separated into independent or slightly dependent regions by shock waves. If the shock wave is stationary, then no perturbations are transformed through it. We will demonstrate this with the numerical solution of the problem with the formation of a stationary shock wave.

4.1.2. Numerical simulation

We wish to analyze properties of the phase space of Eq. (7) in case of appearance of strong shocks and confirm results of the previous subsection. More results can be found in Ref. [55]. We are considering a rectangular domain Ω, sized $8 \times 3 \times 3$. A small rectangular body B is located in the center of the domain. This results in the formation of the front stationary detached shock wave in front of the body in case of the supersonic flow. Gas is homogeneous in $\Omega \backslash B$ at initial conditions with $\rho = 1.24, \mathbf{u} = (1, 0, 0)^T, M_0 = 2$. Wall boundary conditions are set for all $\partial\Omega_W$ in accordance with considering equations, except for inflow and outflow boundaries. No-slip conditions are set for viscous gas flow, and slip conditions are set for Eq. (7) with $\mu = 0$. We impose two unphysical boundary conditions: inflow boundary $\partial\Omega_I$ on the left plane yz for $x = 0$ ($\rho_0 = 1.24, \mathbf{u} = (1, 0, 0)^T, M_\infty = 2$) and outflow boundary $\partial\Omega_O$ on the right plane yz for $x = 8$. These boundary conditions are set with the procedure described in Section 2.2.

We are performing the simulation for quasi-stationary solution on the grid of $500 \times 200 \times 200$ elements. Phase space portraits are constructed using $10 \cdot 10^6$ time points after the flow is quasi-stationary. The viscosity coefficient is set to $1 \cdot 10^{-5}$. We do not perform convergence analysis for viscous flow since this solution is used to show the validity of the phase space separation. Result of the 2D middle section of the numerical Schlieren is presented in **Figure 5**.

One can see point numbers in the section that demonstrate different phase space portraits from these points. (**Figure 6**).

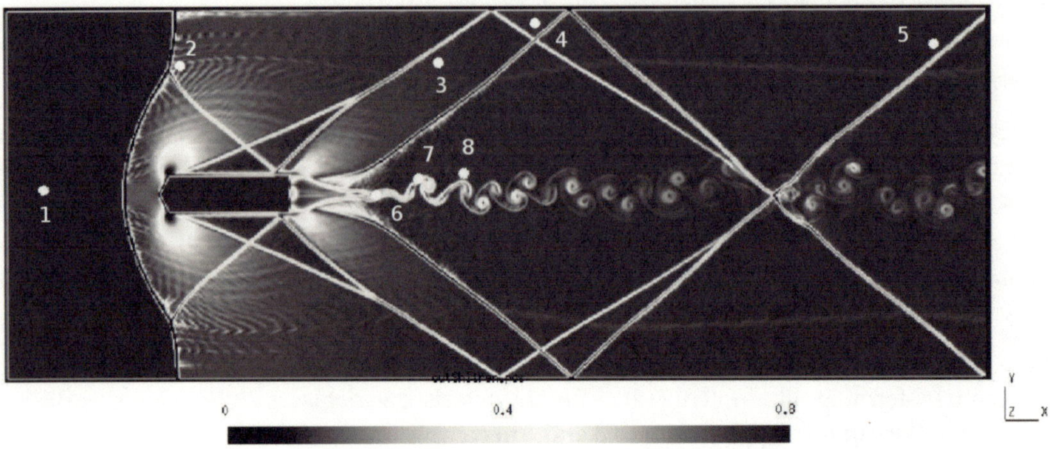

Figure 5. Schlieren picture of the supersonic Mach 2 flow over a body. Points indicate phase space probe points.

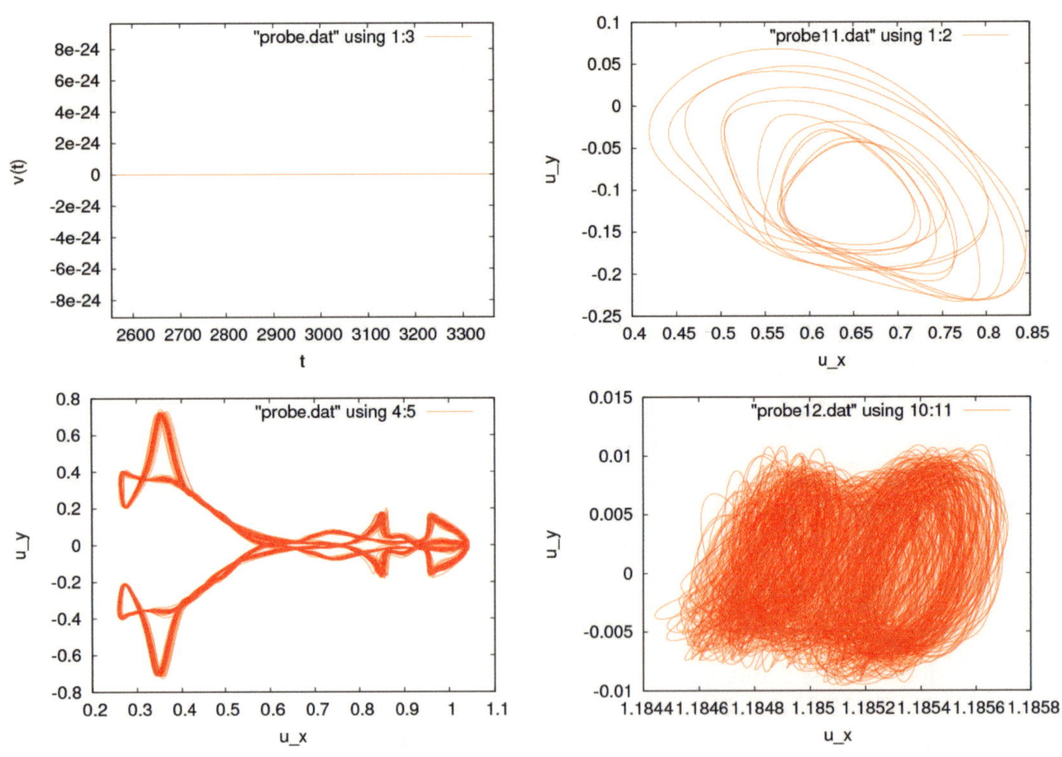

Figure 6. Projections of phase space portraits in different points of the physical domain. Top left to right: point 1, point 2, bottom left to right: point 6, point 4.

We can monitor no perturbations in the inflow part with the formation of the fixed point in the phase space. This confirms the results of the previous subsection. At point 2, near the front shock wave, one can see the formation of the limited cycle of period 11. Perturbations from the body are substantially weakened resulting in a low dimensional process in this point. Please note that the area in which point 2 is located is isolated by shock wave configuration. The low dimensional attractor in this area is formed by the oscillations of the attached shock wave and contact discontinuity. At point 6, we can see an almost symmetrical two-dimensional invariant torus that corresponds to the quasi-periodic solution. This is the type of von-Karman sheet solution generated by the flow over the body in subsonic region of the flow. In this area, we expected location of Andronov-Hopf and secondary Hopf bifurcations. Another separated area with point 4 was tested, where we can monitor chaotic solution. The dimension of the chaos is greater than two, since we are able to monitor chaotic solution in the second Poincaré section. Other points gave qualitatively identical results.

4.2. Kelvin-Helmholtz problem bifurcation scenario

In this subsection, we are considering the bifurcation scenario of the KH problem. The setup of the problem is the following. The domain is $\Omega = [8\times1\times1]$, the flow direction is from left to right. The following boundary conditions are imposed: $\partial\Omega(0, y, z) = \partial\Omega_0$ is the inflow boundary condition, $\partial\Omega(8,y, z) = \partial\Omega_1$ is the outflow boundary condition, $\partial\Omega(x, y, 0) = \partial\Omega(x, y, 1) = \partial\Omega_2$ is the periodic boundary condition in z-direction, and $\partial\Omega(x, 0, z) = \partial\Omega(x, 1, z) = \partial\Omega_3$ is the non-reflecting (symmetry) boundary condition. We use simple ghost cell boundary conditions for $\partial\Omega_2$ and $\partial\Omega_3$. For $\partial\Omega_1$, we use appropriately chosen absorbing boundary conditions with characteristic far-field treatment and we use characteristic boundary conditions for $\partial\Omega_0$. The initial conditions are identical to the boundary conditions on $\partial\Omega_0$ and are given below:

$$\begin{aligned} \rho|_{\partial\Omega_0} &= 1.0, \\ u_x|_{\partial\Omega_0} &= 3/2 + 1/2\tanh\left(\frac{y-1/2}{\delta} + C_z\cos(6\pi z)\right), \\ u_y|_{\partial\Omega_0} &= 1\cdot10^{-5}\sin(6\pi y), \\ u_z|_{\partial\Omega_0} &= 0, \\ p|_{\partial\Omega_0} &= 22.14. \end{aligned} \tag{29}$$

The perturbations are set depending on the problem setup, either 2D (in this case $C_z = 0$) or 3D (in this case $C_z = 1$). The gravitational force is absent. The perturbations are not set in the classical sense, since those must be set along the x-direction. However, we noticed that this kind of initial perturbations is sufficient to trigger instability. For computation, we are using a grid of $1000\times250\times250$ elements totaling $62.5\cdot10^6$ grid points. The sponge zone for $\partial\Omega_1$ is chosen to be 20 elements in x-direction. We use the Reynold number (R) as bifurcation parameter with further details given below.

4.2.1. *Eigenvalue analysis*

First, we analyze linear stability of the main flow for the described problem setup. For $R\leq157$ the flow remained stable. The diffusion is sufficient enough to suppress instability. This solution corresponds to the stable fixed point in the phase space. With the increase of $R > 158$, the flow loses stability with the formation of the initial billows. We are able to show the leading unstable eigenvector for this case; see **Figure 7**. As one can see, the most dangerous eigenvector is a 2D symmetrical one. This corresponds with the conclusions from Orr-Sommerfeld equations for flat-plane viscous flow stability. However, one can see, that the third most dangerous eigenvector has already a 3D structure. The formation of the initial billows corresponds to the Andronov-Hopf bifurcation (two complex conjugate eigenvalues are crossing the imaginary axis). The leading eigenvalues for this problem are presented in **Figure 8**. One can see the monodromy matrix eigenvalues as well. One eigenvalue lies on a unit circle at the point $(+1, 0i)$ on the complex plane. This eigenvalue corresponds to the zero Lyapunov exponent of the dynamical system that represents the cycle direction. This limited cycle is shown in **Figure 9**. It was found that the length of the domain in the x-direction (L_x) is another parameter. With the increase of the domain up to 16, the first critical Reynolds number gradually changes to $R_c = 25$. It can be explained by the wavelength in the x-direction. It is interesting to note that using $L_x = 4$, the inviscid flow is numerically stable (assuming constant perturbation amplitude in Eq. (29)). This leads to the conclusion that the diffusion may pay a positive role in initial instability.

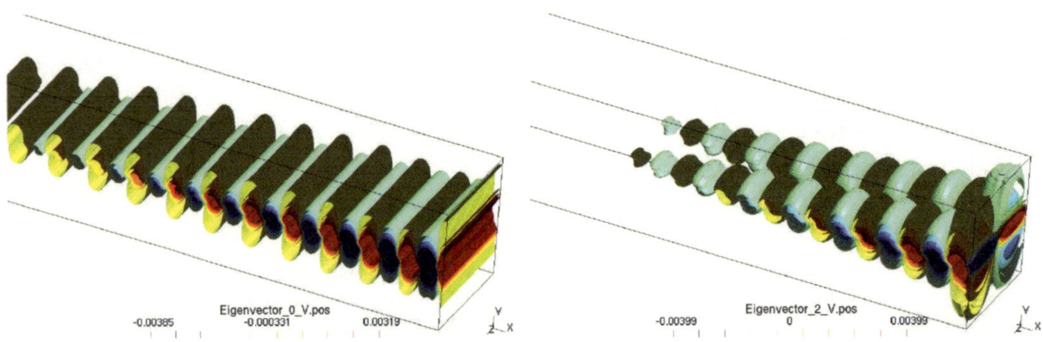

Figure 7. Modulus of the first and the third most dangerous eigenvectors of u_y for the KH coaxial shear flow problem.

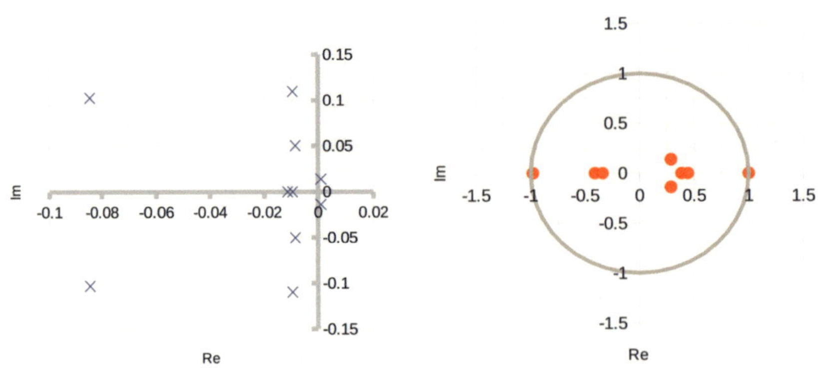

Figure 8. Largest real eigenvalues for the KH problem (left) and monodromy matrix largest magnitude eigenvalues for limited cycle solution, $R = 400$.

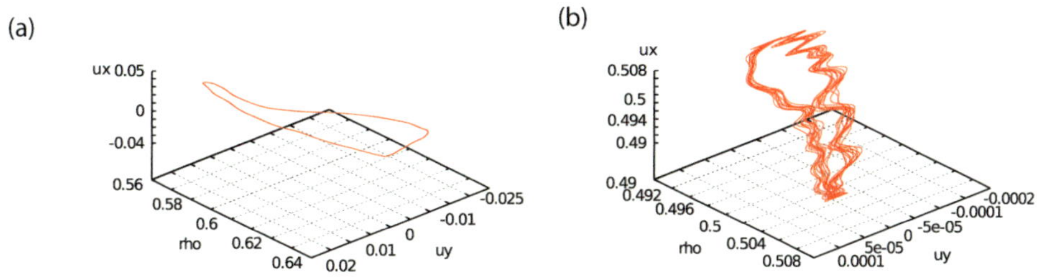

Figure 9. Projection of the limited cycle for $R = 1000$ and cycle period 11 for $R = 1140$.

Let us consider time scales in the problem. The maximum physical process time τ can be roughly estimated as advection time scale $\tau_a \sim L_x/|u_x|$ for advection-dominated flow. Providing that, this time is smaller for $L_x = 4$, initial perturbations for inviscid flow are dominated by advection timescale rather than diffusion timescale $\tau_d \sim L_x^2/\mu = \mathbf{R}\mathbf{L}_x^2$. The timescales are equal if $R \sim \frac{1}{L_x|u_x|}$, so with the increase of L_x, one gets a smaller value of R to have diffusion play a substantial role in the process. For a fixed value of R, the diffusion is decreasing with the decrease of L_x. The mechanism works as follows. Initially, the diffusion is dominated by setting a small value of R with the obvious stabilizing effect to all wavenumbers. One can see that the first unstable eigenvector (**Figure 7**) demonstrates about $\lambda_{critical}$~12π. This wavelength is dominated by diffusion for small R. The destabilizing role of diffusion starts playing a role with the increase of R, resulting in sequential bifurcations. The further route to chaos depends on the initial perturbations and bifurcation parameters.

4.2.2. 2D perturbations

With the further increase of the Reynolds number, we can monitor the formation of singular limited cycles. These cycles are low dimensional attractors that are formed by a continuous set of period doubling bifurcations. All these cycles lie in the Feigenbaum cascade. We are able to monitor the formation of cycles until cycle period 11, in **Figure 9** at around $R = 1140$. This cycle originates from the Sharkovskii order and is one of the key cycles in Feigenbaum-Sharkovskiy-Magnitskii (FShM) cascade [51]. Unfortunately, we are unable to trace the cycle period three which is the last cycle in Sharkovskii direct order. It is known that chaos emerges after this cycle. With further increase of R value, we witnessed the reversed cascade up to the solution of a limited cycle. All these cycles are corresponding to the 2D instabilities of the primary flow resulting in low dimensional chaos and correspond to the oscillation of billows in physical space in xy plane. It is noted that the increase of domain length in x-direction, or decrease of base velocity u_x results in the change of the bifurcation scenario. This again can be explained by the domination of τ_a with the increase of R. We confirmed this proposition by increasing the L_x up to 16 for the fixed value of $R = 2500$ thus giving the diffusion scale more time to develop. While cycle was formed for $L_x = 8, R = 2500$, the chaotic solution emerges for $L_x = 16, R = 2500$, so the value of L_x becomes another bifurcation parameter. Moreover, in this case a two dimensionality of billows is destroyed with the formation of chaotic flow. We are unable to trace any regular attractors after that. This can be the result of disturbing diffusion action on the flow through screw symmetric viscous tensor components.

4.2.3. *3D perturbations*

We found that the initial 2D instability is reformed for the 3D perturbations. This can be explained since the flow is no longer flat parallel. The change in the inflow conditions resulted in the formation of chaotic solution right after the flow is developing. It was difficult to trace any bifurcations further with only one regular structure found being a cycle, two-dimensional and three-dimensional invariant tori for $100 < R < 950$; see **Figure 10**.

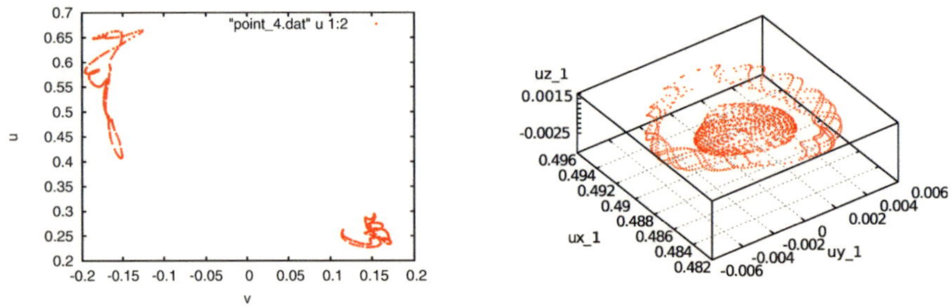

Figure 10. 2D invariant torus ($R = 760, L_x = 8$) and 3D invariant torus ($R = 950, L_x = 8$) first Poincaré sections.

4.3. Coupled (Kelvin-Helmholtz and Rayleigh-Taylor) problem bifurcation scenario

We are considering the problem of stratified gas flow with velocity shear layer. This is a coupled KH and RT problem that can be characterized by the density difference and application of the gravitational force. We set up the problem analogously to the previous one but we have one more parameter that is the Richardson number. The boundary on $\partial\Omega_0$ and initial values are set as follows:

$$\begin{aligned} \rho|_{\partial\Omega_0} &= 1.1 + \frac{1}{10}\mathrm{erf}\left(\frac{y - 1/2}{\delta} + C_z \cos(6\pi z)\right), \\ u_x|_{\partial\Omega_0} &= 2.0 + \mathrm{erf}\left(\frac{y - 1/2}{\delta} + C_z \cos(6\pi z)\right), \\ p|_{\partial\Omega_0} &= 22.14 + \rho g y + (\rho_{max} - \rho)g. \end{aligned} \tag{30}$$

The Richardson number (Eq. (3)) is set depending on the problem in range of $1/5 - 1/8$, where $H = 4\delta$ is the length of the mixing layer and $\rho_{max} = 1.2$. We also consider two options—$C_z = 0$ or $C_z = 1$. The results of the 2D perturbations are analogous to the previous subsection with the formation of limited cycles period five and three presented in **Figure 11**. Interesting to note that this cycle was found in the reverse cascade due to the limited length, $L_x = 6$. The increase of L_x up to 10 leads to the formation of chaotic solution. It was noted that the process was more related to the shear flow of gases with nonzero density ratio. The development of RT instability was not detected; thus, these series can be qualified as pure KH instability problem. The initial stage of instability is formed by low dimensional attractors with confirmed inverse Feigenbaum-Sharkovskiy cascade.

http://dx.doi.org/10.5772/67918

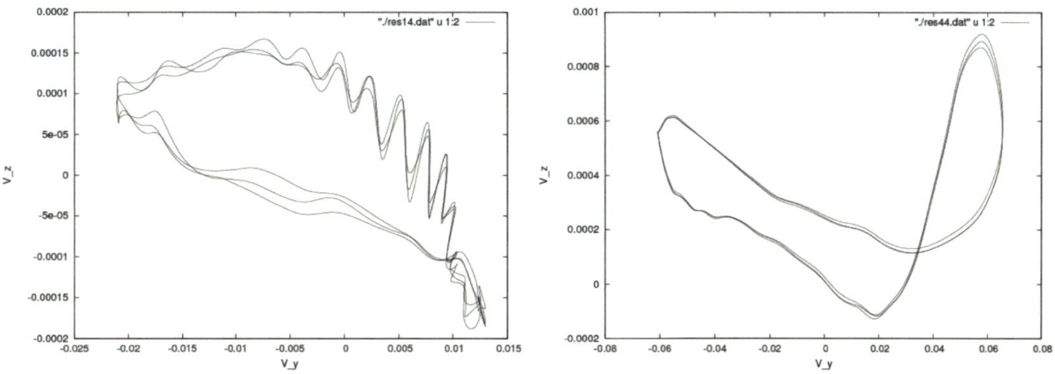

Figure 11. Projection of limited cycles of period three into v_z,v_y phase subspace for 2D perturbation problem for $R = 2417$,$Ri = 1/6$,$L_x = 6$.

4.3.1. 3D perturbations

Results for 3D perturbations are presented in **Figures 12** and **13**. One can clearly see the formation of initial billows of KH instability, followed by RT instability. The latter can be clearly noticed in later stages of the flow; see **Figure 13**. With L_x acting as bifurcation parameter in mind, we fix the length and increase value of R until the solution is a regular attractor. Please mind that the initial flow is linearly unstable, and small values of R may give a destabilizing effect.

The starting point is $R = 1$ for which the solution is stationary and totally diffusion dominated. This solution corresponds to the stable point in the physical space. At around $R = 10.5$, the first bifurcation of the stationary flow is detected with the formation of the limited cycle; see **Figure 14**.

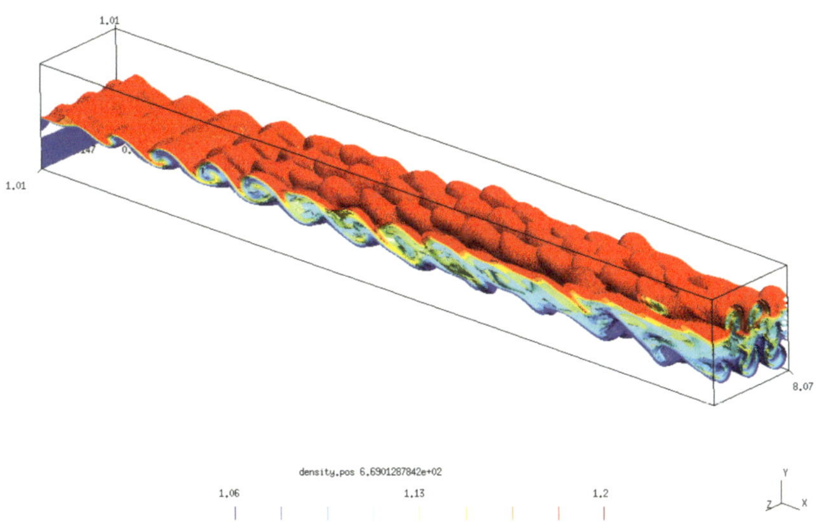

Figure 12. Coupled problem quasi-stationary solution. Density isosurfaces.

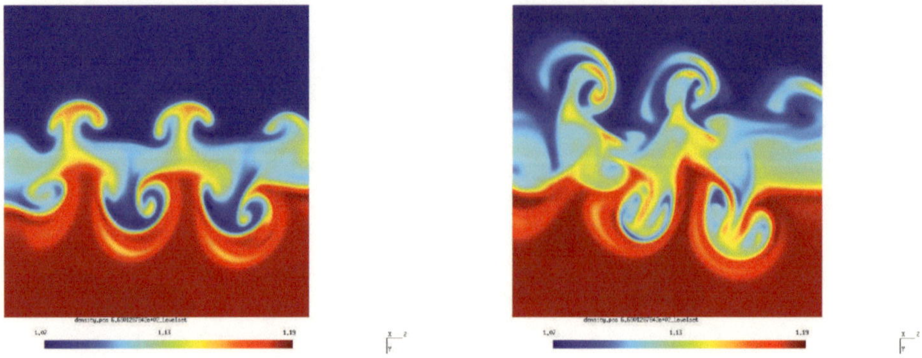

Figure 13. Coupled problem quasi-stationary solution. Density sections in yz planes with $x = 3$ and $x = 5$.

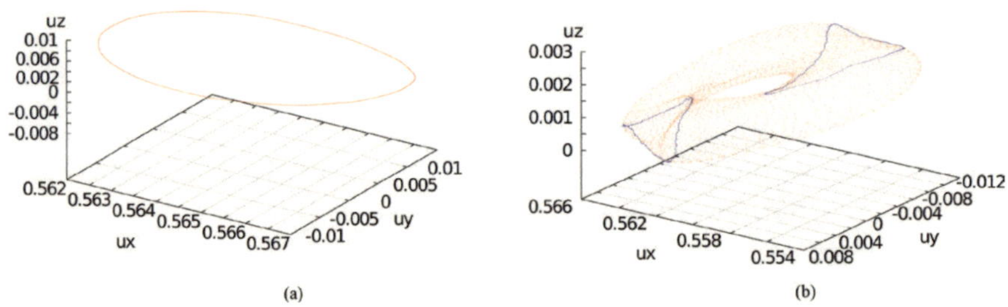

Figure 14. Projection of the limited cycle $R = 10.5$ and invariant torus (with Poincaré section) $R = 510$ to three-dimensional phase subspace.

After that the influence of viscosity compressed the region of parameters forming chaotic attractors around the cycle with small amplitude. These attractors are, supposedly, low dimensional singular cycles. But the small amplitude of the solution over the cycle made it impossible to analyze. The next attractor that we are able to detect is the limited torus, presented in **Figure 14** with the Poincaré section indicated. Close resemblance to the cycle may indicate that the attractor was formed as the result of multistability from the cycle by Hopf bifurcation. This indicates the existence of two irrational frequencies in the system. Further increase of the Reynolds number up to $R = 516$ resulted in the other Hopf bifurcation with the formation of the three-dimensional invariant torus, presented in **Figure 15**. This torus becomes singular (by

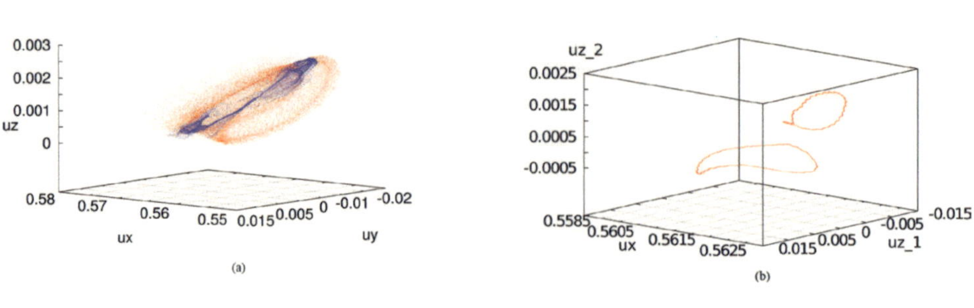

Figure 15. Projection of the invariant three-dimensional torus and first Poincaré section for $R = 516$ to the three-dimensional phase subspace and second section for $R = 518$.

period doubling bifurcations on one of the frequencies). This is shown in **Figure 16** for $R = 518$. However, this cascade of period doubling bifurcations is reversed to the original 3D torus. The next bifurcation that could be traced at $R = 520.5$ is another Hopf bifurcation leading to the formation of the four-dimensional invariant torus.

Further increase of the Reynolds number leads to the chaotic solution that corresponds to the dense field of points in phase subspace projections up to about $R = 2100$. With further increase of R, we observed the formation of inverse cascades. For verification purposes, the domain length was increased again up to $L_x = 16$ for fixed $R = 2500$. This resulted in the chaotic solution.

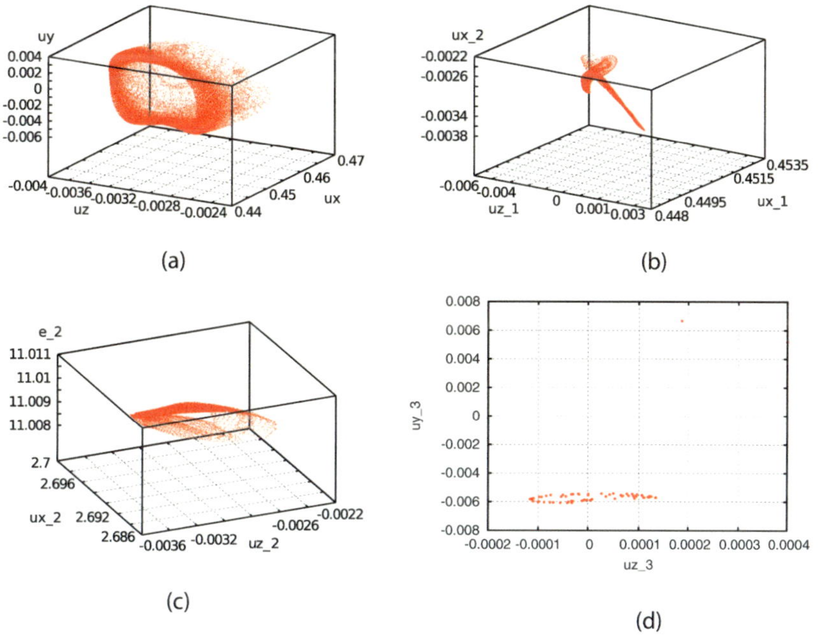

Figure 16. Projection of the invariant four-dimensional torus into three-dimensional phase subspace (top left) and sequential first, second, and third sections in the phase space for $R = 520.5$ (left to right, up to down). Only parts of sections are shown.

5. Discussion and conclusion

5.1. Discussion

This research for the bifurcation scenarios in RT and KH instabilities started in 2013 and continues still. From this research, we obtained information that raises some questions that must be addressed. The first question is the stability of the KH and RT problems for low Reynolds number. It is known [38] that viscosity may result in instability. However, we found two ways of this instability to occur. Starting with very low Reynolds number results in linear stability of the flow. When linear stability is violated, we can witness a rapidly growing instability that results in highly suppressed cascade of bifurcations. We are still unable to deduce this cascade. However, from a certain value of R, the solution becomes regular with the formation of regular attractor (either invariant cycle or torus). This indicates that the cascade of bifurcations that is induced by diffusion is reversed at some point resulting in the final return to the attractor with the maximum basin of attraction. Those are the bifurcations that were traced by other authors [48–50].

Another topic of discussion is the Landau-Hopf scenario that was found in coupled problem. This means that the system triggers secondary instabilities that have irrational frequencies. It is known, say from [68], that the existence of secondary oscillations on KH instability is triggered by density stratification. This results in the formation of instabilities of outer filament or inner vortex type. These instabilities might have irrational frequencies resulting in the formation of high dimensional attractor. That is clearly the case for coupled problem; see **Figure 12**.

Another difficulty is related to the limited length of the domain during presented bifurcation analysis. We show in Section 4.1, that the information for subsonic flow is translated back to the upwind direction. This results in the global phase space formation. On the other hand, the length of the domain becomes a bifurcation parameter since physical process evolves with the background flow and is cut at the point of flow leaving the domain. The situation for the initial valued problem is different since the process will eventually relax (e.g., densities are swapped for RT instability), and phase space solution will be represented by a fixed stable point. It means that the process consists of direct and inverse cascades of bifurcations starting and finishing at the point. In most cases, the initial point is linearly unstable and the final point is stable.

5.2. Conclusion

In this paper, we show some results regarding RT and KH instabilities and phase space properties for supersonic solution with shock waves. We show that the phase space is separated for inviscid flow and give proof of this fact for 1D gas flow. This fact is numerically demonstrated, and some results are obtained that confirm the validity of this proposition for multidimensional case. We also considered linear stability and bifurcation analysis for KH and RT instabilities in the setup with the coaxial flow.

We introduce the following notations of regular attractors to write down bifurcation scenarios: P is a point, Cn is a limited cycle of period n, nTM is an invariant torus of dimension M and period n (on any frequency). For fixed values of domain length, we obtained the following bifurcation scenarios.

KH instability with 2D perturbations:

$$P \to C \to \ldots \to C5 \to \ldots \to C11 \to \ldots \to C \tag{31}$$

KH instability with 3D perturbations:

$$P \to C \to T2 \to T3 \to Chaos.$$

Coupled problem (KH and RT instability) with 2D perturbations:

$$P \to C \to C2 \to C4 \to Cn \to .. \to C5 \to C3 \to Chaos. \tag{32}$$

Coupled problem (KH and RT instability) with 3D perturbations:

$$P \to C \to nC \to \ldots \to C \to T2 \to T3 \to \ldots \to nT3 \to \ldots \to T4 \to Chaos. \tag{33}$$

The following scenarios show the following. First, the existence of FShM scenario in KH instability is confirmed by Eqs. (31) and (32). In Eq. (31), we notice the direct and inverse cascades with the Sharkovskiy sequence are not fully developed. In Eq (32), we are able to show the full FShM cascade up to limited cycle of period three. However, these low dimensional attractors are only the initial trigger mechanism of laminar-turbulent transition in such complicated problems. It can be clearly seen in Eq. (33) where a Landau-Hopf scenario is developed up to a regular attractor represented by the 4D invariant torus. The existence of computationally stable 4D invariant torus is a remarkable fact. It took $2.6 \cdot 10^9$ time samples to analyze and about 3.5 months to calculate. We can notice the appearance of singular 3D tori in the cascade. Most likely, some part of incomplete FShM scenario with direct and inverse cascades is involved in these singular tori. It is interesting to note that we found no resonance tori during the bifurcation analysis that are typical for other problems.

Author details

Nikolay Mihaylovitch Evstigneev and Nikolai Alexandrovitch Magnitskii*

*Address all correspondence to: nikmagn@gmail.com

Federal Research Center "Informatics and Control", Institute for System Analysis of Russian Academy of Science, Russia

References

[1] Evstigneev N.M., Magnitskii N.A. (2012) Chapter 10: FSM scenarios of laminar-turbulent transition in incompressible fluids. In Nonlinearity, bifurcation and chaos – theory and applications. Intech, Rijeka, pp. 251–280. http://dx.doi.org/10.5772/48811

[2] Charles Li Y. (2007) Chaos in partial differential equations, Navier-Stokes equations and turbulence. ICCM. III, 110–122

[3] Arnold V.I., Meshalkin L.D. (1958–1959) The seminar of A.N. Kolmogorov on selected topics in analysis. Usp. Mat. Nauk. 15, 247–250

[4] Meshalkin L.D., Sinai I.G. (1961) Investigation of the stability of a stationary solution of a system of equations for the plane movement of an incompressible viscous liquid. J. Appl. Math. Mech. 25(4), 1700–1705

[5] Yudovich V.I. (1973) Natural oscillations arising from loss of stability in parallel flows of a viscous liquid under long-wavelength periodic disturbances. Fluid Dyn. 8, 26. doi:10.1007/BF01017632

[6] Sivashinsky G. (1985) Weak turbulence in periodic flows. Phys. D. 17(2), 243–255

[7] Lucas D., Kerswell R. (2014) Spatiotemporal dynamics in two-dimensional Kolmogorov flow over large domains. J. Fluid Mech. 750, 518–554. doi:10.1017/jfm.2014.270

[8] Evstigneev N.M., Magnitskii N.A., Silaev D.A. (2015) Qualitative analysis of dynamics in Kolmogorov's problem on a flow of a viscous incompressible fluid. Differ. Eqn. 51(10), 1292–1305

[9] Lord Rayleigh O.M. F.R.S. (1916) On convection currents in a horizontal layer of fluid, when the higher temperature is on the under side. Philos. Mag. Ser. 6(32), 192, 529–546

[10] Manneville P. (2010) Rayleigh-Benard convection: thirty years of experimental, theoretical, and modeling work. Springer Tracts Modern Phys. 207, 41–65

[11] Paul Supriyo, Verma Mahendra K., Wahi Pankaj, Reddy Sandeep K., Kumar Krishna (2012) Bifurcation analysis of the flow patterns in two-dimensional Rayleigh-Benard convection. Int. J. Bifurc. Chaos. 22(5), 1230018

[12] Evstigneev N.M. (2016) Laminar-turbulent bifurcation scenario in 3D Rayleigh-Benard convection problem. Open J. Fluid Dyn. 6, 496–539. doi:10.4236/ojfd.2016.64035

[13] Getling A.V. (1998) Rayleigh-Benard convection: structures and dynamics. World Scientific, Singapore. ISBN 978-981-02-2657-2

[14] Ashwin P., Podvigina O. (2003) Hopf bifurcation with cubic symmetry and instability of ABC flow. Proc. R. Soc Lond. A. 459(2035), 1801–1827

[15] Pfister G., Schmidt H., Cliffet K.A., Mullin T. (1988) Bifurcation phenomena in Taylor-Couette flow in a very short annulus. J. Fluid Mech. 191, 1–18

[16] Mamun C.K., Tuckerman L.S. (1994) Asymmetry and Hopf bifurcation in spherical Couette Flow. Phys. Fluids. 7(1). doi:10.1063/1.868730

[17] Lopez J.M., Marques F. (2005) Finite aspect ratio Taylor–Couette flow: Shil'nikov dynamics of 2-tori. Phys. D. 211, 168–191

[18] Ma Tian, Wang Shouhong (2006) Stability and bifurcation of the Taylor problem. Arch. Ration. Mech. Anal. 181, 149–176

[19] Furukawa H., Horikoshi H., Ohazama N., Watanabe T. (2012, June 25–28) PIV analysis of mode bifurcation in Taylor vortex flow. In 15th International Symposium on Flow Visualization, Minsk, Belarus

[20] Barkley D., Gomes M.G.M., Henderson R.D. (2002) Three-dimensional instability in flow over a backward-facing step. J. Fluid Mech. 473, 167–190. doi:10.1017/S002211200200232X

[21] Rani H.P., Sheu Tony W.H. (2006) Nonlinear dynamics in a backward-facing step flow. Phys. Fluids. 18, 084101

[22] Evstigneev N.M., Magnitskii N.A., Sidorov S.V. (2009) On the nature of turbulence in a problem on the motion of a fluid behind a ledge. Differ. Eqn. 45(1), 68–72

[23] Soibelman I., Meiron D.I. (1991) Finite-amplitude bifurcations in plane Poiseuille flow: two-dimensional Hopf bifurcation. J. Fluid Mech. 229, 389–416

[24] Casas P.S., Jorba A. (2012) Hopf bifurcations to quasi-periodic solutions for the two-dimensional plane Poiseuille flow. Commun. Nonlinear Sci. Numer. Simul. 17(7), 2864

[25] Lord Rayleigh O.M. F.R.S. (1883) Investigation of the character of the equilibrium of an incompressible heavy fluid of variable density. Proc. Lond. Math. Soc. XIV, 70–177

[26] Taylor G.I. (1950) The instability of liquid surfaces when accelerated in a direction perpendicular to their planes. Proc. R. Soc. Lond. 201, 192–196

[27] von Helmholtz Hermann (1868) Uber discontinuierliche Flussigkeits-Bewegungen. Monatsberichte der Koniglichen Preussische Akademie der Wissenschaften zu Berlin. 23, 215–228

[28] Lord Kelvin (William Thomson). (1871) Hydrokinetic solutions and observations. Philos Mag. 42, 362–377

[29] Drazin P.G. (2002) Introduction to hydrodynamic stability. Cambridge University Press, Cambridge, United Kingdom

[30] Kull H.J. (1991) Theory of the Rayleigh-Taylor instability. Phys. Rep. 206(5), 197–325. North-Holland

[31] Book D.L. (1984) Rayleigh-Taylor instability in compressible media. NRL memorandum report no. 5373. Washington, DC: Naval Research Laboratory

[32] Schmid P.J., Henningson D.S. (2001) Stability and transition in shear flows. Springer, Springer-Verlag New York, lnc.

[33] Blumen W. (1970) Shear layer instability of an inviscid compressible fluid. J. Fluid Mech. 40(Part 4), 769–781

[34] Hesla Todd I., Pranckh Ferdinand R., Preziosi Luigi (1986) Squire's theorem for two stratified fluids. Phys. Fluids 29(9), 2808–2811. doi:10.1063/1.865478

[35] Cushman-Roisin B. (2005) Kelvin-Helmholtz instability as a boundary-value problem. Environ Fluid Mech. 5, 507–525

[36] Richardson L.F. (1920) The supply of energy from and to atmospheric eddies. Proc. R. Soc. London. A97, 354–373

[37] Kumar K., Tuckerman L.S. (1994) Parametric instability of the interface between two fluids. J. Fluid Mech. 279, 49–68

[38] Funada T., Joseph D.D. (2001) Viscous potential flow analysis of Kelvin-Helmholtz instability in a channel. J. Fluid Mech. 445, 263–283

[39] Yoshikawa H., Wesfreid J.E. (2011) Oscillatory Kelvin-Helmholtz instability. Part 1. A viscous theory. J. Fluid Mech. 675, 223–248

[40] Fjørtoft T. (1950) Application of integral theorems in deriving criteria of instability of laminar flow and for the baroclinic circular vortex. Geophys. Publ. 17, 1–52

[41] Mack C. (2009) Global stability of compressible flow about a swept parabolic body. PhD thesis, Ecole Polytechnique

[42] Henze O., Lemke M., Sesterhenn J. (2015) A parallel and matrix free framework for global stability analysis of compressible flows. arXiv:1502.03701

[43] Bernstein I.B., Book D.L. (1983) Effect of compressibility on the Rayleigh-Taylor instability. Phys. Fluids 26, 453

[44] Lezzi A.M., Prosperetti A. (1989) Rayleigh-Taylor instability for adiabatically stratified fluids. Phys. Fluids A. 1, 1784

[45] Turner L. (2002) Rayleigh-Taylor instabilities and gravity waves in compressible fluids. Los Alamos National Laboratory Report LA-UR-02-6439, Los Alamos, USA.

[46] Livescu D. (2004) Compressibility effects on the Rayleigh-Taylor instability growth between immiscible fluids. Phys. Fluids. 16(1), 118–127

[47] Guo Yan, Tice Ian (2009) Compressible, inviscid Rayleigh-Taylor instability. Indiana Univ. Math. J. 60(2). doi:10.1512/iumj.2011.60.4193

[48] Poehlmann A., Richter R., Rehberg I. (2013) Unravelling the Rayleigh-Taylor instability by stabilization. J. Fluid Mech. 732, R3. doi:10.1017/jfm.2013.424

[49] Ilak M., Schlatter P., Bagheri S., Henningson D.S. (2012) Bifurcation and stability analysis of a jet in cross-flow: onset of global instability at a low velocity ratio. J. Fluid Mech. 686, 94–121

[50] Hunter J.K., Thoo J.B. (2011) On the weakly nonlinear Kelvin–Helmholtz instability of tangential discontinuities in MHD. J. Hyperbol. Differ. Eqn. 8(4), 691–726

[51] Magnitskii N.A., Sidorov S.V. (2006) New methods for chaotic dynamics. Singapore: World Scientific

[52] Magnitskii N.A. (2012) Chapter 6: Universality of transition to chaos in all kinds of nonlinear differential equations. In Nonlinearity, bifurcation and chaos – theory and applications. Intech, pp. 133–174. http://dx.doi.org/10.5772/48811

[53] Toro E.F. (1999) Riemann solvers and numerical methods for fluid dynamics. Springer-Verlag, Berlin-Heidelberg.

[54] Evstigneev N.M. (2016) On the construction and properties of WENO schemes order five, seven, nine, eleven and thirteen. Part 1. Construction and stability. Comput. Res. Model. 8 (5), 721–753. (in Russian)

[55] Evstigneev N.M., Magnirskii N.A. (2012) On phase space peculiarities of gas dynamics equations for a supersonic initial-boundary value problem. Proc. ISA RAS. 62(4), 85–102. (in Russian)

[56] Evstigneev N.M., Ryabkov O.I. (2016) Application of multi GPU+CPU architecture for the direct numerical simulation of laminar-turbulent transition. Numer. Methods Program. 17, 55–64. (in Russian)

[57] Evstigneev N.M., Magnitskii N.A. (2014) On the initial stage of Kelvin-Helmholtz instability for laminar-turbulent transition in viscous gas flow. Proc. ISA RAS. 64(3), 41–52. (in Russian)

[58] Evstigneev N.M. (2016) On the construction and properties of WENO schemes order five, seven, nine, eleven and thirteen. Part 2. Comput. Res. Model. 8(6), 885–910. (in Russian)

[59] Evstigneev N.M. (2017) Implementation of implicitly restarted Arnoldi method on multiGPU architecture with application to fluid dynamics problems. Commun. Comput. Inform. Sci. (accepted for publication)

[60] McNally C.P., Lyra W., Passy J.C. (2012) A well-posed Kelvin-Helmholtz instability test and comparison. Astrophys. J. Suppl. Ser. 201(2), 1–17

[61] Youngs D.L. (1984) Numerical simulation of turbulent mixing by Rayleigh-Taylor instability. Phys. D. 12, 32–44

[62] Dimonte G, Youngs D.L., et al. (2004) A comparative study of the turbulent Rayleigh-Taylor instability using high-resolution three-dimensional numerical simulations: the alpha-group collaboration. Phys. Fluids. 16, 1668–1693

[63] Kucherenko, Yu. A. et al. (2003) Experimental investigation into the self-similar mode of mixing of different density gases in the Earth's gravitational field. Laser Part. Beams. 21, 385–388

[64] Cabot W.H., Cook A.W. (2006) Reynolds number effects on Rayleigh-Taylor instability with possible implications for type Ia supernovae. Nat. Phys. 2(8), 562–568

[65] Belotserkovskii O.M., Oparin A.M. (2000) Numerical experiment on turbulence: from order to chaos. Moscow: Nauka. (in Russian)

[66] Sin'kova O.G., et al. (1997) Three-dimensional DNS of gravitational turublence mixing. In 6th International Workshop on the Physics of Compressible Turbulent Mixing, Marseille, France, pp. 470–479

[67] Yanilkin Y.V., Statsenko V.P., Kozlov V.I. (2009) Mathematical simulation of turbulent mixing in compressible fluids. Nauka, Sarov, Russia.

[68] Lecoanet D., McCourt M., Quataert E., Burns K.J., Vasil G.M., Oishi J.S., Brown B.P., Stone J.M., O'Leary R.M. (2015) A validated nonlinear Kelvin-Helmholtz benchmark for numerical hydrodynamics. Mon. Not. R. Astron. Soc. 455(4), 4274–4288

Chapter 3

Interface Instability and Turbulent Mixing

Jingsong Bai and Tao Wang

Additional information is available at the end of the chapter

http://dx.doi.org/10.5772/67750

Abstract

Richtmyer-Meshkov instability and turbulent mixing are fundamental problems of multi-materials interface dynamics, which mainly focuses on the growth of perturbation on the interface and mixing of different materials. It is very important in many applications such as inertial confinement fusion, high-speed combustion, supernova, etc. In this chapter, we will gain advances in understanding this problem by numerical investigations, including the numerical method and program we used, the verification and validation of numerical method and program, the growth laws and mechanics of turbulent mixing, the effects of initial conditions, the dynamic behavior, and some new phenomenon for Richtmyer-Meshkov instability and turbulent mixing.

Keywords: Richtmyer-Meshkov instability, turbulent mixing, interface dynamics, dynamic behavior

1. Introduction

When a shock wave accelerates a perturbed interface separating two different fluids, the Richtmyer-Meshkov (RM) instability will occur. This phenomenon was theoretically analyzed by Richtmyer [1] for the first time in 1960, which was confirmed in experiment by Meshkov [2] in 1969. The main mechanism of the Richtmyer-Meshkov instability (RMI) is the baroclinic vorticity deposition at the interface due to the misalignment of the pressure gradient across the shock and the local density gradient at the interface ($\nabla\rho \times \nabla p \neq 0$). At the beginning, the perturbations grow linearly. When entering the nonlinear stage, the perturbations develop into complex structures formed by bubbles (the portion of the light fluid penetrating into the heavy fluid) and spikes shaped with "mushrooms" (the portion of the heavy fluid penetrating into the light fluid), see **Figure 1**. Afterward, the mushroom structures are eroded and broke up which results in the turbulent mixing eventually. When

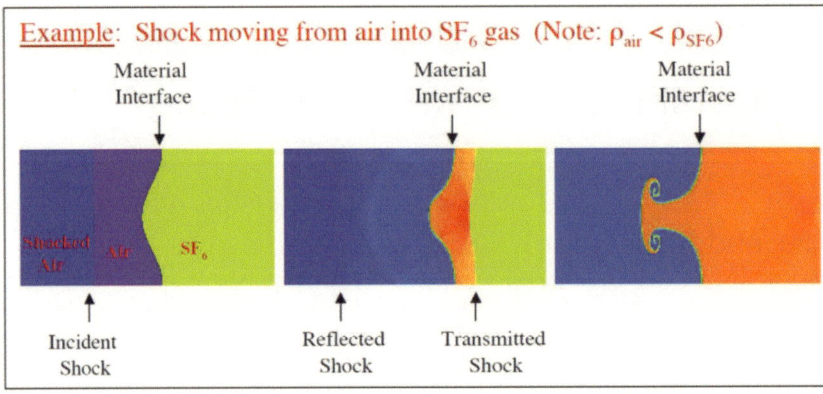

Figure 1. Development of the Richtmyer-Meshkov perturbed interface between air and SF_6 gases [3].

the incident shock wave impacts the initial perturbed interface, it bifurcates into a transmitted shock wave and a reflected wave. Depending on the material properties of the fluid on both sides of the interface, the reflected wave can be either a shock or a rarefaction wave. The criterion is that when the incident shock wave interacts with the interface from light fluid to heavy fluid, the reflected wave is a shock wave; otherwise, it is a rarefaction wave. If the transmitted shock wave meets a rigid wall and is reflected back to collide with the interface once again, this process is called reshock which can advance the transition to turbulent mixing.

The Richtmyer-Meshkov instability and induced turbulent mixing are very important in a variety of man-made applications and natural phenomena such as inertial confinement fusion (ICF) [4], deflagration-to-detonation transition (DDT) [5], high-speed combustion [6], and astrophysics (i.e., supernova explosions) [7]. For ICF, the ablative shell that encapsulates the deuterium-tritium fuel becomes RM unstable as it is accelerated inward by the ablation of its outer surface by laser or secondary X-ray radiation. The degree of compression achievable in laser fusion experiments is ultimately limited by Richtmyer-Meshkov and Rayleigh-Taylor instabilities. Thus, these instabilities represent the most significant barriers to attaining positive-net-yield fusion reactions in laser fusion facilities. For the supersonic combustion, the Richtmyer-Meshkov instability caused by the interaction of a shock wave with a flame front can greatly promote the mixing of fuel and oxidant and enhance the burning rate. For the supernova explosions, the Richtmyer-Meshkov instability was believed to occur when the outward propagating shock wave generated by the collapsing core of a dying star passes over the helium-hydrogen interface. Observations of the optical output of the supernova 1987A suggest that the outer regions of the supernova were much more uniformly mixed than expected, and indicating significant Richtmyer-Meshkov mixing had occurred [8]. The Richtmyer-Meshkov instability and turbulent mixing are so important and have gained significant attention. However, the turbulent mixing is a complicated three-dimensional unstable flow, which spans a wide range of time-space scales.

http://dx.doi.org/10.5772/67750

2. Numerical method and program

By applying the piecewise parabolic method (PPM), volume of fluid (VOF), parallel technique, and so on, we have developed a series of Euler compressible multi-fluid dynamic programs with three orders precision, such as MFPPP [9], MVPPM [10], and multi-viscous-flow and turbulence (MVFT) [11, 12]. MFPPM program is not considering the fluid viscosity, which only solves the Euler equations. MVPPM program is considering the molecular viscosity but it is not changed with temperature. Multi-viscous-flow and turbulence (MVFT) is a large-eddy simulation (LES) program that has four choices of subgrid-scale (SGS) stress models including the Smagorinsky model [13], Vreman model (VM) [14], dynamic Smagorinsky model (DSM) [15], and stretched-vortex model (SVM) [16].

2.1. Governing equations

The governing equations of large-eddy simulation are the Favre-filtered compressible multi-viscous-flow Navier-Stokes (NS) equations and are written as in tensor form:

$$
\begin{aligned}
&\frac{\partial \overline{\rho}}{\partial t}+\frac{\partial \overline{\rho}\tilde{u}j}{\partial x_j}=0\\
&\frac{\partial \overline{\rho}\tilde{u}i}{\partial t}+\frac{\partial(\overline{\rho}\tilde{u}_i\tilde{u}_j+\overline{p}\delta_{ij})}{\partial x_j}=\frac{\partial \sigma_{ij}}{\partial x_j}-\frac{\partial \tau_{ij}}{\partial x_j}\\
&\frac{\partial \overline{\rho}\overline{E}}{\partial t}+\frac{\partial(\overline{\rho}\tilde{u}_j\overline{E}+\overline{p}\tilde{u}_j)}{\partial x_j}=\frac{\partial(\sigma_{ij}-\tau_{ij})\tilde{u}_i}{\partial x_j}-\frac{\partial\left(q_j^l+Q_j^T\right)}{\partial x_j}\\
&\frac{\partial \tilde{Y}^{(s)}}{\partial t}+\tilde{u}_j\frac{\partial \tilde{Y}^{(s)}}{\partial x_j}=\frac{\partial}{\partial x_j}\left(\tilde{D}\frac{\partial \tilde{Y}^{(s)}}{\partial x_j}\right)-\frac{\partial Q_j^Y}{\partial x_j}\qquad s=1,\ 2,\ \cdots,\ N-1
\end{aligned}
\tag{1}
$$

here i and j represent the directions of x, y, and z, and the same subscripts denote the tensor summation; $\overline{\rho}$, $\tilde{u}_k(k=i,j)$, $\overline{p}$, and $\overline{E}$ are the resolved-scale density, velocity, pressure, and total energy per unit mass; N is the number of species; $\tilde{Y}(s)$ is the volume fraction of the sth fluid and satisfies $\sum_1^N \tilde{Y}^{(s)}=1$; $\tilde{D}$ is the diffusion coefficient set to $\tilde{D}=\nu/\mathrm{Sc}$, where ν is the kinematic viscosity of the fluid, Sc is the Schmidt number. σ_{ij} is the deviatoric Newtonian stress tensor, i.e.

$$
\sigma_{ij}=\mu_l\left(\frac{\partial \tilde{u}_i}{\partial x_j}+\frac{\partial \tilde{u}_j}{\partial x_i}-\frac{2}{3}\delta_{ij}\left(\frac{\partial \tilde{u}_k}{\partial x_k}\right)\right)
\tag{2}
$$

where μ_l is the dynamic viscosity. $q_j^l=-\lambda_l\partial\tilde{T}/\partial x_j$ is the resolved heat transport flux per unit time and space, $\lambda_l=\mu_l c_p/\mathrm{Pr}_l$ is the resolved heat conduction coefficient, c_p is the constant pressure specific heat, Pr_l is the Prandtl number, and $\tilde{T}$ is the fluid temperature. The equation of state (EOS) has choices of the ideal gas state form, stiffen gas state form, reduction form of Gruneisen EOS for condensed matter.

For a multi-material mixture case, the average quantities and physical properties of a mixed phase are supposed as a volume-weighted sum over the set of components $(\overline{\rho})^{(s)}$, $(\overline{\rho}\tilde{u}_i)^{(s)}$, and $\left(\overline{\rho}\tilde{E}\right)^{(s)}$ used at the initial time, and so on for each, respectively,

$$\overline{\rho} = \sum_{s=1}^{N} \tilde{Y}^{(s)} \overline{\rho}^{(s)} \tag{3}$$

$$\overline{\rho}\tilde{u}_i = \sum_{s=1}^{N} \tilde{Y}^{(s)} (\overline{\rho}\tilde{u}_i)^{(s)} \tag{4}$$

$$\overline{\rho}\tilde{E} = \sum_{s=1}^{N} \tilde{Y}^{(s)} \left[(\overline{\rho e})^{(s)} + \frac{1}{2}\left(\overline{\rho}\tilde{u}_i^2\right)^{(s)} \right] \tag{5}$$

$$\frac{1}{\gamma - 1} = \sum_{s=1}^{N} \frac{\tilde{Y}^{(s)}}{\gamma^{(s)} - 1} \tag{6}$$

$$\mu = \sum_{s=1}^{N} \mu^{(s)} \tilde{Y}^{(s)} \tag{7}$$

$$\overline{D} = \sum_{s=1}^{N} \overline{D}^{(s)} \tilde{Y}^{(s)} \tag{8}$$

For large-eddy simulation, in the filtering operation, the unresolved-scale motions identified as the "subgrid-scale" are filtered, but the effects of subgrid-scale motions on the resolved-scale motions are retained in the governing equations in the form of SGS turbulence transport fluxes, which must be modeled to complete the closure of LES equations. The SGS stress tensor, the heat, and scalar transport flux are defined as

$$\tau_{ij} = \overline{\rho}\left(\widetilde{u_i u_j} - \tilde{u}_i \tilde{u}_j\right) \tag{9}$$

$$Q_j^T = \overline{\rho}\left(\widetilde{c_p T u_j} - \tilde{c}_p \tilde{T} \tilde{u}_j\right) \tag{10}$$

$$Q_j^Y = \left(\widetilde{Y u_j} - \tilde{Y} \tilde{u}_j\right) \tag{11}$$

2.2. Subgrid-scale stress models for LES

2.2.1. Smagorinsky model

The SGS turbulence behavior is assumed to be analogous to the molecular dissipative mechanism, so that the molecular viscosity and diffusion models can be used to simulate the SGS fluxes, and the SGS stress tensor, the heat, and the scalar transport flux are calculated as follows [13]

http://dx.doi.org/10.5772/67750

$$\tau_{ij} = -\mu_t\left(\frac{\partial \tilde{u}_i}{\partial x_j} + \frac{\partial \tilde{u}_j}{\partial x_i} - \frac{2}{3}\delta_{ij}\left(\frac{\partial \tilde{u}_k}{\partial x_k}\right)\right) \tag{12}$$

$$Q_j^T = -\lambda_t \frac{\partial \tilde{T}}{\partial x_j} = -\frac{\mu_t c_p}{\mathrm{Pr}_t}\frac{\partial \tilde{T}}{\partial x_j} \tag{13}$$

$$Q_j^Y = -D_t \frac{\partial \tilde{Y}^{(s)}}{\partial x_j} = -\frac{\mu_t}{\overline{\rho} Sc_t}\frac{\partial \tilde{Y}^{(s)}}{\partial x_j} \tag{14}$$

where the SGS turbulent viscosity μ_t is calculated by the Smagorinsky eddy viscosity model,

$$\mu_t = 2C\rho\Delta^2|\overline{S}| \tag{15}$$

where the dimensionless coefficient $C = C_S^2$, for the isotropic turbulence, the model constant is $C_S = 0.17$, Δ is the grid-filter width, and $|\overline{S}| = (2\overline{S}_{ij}\overline{S}_{ij})^{1/2}$ is the magnitude of the resolved strain-rate tensor,

$$\overline{S}_{ij} = \frac{1}{2}\left(\frac{\partial \tilde{u}_i}{\partial x_j} + \frac{\partial \tilde{u}_j}{\partial x_i}\right) \tag{16}$$

2.2.2. Vreman model

The Vreman SGS model is also an eddy viscosity model and is as follows [14]:

$$\mu_t = c\overline{\rho}\sqrt{\frac{B_\beta}{\alpha_{ij}\alpha_{ij}}} \tag{17}$$

with

$$\begin{aligned} \alpha_{ij} &= \partial_i \tilde{u}_j = \frac{\partial \tilde{u}_j}{\partial x_i} \\ \beta_{ij} &= \Delta_m^2 \alpha_{mi}\alpha_{mj} \\ B_\beta &= \beta_{11}\beta_{22} - \beta_{12}^2 + \beta_{11}\beta_{33} - \beta_{13}^2 + \beta_{22}\beta_{33} - \beta_{23}^2 \end{aligned} \tag{18}$$

The model constant c is related to the Smagorinsky SGS model constant C_S by $c \approx 2.5C_S^2$. The symbol α represents the 3×3 matrix of derivatives of the filtered velocity $\tilde{u}$. If $\alpha_{ij}\alpha_{ij}$ equals zero, μ_t is consistently defined to be zero. In fact, B_β is an invariant of the matrix β, while $\alpha_{ij}\alpha_{ij}$ is an invariant of $\alpha^T\alpha$. If the filter width is the same in each direction, then $\Delta_i = \Delta$ and $\beta = \Delta^2\alpha^T\alpha$.

2.2.3. Dynamic Smagorinsky model

In general, the turbulent kinetic energy is transferred from large scales to small scales in turbulent flows, which is called forward scatter of energy, and then it is dissipated by the viscous action. However, the reversed energy flow from small scales to large scales (the process called backscatter) may also occur. Although the backscatter is just a small range of local phenomenon, it has been shown to be of importance, especially in the transition regime. The physical mechanism of most of the widely used SGS models is the forward scatter of energy; in other words, the SGS models are absolutely dissipative, such as the eddy viscosity models (Smagorinsky and Vreman SGS models). In order to account for the backscatter of energy, several modifications to the eddy viscosity models have been proposed. An improvement is to calculate the eddy viscosity coefficient dynamically which is a function of space and time and can be negative in some regions [15].

The dynamic Smagorinsky model for compressible turbulence is as follows, the model coefficient is

$$C_D = \frac{\langle L_{ij} M_{ij} \rangle}{\langle M_{ij} M_{ij} \rangle} \tag{19}$$

where L_{ij} is the Leonard stress, $\langle\ \rangle$ indicates the statistics averaging.

$$L_{ij} = \widehat{\overline{\rho}\tilde{u}_i\tilde{u}_j} - \frac{1}{\hat{\overline{\rho}}}\left(\widehat{\overline{\rho}\tilde{u}_i}\widehat{\overline{\rho}\tilde{u}_j}\right) \tag{20}$$

$$M_{ij} = -\left[2\hat{\overline{\Delta}}^2\hat{\overline{\rho}}\left|\hat{\overline{S}}\right|\left(\hat{\overline{S}}_{ij} - \frac{\delta_{ij}}{3}\hat{\overline{S}}_{kk}\right) - 2\overline{\Delta}^2\overline{\rho}\left|\overline{S}\right|\widehat{\left(\overline{S}_{ij} - \frac{\delta_{ij}}{3}\overline{S}_{kk}\right)}\right] \tag{21}$$

the overbars of "‾" and "^" denote the grid filter and test filter, respectively.

2.2.4. Stretched-vortex model

The stretched-vortex model is based on an explicit structural modeling of small-scale dynamics. It can simulate the multiscale compressible turbulence and allows the anisotropy of the flow to be extended to the dissipation scale. In the stretched-vortex model, the flow within a computational grid cell is assumed to result from an ensemble of straight, nearly axisymmetric vortices aligned with the local resolved scale strain or vorticity. The resulting SGS flux terms are [16]

$$\tau_{ij} = \overline{\rho}\tilde{k}\left(\delta_{ij} - e_i^v e_j^v\right) \tag{22}$$

$$Q_i^T = -\overline{\rho}\frac{\Delta_c}{2}\tilde{k}^{1/2}\left(\delta_{ij} - e_i^v e_j^v\right)\frac{\partial(\tilde{c}_p\tilde{T})}{\partial x_j} \tag{23}$$

$$Q_i^Y = -\frac{\Delta_c}{2}\tilde{k}^{1/2}\left(\delta_{ij} - e_i^v e_j^v\right)\frac{\partial\tilde{Y}}{\partial x_j} \tag{24}$$

where $\tilde{k} = \int_{k_c}^{\infty} E(k)dk$ is the subgrid turbulent kinetic energy, e^v is the unit vector aligned with the subgrid vortex axis, $k_c = \pi/\Delta_c$ is the cutoff wave number, $E(k) = K_0\varepsilon^{2/3}k^{-5/3}\exp\left[-2k^2\nu/(3|\tilde{a}|)\right]$ represents the energy spectrum of subgrid vortices, and K_0 is the Kolmogorov prefactor, ε is the

http://dx.doi.org/10.5772/67750

local cell-averaged dissipation, $\tilde{a} = \tilde{S}_{ij} e_i^v e_j^v$ is the axial strain along the subgrid vortex axis, $\tilde{S}_{ij}$ denotes the resolved rate-of-strain tensor.

2.3. Numerical algorithm

For the convenience, in the following sections, the overbar and overtilde of variables are omitted. Operator splitting technique is used to solve the physical problems, described by Eq. (1), into three sub-processes, i.e., the computation of inviscid flux, viscous flux, and heat flux. Eq. (1) can be split into two equations as follows:

$$\begin{aligned} &\frac{\partial \rho}{\partial t} + \frac{\partial \rho u_j}{\partial x_j} = 0 \\ &\frac{\partial \rho u_i}{\partial t} + \frac{\partial \left(\rho u_i u_j + p \delta_{ij}\right)}{\partial x_j} = 0 \\ &\frac{\partial \rho E}{\partial t} + \frac{\partial \left(\rho u_j E + p u_j\right)}{\partial x_j} = 0 \\ &\frac{\partial Y^{(s)}}{\partial t} + u_j \frac{\partial Y^{(s)}}{\partial x_j} = 0 \quad s = 1,\ 2,\ \ldots,\ N-1 \end{aligned} \tag{25}$$

and

$$\begin{aligned} &\frac{\partial \rho}{\partial t} = 0 \\ &\frac{\partial \rho u_i}{\partial t} = \frac{\partial \sigma_{ij}}{\partial x_j} - \frac{\partial \tau_{ij}}{\partial x_j} \\ &\frac{\partial \rho E}{\partial t} = \frac{\partial \left(\sigma_{ij} - \tau_{ij}\right) u_i}{\partial x_j} - \frac{\partial \left(q_j^l + Q_j^T\right)}{\partial x_j} \\ &\frac{\partial Y^{(s)}}{\partial t} = \frac{\partial}{\partial x_j}\left(D \frac{\partial Y^{(s)}}{\partial x_j}\right) - \frac{\partial Q_j^Y}{\partial x_j} \quad s = 1,\ 2,\ \ldots,\ N-1 \end{aligned} \tag{26}$$

For the inviscid flux, the three-dimensional problem can be simplified into one-dimensional (1D) problem in three directions of x, y, and z by the dimension splitting technique,

$$\begin{aligned} &\frac{\partial \rho}{\partial t} + \frac{\partial (\rho u)}{\partial x} = 0 \\ &\frac{\partial (\rho u)}{\partial t} + \frac{\partial \left(\rho u^2 + p\right)}{\partial x} = 0 \\ &\frac{\partial (\rho E)}{\partial t} + \frac{\partial (\rho u E + p u)}{\partial x} = 0 \\ &\frac{\partial Y^{(s)}}{\partial t} + u \frac{\partial Y^{(s)}}{\partial x} = 0 \qquad s = 1,\ 2,\ \ldots,\ N-1 \end{aligned} \tag{27}$$

And then the one-dimensional Eq. (27) in each direction is resolved by the two-step Lagrange-Remapping algorithm. Also, one time step is divided into four substeps: (i) the piecewise parabolic interpolating of physical quantities, (ii) solving the Riemann problems approximately, (iii) marching of the Lagrange equations, and (iv) remapping the physical quantities back to the stationary Euler meshes. The orders of accuracy of the spatial and temporal schemes are the third and second, respectively, for smooth flows.

2.3.1. Lagrange step of finite volume method

The Lagrange equations in 1D for multi-fluid can be written as

$$\begin{cases} \dfrac{\partial \tau}{\partial t} - \dfrac{\partial (r^a u)}{\partial m} = 0 \\ \dfrac{\partial u}{\partial t} + r^a \dfrac{\partial p}{\partial m} = 0 \\ \dfrac{\partial E}{\partial t} + \dfrac{\partial (r^a u p)}{\partial m} = 0 \\ \dfrac{\partial Y^{(s)}}{\partial t} = 0 \qquad s = 1,\ 2,\ \ldots,\ N-1 \end{cases} \tag{28}$$

where τ is the specific volume, α is 0, 1, and 2 corresponding to the plane, axial symmetry, and spherical symmetry problems, r is the Lagrange spatial coordinates, m is the Lagrange mass coordinates $m = \int_{r_0}^{r} \rho r^{\alpha} dr$, $u = \frac{\partial r}{\partial t}$, $m_{j-1/2}$ and $m_{j+1/2}$ are the mass coordinates of both sides of the jth grid, $\Delta m = m_{j+1/2} - m_{j-1/2}$. The mass average of physical quantities $\left(\tau,\ u,\ E,\ Y^{(s)}\right)_j^n$ for the jth grid can be defined as

$$\begin{pmatrix} \tau \\ u \\ E \\ Y^{(s)} \end{pmatrix}_j^n = \frac{1}{\Delta m_j} \int_{m_{j-1/2}}^{m_{j+1/2}} \begin{pmatrix} \tau(m,\ t_n) \\ u(m,\ t_n) \\ E(m,\ t_n) \\ Y^{(s)}(m,\ t_n) \end{pmatrix} dm \tag{29}$$

Because of the calculations are referred to as scalars, here the averaged physical quantities in a cell are written as a uniformity Q_j^n at time t. The piecewise parabolic function $Q_j^n(m)$ in cells are constructed to compute the time average of physical quantity Q on both sides of the grid edge $m_{j+1/2}, \tilde{Q}_{j+1/2,L}$ and $\tilde{Q}_{j+1/2,R}$. Then the Riemann problem at the grid edge $m_{j+1/2}$ is solved by using double shock wave approximation. After Lagrange marching step, we can obtain distribution of the physical quantities $\left\{\tau_j^{n+1}\right\}$, $\left\{u_j^{n+1}\right\}$, $\left\{E_j^{n+1}\right\}$, and the position of new grids $\left\{r_{j\pm 1/2}^{n+1}\right\}$ at time t_{n+1}, the pressure $\left\{p_j^{n+1}\right\}$ can be computed by equation of state, the $\left\{\left(Y^{(s)}\right)_j^{n+1}\right\}$ will not change. The marching formula is as follows

http://dx.doi.org/10.5772/67750

$$
\begin{cases}
r_{j+1/2}^{n+1} = r_{j+1/2}^{n} + \tilde{u}_{j+1/2}\Delta t \\
\tau_j^{n+1} = \dfrac{\left(r_{j+1/2}^{n+1}\right)^{\alpha+1} - \left(r_{j-1/2}^{n+1}\right)^{\alpha+1}}{(\alpha+1)\Delta m_j} \\
u_j^{n+1} = u_j^n - \dfrac{1}{2}\dfrac{\Delta t}{\Delta m_j}\left(\sigma_{j+1/2} + \sigma_{j-1/2}\right)\left(\tilde{p}_{j+1/2} - \tilde{p}_{j-1/2}\right) \\
E_j^{n+1} = E_j^n - \dfrac{\Delta t}{\Delta m_j}\left(\sigma_{j+1/2}\tilde{u}_{j+1/2}\tilde{p}_{j+1/2} - \sigma_{j-1/2}\tilde{u}_{j-1/2}\tilde{p}_{j-1/2}\right)
\end{cases}
\tag{30}
$$

$$
\sigma_{j+1/2} = \frac{\left(r_{j+1/2}^{n+1}\right)^{\alpha+1} - \left(r_{j+1/2}^{n}\right)^{\alpha+1}}{(\alpha+1)\left(r_{j+1/2}^{n+1} - r_{j+1/2}^{n}\right)}
\tag{31}
$$

2.3.2. Remap step

In Lagrange step, the computational cells distort to follow the material motion, and in Remap step, the averaged physical quantities at the distorted Lagrange cells are remapped back to the stationary Euler meshes. The piecewise parabolic interpolation and integral methods in this step are the same as the ones in Lagrange step.

After Remap step, we define the physical quantities in Euler mesh as $\left(\rho_{\text{Euler}}\right)_j^{n+1}$, $\left(u_{\text{Euler}}\right)_j^{n+1}$, $\left(E_{\text{Euler}}\right)_j^{n+1}$, $\left(\left(Y^{(s)}\right)_{\text{Euler}}\right)_j^{n+1}$, the volume of fluid through the grid boundary $j+1/2$ as $\Delta V_{j+1/2}^*$, and the average of physical quantities as $\left(\rho^*, u^*, p^*, \left(Y^{(s)}\right)^*\right)_{j+1/2}$. The density after Lagrange step is $\rho_j^{n+1} = 1/\tau_j^{n+1}$. In Remap step

$$
\Delta m_j^{n+1} = \rho_j^{n+1} V_j^{n+1}
\tag{32}
$$

$$
\Delta m_j^n = \left(\rho_{\text{Euler}}\right)_j^n V_j^n
\tag{33}
$$

$$
\Delta m_{j+1/2}^* = \rho_{j+1/2}^* \Delta V_{j+1/2}^*
\tag{34}
$$

$$
\Delta E_{j+1/2}^* = \frac{1}{2}\left(u_{j+1/2}^*\right)^2 \Delta m_{j+1/2}^* + \Delta e_{j+1/2}^*
\tag{35}
$$

$$
\left(\rho_{\text{Euler}}\right)_j^{n+1} = \left[\Delta m_j^{n+1} - \left(\rho_{j+1/2}^* \Delta V_{j+1/2}^* - \rho_{j-1/2}^* \Delta V_{j-1/2}^*\right)\right] \Big/ V_j^n
\tag{36}
$$

$$
\left(u_{\text{Euler}}\right)_j^{n+1} = \left[u_j^{n+1}\Delta m_j^{n+1} - \left(u_{j+1/2}^*\Delta m_{j+1/2}^* - u_{j-1/2}^*\Delta m_{j-1/2}^*\right)\right] \Big/ \Delta m_j^n
\tag{37}
$$

$$
\left(\left(Y^{(s)}\right)_{\text{Euler}}\right)_j^{n+1} = \left[\left(Y^{(s)}\right)_j^{n+1} V_j^{n+1} - \left(\left(Y^{(s)}\right)_{j+1/2}^* \Delta V_{j+1/2}^* - \left(Y^{(s)}\right)_{j-1/2}^* \Delta V_{j-1/2}^*\right)\right] \Big/ V_j^n
\tag{38}
$$

$$\left(E_{\text{Euler}}\right)_j^{n+1} = \left[E_j^{n+1}\Delta m_j^{n+1} - \left(\Delta E_{j+1/2}^* - \Delta E_{j-1/2}^*\right)\right]/\Delta m_j^n \tag{39}$$

where $\Delta e_{j+1/2}^*$ is the advection of specific energy through the grid boundary $j + 1/2$

2.3.3. Viscous flux, heat flux, and scalar flux

The viscous flux, heat flux, and scalar flux of Eq. (26) are calculated based on the computed inviscid flux by using second-order spatial center difference and two-step Runge-Kutta time marching. The first equation of Eq. (26) can be neglected, and which is written in conserved form

$$\frac{\partial \overline{\mathbf{U}}}{\partial t} + \frac{\partial \overline{\mathbf{F}}}{\partial x} + \frac{\partial \overline{\mathbf{G}}}{\partial y} + \frac{\partial \overline{\mathbf{H}}}{\partial z} = 0 \tag{40}$$

and

$$\begin{aligned}
\overline{\mathbf{U}} &= \left(\rho u, \rho v, \rho w, \rho E\right)^T \\
\overline{\mathbf{F}} &= \left(-\sigma_{xx}, -\sigma_{yx}, -\sigma_{zx}, -u\sigma_{xx} - v\sigma_{yx} - w\sigma_{zx} + q_x\right)^T \\
\overline{\mathbf{G}} &= \left(-\sigma_{xy}, -\sigma_{yy}, -\sigma_{zy}, -u\sigma_{xy} - v\sigma_{yy} - w\sigma_{zy} + q_y\right)^T \\
\overline{\mathbf{H}} &= \left(-\sigma_{xz}, -\sigma_{yz}, -\sigma_{zz}, -u\sigma_{xz} - v\sigma_{yz} - w\sigma_{zz} + q_z\right)^T
\end{aligned} \tag{41}$$

In the Cartesian coordinate system, the spatial derivative terms of Eq. (40) can be dispersed as follows:

$$\mathbf{L}_h\left(\overline{\mathbf{U}}_{i,j,k}^n\right) = \frac{\overline{\mathbf{F}}_{i-1/2,j,k}^n - \overline{\mathbf{F}}_{i+1/2,j,k}^n}{\Delta x} + \frac{\overline{\mathbf{G}}_{i,j-1/2,k}^n - \overline{\mathbf{G}}_{i,j+1/2,k}^n}{\Delta y} + \frac{\overline{\mathbf{H}}_{i,j,k-1/2}^n - \overline{\mathbf{H}}_{i,j,k+1/2}^n}{\Delta z} \tag{42}$$

Then, Eq. (40) is solved by two Runge-Kutta time marching,

$$\begin{cases}
\overline{\mathbf{U}}_{i,j,k}^{(l)} = \overline{\mathbf{U}}_{i,j,k}^{E} + \Delta t \mathbf{L}_h\left(\mathbf{U}_{i,j,k}^n\right) \\
\mathbf{U}_{i,j,k}^{n+1} = \dfrac{1}{2}\left[\overline{\mathbf{U}}_{i,j,k}^{E} + \overline{\mathbf{U}}_{i,j,k}^{(l)} + \Delta t \mathbf{L}_h\left(\overline{\mathbf{U}}_{i,j,k}^{(l)}\right)\right]
\end{cases} \tag{43}$$

The detailed descriptions of numerical algorithm are referred to Ref. [17].

3. Verification and validation

In this section, the validity and reliability of our compressible multi-fluid dynamic programs are to be verified and validated by comparisons with analytical solutions and hydrodynamic interface instability experiments in a shock tube.

http://dx.doi.org/10.5772/67750

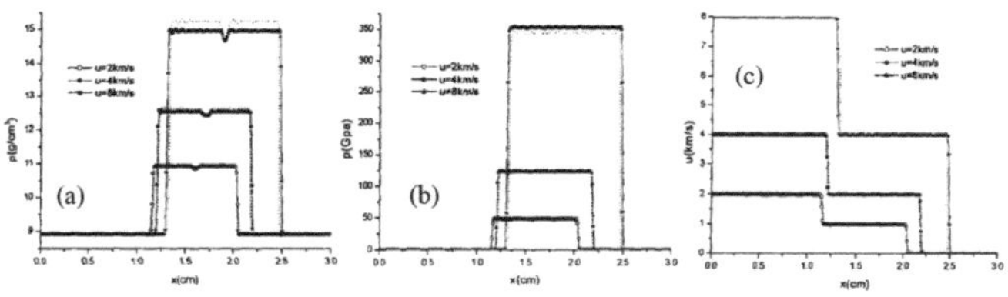

Figure 2. Comparison of postshock density (a), pressure (b), and velocity (c) in copper at $t = 1.0$ μs when a copper pellet collides with a copper target at three different velocities.

3.1. Riemann problem of condensed matter

A copper pellet collides with a copper target with three velocities of 2, 4, and 8 km/s. Using the reduction form of Gruneisen EOS for copper in simulations, the one-dimensional numerical results of postshock density (a), pressure (b), and velocity (c) compare with theoretical solutions, as shown in **Figure 2**, the solid lines corresponding to numerical results and the dot lines corresponding to the theoretical solutions.

3.2. Riemann problem of one-dimensional gas/liquid

At initial time, the region [0, 1.0 cm] is filled with gas with high pressure 1.0×10^8 Pa and density 1.29 g/cm^3. The region [1.0 cm, 5.0 cm] is filled with water with pressure 1.0×10^5 Pa and density 1.0 g/cm^3. The gas and water are all described by Stiffen gas EOS. The left and right boundaries are flow. **Figure 3** shows the distributions of the density (a), pressure (b), and velocity (c) at 20 μs. There are a forward shock wave and a backward rarefaction wave after interaction. The pressure and velocity around the interface are well continuous and have no nonphysical oscillation.

3.3. Single-mode Richtmyer-Meshkov instability

The two- and three-dimensional single-mode Richtmyer-Meshkov instabilities are numerical simulated by MVPPM program, which also compare with the theoretical model [18]. The initial small perturbation is a sinusoidal one with wavelength 60 mm (global wavelength for a three-dimensional case) and amplitude 1.0 mm. The incident air shock wave with Mach number 1.2 impacts the air/SF_6 single-mode interface. **Figure 4** shows the comparisons of amplitude of single-mode perturbed interface with the linear impulsive and nonlinear models, the left one corresponding to the two-dimensional (2D) results, and the right one corresponding to the three-dimensional results.

3.4. Simulations of shock tube experiments

The compressible multi-fluid dynamic programs are used to simulated several shock tube experiments of interface instability for validation. These experiments include planar interface and gas cylinder shock tube experiment and planar and cylindrical jelly experiments.

Figure 5 shows the comparison of 2D calculated width of turbulent mixing zone (TMZ) [19] of Leinov's shock tube experiment with reshock with experiment [20] in which the Mach number

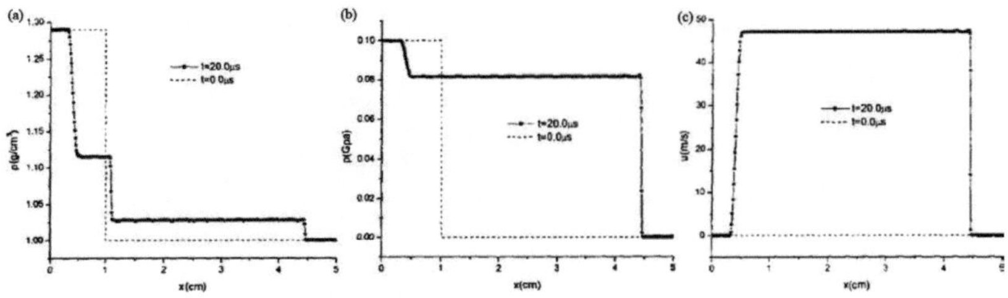

Figure 3. Distributions of the density (a), pressure (b), and velocity (c) at 20 μs for one-dimensional Riemann problem of gas/liquid interface.

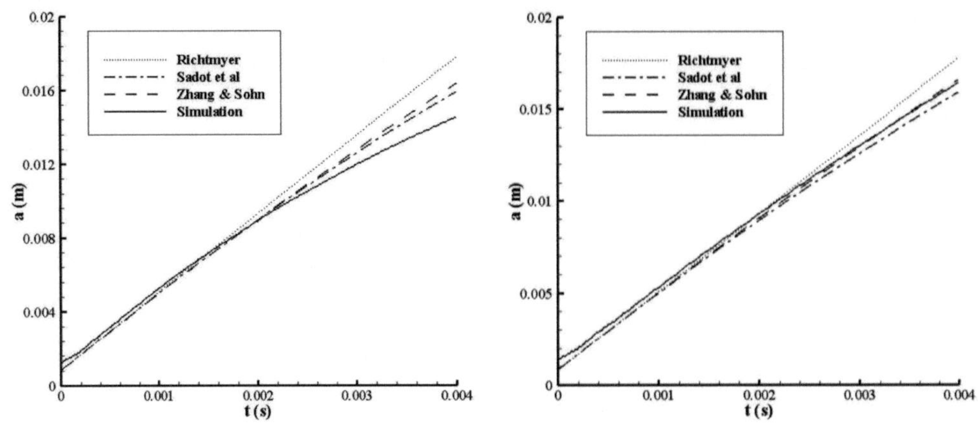

Figure 4. Comparisons of amplitude of single-mode perturbed interface with the linear impulsive model and nonlinear models, left: two-dimension and right: three-dimension.

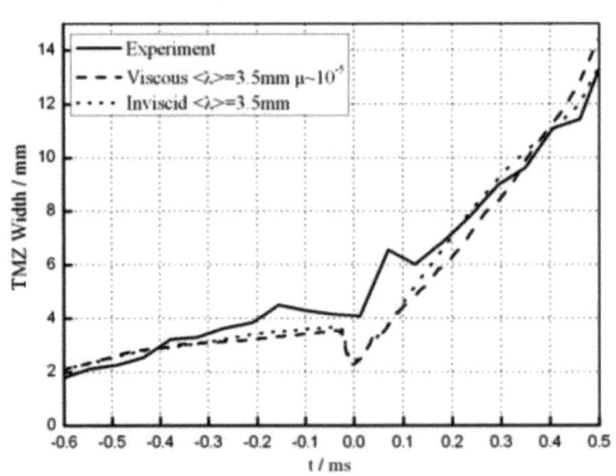

Figure 5. TMZ width versus time ($t = 0$ denotes the reshock arrival to the interface).

http://dx.doi.org/10.5772/67750

of incident air shock wave is 1.2, which impacts the air/SF_6 interface, the grid size is 50 μm. **Figure 6** shows the 3D calculated TMZ width [21] of Poggi's multi-mode shock tube experiment [22] with reshock in which the Mach number of incident SF_6 shock wave is 1.453, which impacts the SF_6/air interface. **Figures 7** [10] and **8** [23] show the calculated and experimental interface images of AWE's SF_6 half-height and double-bump shock tube experiment [24, 25], the Mach number of incident air shock wave is 1.26, respectively. **Figure 9** [26] shows the SF_6 gas cylinder evolution at different times under the initial air shock wave with the Mach number 1.2, **Figure 10** shows the width and height of gas cylinder at different times for experiment and numerical simulations.

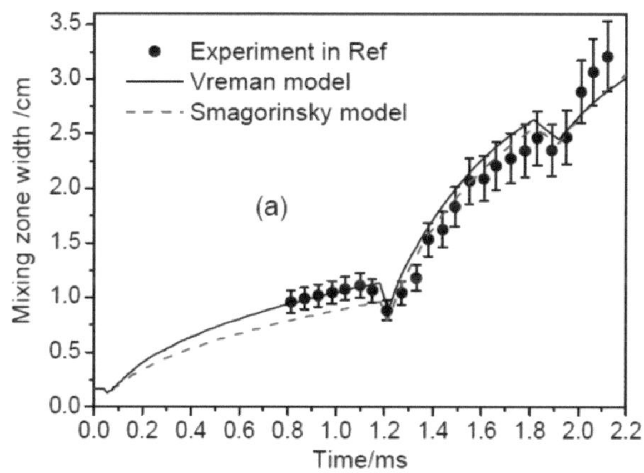

Figure 6. Time history of TMZ width.

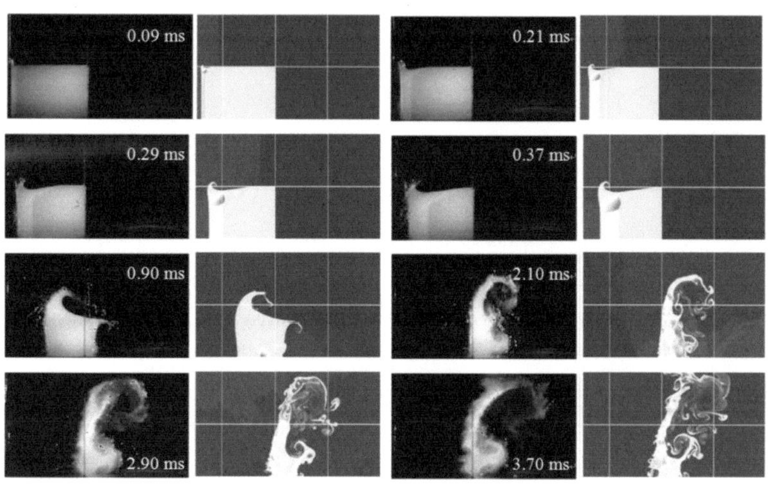

Figure 7. Two dimension calculated from MVFT and experimental half-height interface at different times.

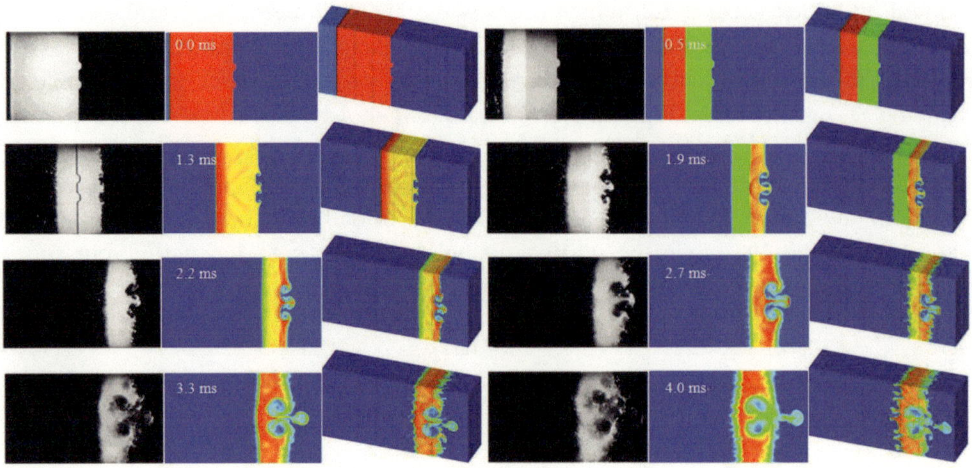

Figure 8. Three dimension calculated from MVFT and experimental double-bump interface at different times, left column corresponding to experimental images, right column corresponding to 3D calculated images, and middle column corresponding to span-wise average images of 3D calculated results.

Figure 9. Evolutions of SF_6 gas cylinder, upper row corresponding to experiment, lower row corresponding to numerical simulation by MVFT program, the time sequences are 0, 50, 190, 330, 470, 610, and 750 μs severally.

For the problems of interface instability with a high density ratio, our hydrodynamic programs are also applicable. For the LLNL's jelly ring experiments [27], the jelly ring only has a sinusoidal periodic initial perturbation at the outer interface and is driven by expansion of the explosion products of a gaseous mixture of C_2H_2 and O_2. The jelly is mainly made of water. The thickness of ring is 15 mm, the mode number of perturbation is 6 and 36, respectively, and the initial amplitude is 1.0 mm. **Figure 11** shows the evolutions of the jelly ring at different times from simulated from LLNL' CALE program (a) and our MVPPM program (b), and the experimental image (c) at 600 μs [28]. **Figure 12** shows the time histories of radius (a) and the amplitude (b) of the outer and inner interfaces of the jelly ring simulated from CALE and

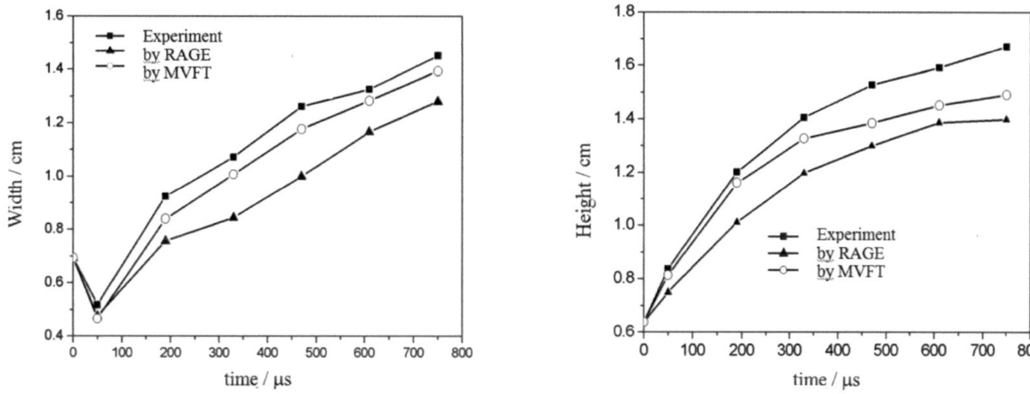

Figure 10. Width and height of gas cylinder at different times for experimental and numerical simulations.

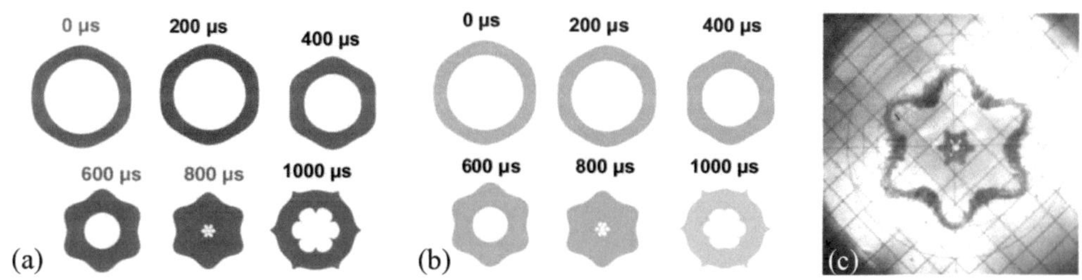

Figure 11. Evolutions of jelly ring at different simulated from CALE (a: left group) and MVPPM (b: middle group) programs, and the experimental image at 600 μs (c).

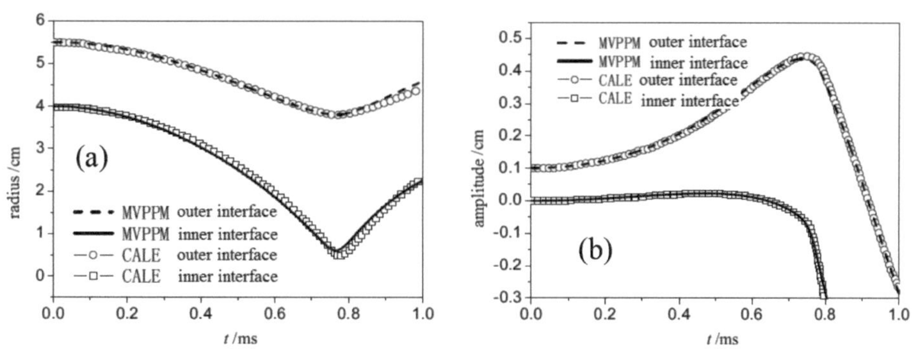

Figure 12. Time histories of radius (a) and the amplitude (b) of the outer and inner interfaces of jelly ring simulated from CALE and MVPPM programs.

MVPPM programs [28]. **Figure 13** shows the images of jelly ring at 600 μs for mode number 36 from experiment (a), CALE (b), and MVPPM (c) programs [28].

From the above comparisons between our numerical simulations and theoretical model, experiments for the problems of hydrodynamic interface instability from the density ratio low

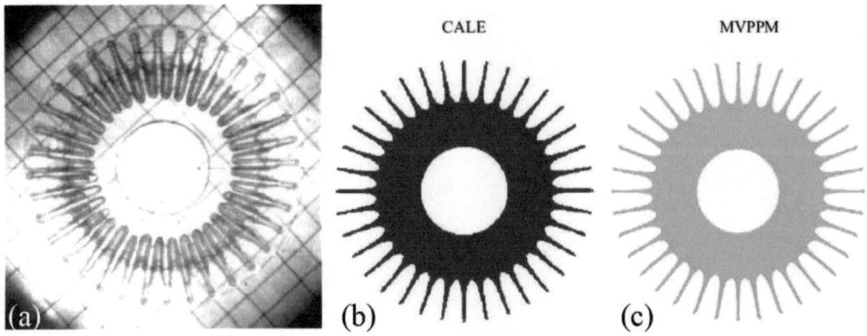

Figure 13. Images of jelly ring at 600 μs for mode number 36 from experiment (a), CALE program (b), and MVPPM program (c).

to the high density ratio, the numerical results agree well with theory and experiments, so the validity and reliability of our compressible multi-fluid dynamic programs have been verified and validated.

4. Interface instability and turbulent mixing

As we know, the physical mechanism for the occurrence of Richtmyer-Meshkov instability is the baroclinic vorticity deposition at the interface resulting from the misalignment of the pressure gradient across the shock front and the local density gradient across the interface. The evolution equation of vorticity is as follows:

$$\frac{d\boldsymbol{\omega}}{dt} = \frac{\nabla\rho \times \nabla p}{\rho^2} + \boldsymbol{\omega} \cdot \nabla \mathbf{u} - \boldsymbol{\omega} \nabla \cdot \mathbf{u} \tag{44}$$

where $\boldsymbol{\omega} = \nabla \times \mathbf{u}$ is the vorticity and viscous terms are neglected. The first term on the right side of Eq. (44) is the baroclinic vorticity production term. The second term on the right side of Eq. (44) is the vortex-stretching term, which is zero in the two-dimensional case, as the vorticity and velocity fields are orthogonal. This term enhances dissipation, resulting in more diffuse and smaller scale structures in the turbulent mixing zone. The third term on the right side is the compression term and does not contribute to the vorticity evolution significantly. The baroclinic vorticity production is much larger when the shock wave impacts the interface and pass through it and constitutes the principal mechanism of the Richtmyer-Meshkov instability.

In this section, we will introduce the growth laws of the Richtmyer-Meshkov instability and the induced turbulent mixing and its dynamic behavior by numerical simulations.

4.1. Growth laws of Richtmyer-Meshkov instability and turbulent mixing

Figure 14 shows the amplitude (a) and growth rate (b) of two-dimensional single-mode Richtmyer-Meshkov instability without reshock for initial perturbation amplitude 1 and 12 mm

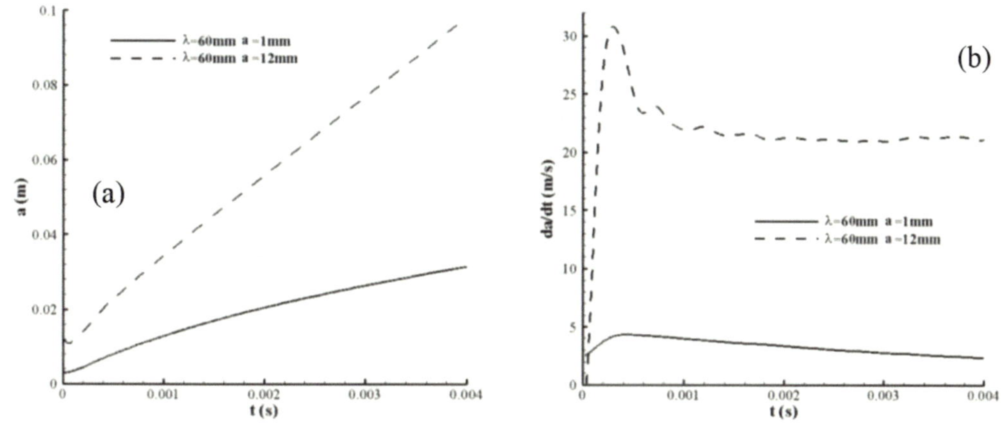

Figure 14. Amplitude (a) and growth rate (b) of two-dimensional single-mode Richtmyer-Meshkov instability without reshock.

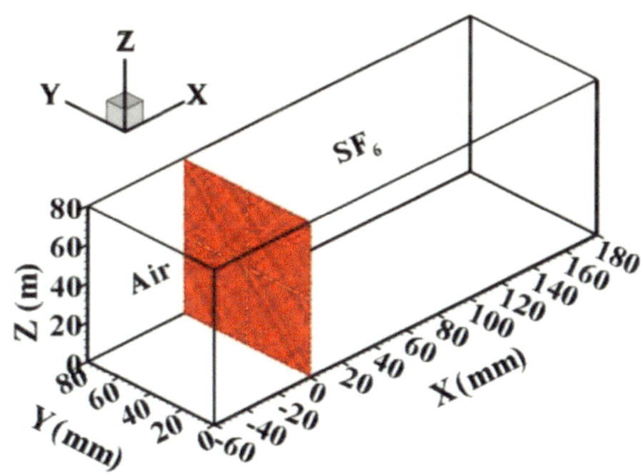

Figure 15. Schematic of shock tube and computational model.

and wavelength 60 mm. After initial shock, the perturbation enters the nonlinear stage quickly. The growth rate increases fast and reaches the highest peak, then will reduce owing to the reduction of the effect of compressibility and the dominant role of flow nonlinearity. At the late times, the amplitude is increasing linearly, and the growth rate remains a constant. For the larger initial amplitude, the perturbation enters the later linear growth stage earlier.

For the multi-mode Richtmyer-Meshkov instability and the induced turbulent mixing with reshock, the initial air shock wave with Mach number 1.2 impacts the air/SF_6 interface. The multi-mode initial perturbation is composed of eight dominant mode wavelengths of 0.8, 1.0, 1.25, 1.6, 2.0, 2.5, 3.2, and 4.0 mm superimposed with a very small random disturbance. The shock tube experiment can be referred to Ref. [29]. **Figure 15** shows the schematic of shock tube and computational model. The transmitted wave rebounds between the interface and the end-wall of shock tube and produces multiple shock-interface interactions. At about 1.7 ms,

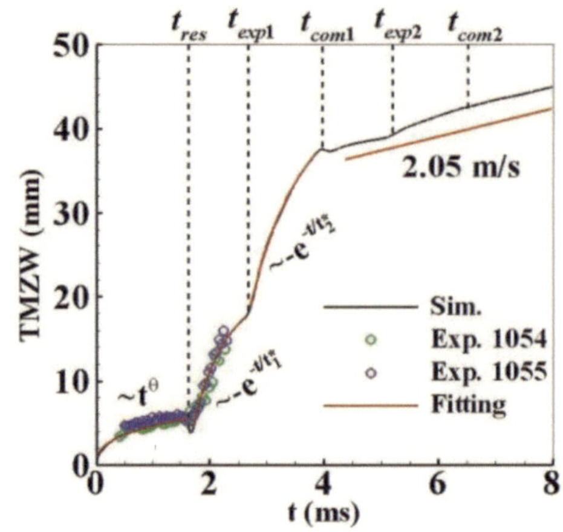

Figure 16. Time history of the TMZ width.

the transmitted shock wave reflected back off the end-wall impacts the interface. **Figure 16** shows the time history of TMZ width, the black line denotes the numerically calculated width and the red line is obtained from fitting of the numerical results. As can be seen, after the initial shock, the TMZ width starts to grow as a power law ~t^{θ} with the value θ to be determined as 0.352. After the reshock, more energy is deposited onto the interface to promote the development of TMZ, and the TMZ width evolves in time as a negative exponential law ~$-e^{-t/t^*}$ where the value of t^* is 0.519. Then after the following interaction of the reflected rarefaction wave with the interface, the TMZ also evolves as a negative exponential law but with a different factor $t^* = 0.875$. Under the subsequent impingements with lower and lower intensity, the TMZ width, after a slight reduction caused by the reflected compression wave, evolves in an approximate linear fashion with a velocity of 2.05 m/s. **Figure 17** shows the instantaneous images of the TMZ visualized by the volume fraction isosurface $Y_{SF6} = 0.1, 0.3, 0.5, 0.7$, and 0.9 from blue to orange at different times, the TMZ exhibits a very complex spatial structure.

4.2. Evaluation of different subgrid-scale stress models

In large-eddy simulation, the effect of small scales on large-scale motions is represented by the SGS stress model. Most of the commonly used SGS models assume that the main function of subgrid scales is to remove energy from the large scales and dissipate it through the action of the viscous forces. But, as we know, in fact the energy is also transferred from the small scales to the large scales (backscatter) in a small and local range. The SGS turbulent dissipation, which is the work of SGS stress, represents the energy transfer between resolved and subgrid scales,

$$\varepsilon_{SGS} = \tau_{ij}\overline{S}_{ij} \tag{45}$$

If it is negative, the subgrid scales remove energy from the resolved scales (forward scatter); if it is positive, they release energy to the resolved scales (backscatter). It is easy to see that the eddy viscosity models such as Smagorinsky model, Vreman model, etc. are absolutely dissipative.

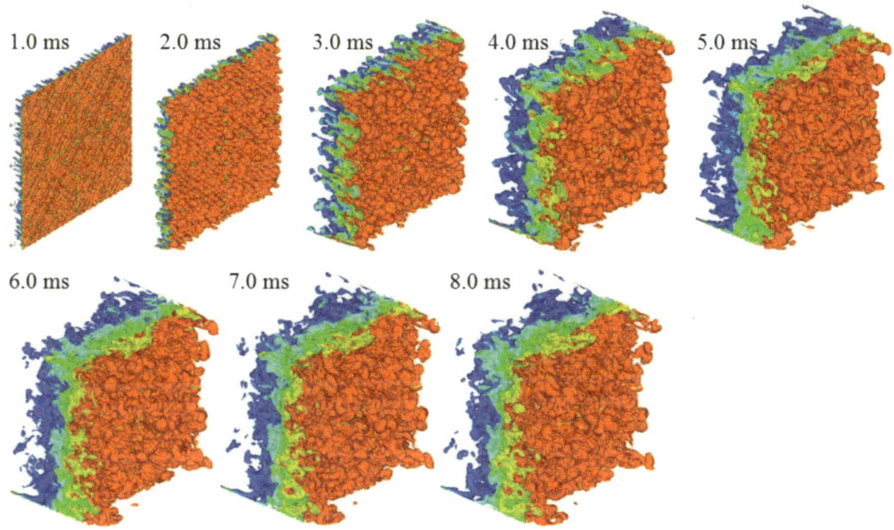

Figure 17. Instantaneous images of TMZ visualized by the volume fraction isosurface $Y_{SF6} = 0.1, 0.3, 0.5, 0.7$, and 0.9 from blue to orange at different times.

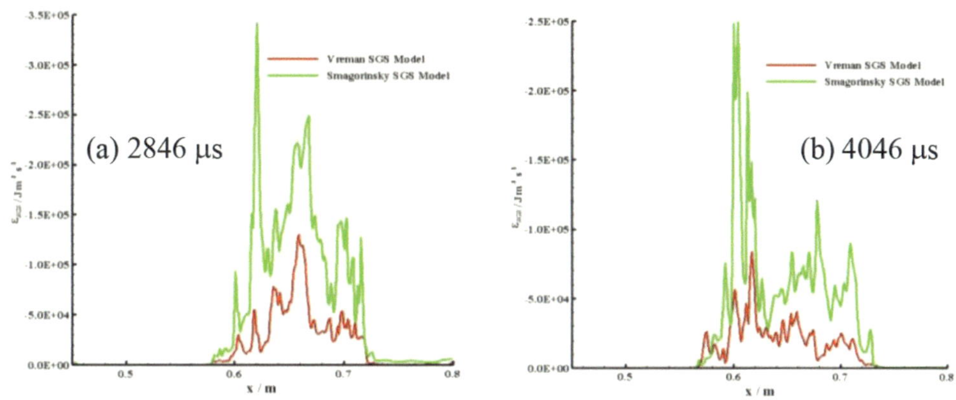

Figure 18. Distribution of calculated SGS turbulent dissipation in a streamwise direction for Smagorinsky and Vreman models and for AWE's SF_6 half-height shock tube experiment [24] at two times. Red line corresponding to the Vreman model, green line corresponding to the Smagorinsky model.

Figure 18 shows the distribution of calculated SGS turbulent dissipation in streamwise direction for Smagorinsky and Vreman models and for AWE's SF_6 half-height shock tube experiment [24] at two times [11]. The SGS dissipation of the Smagorinsky model is much greater than the Vreman model over 1.5 times; therefore, the dissipation is too great for the Smagorinsky SGS model. The SGS turbulent dissipations of the Vreman model (VM), the dynamic Smagorinsky model (DSM), and the stretched-vortex model (SVM) based on a planar Richtmyer-Meshkov instability with incident Mach number 1.2 are shown in **Figure 19** [30]. Before the interfacial flow has developed to be turbulent completely, the dynamic and stretched-vortex models have all predicted the energy backscatter, but the energy backscatter predicted by the dynamic model is

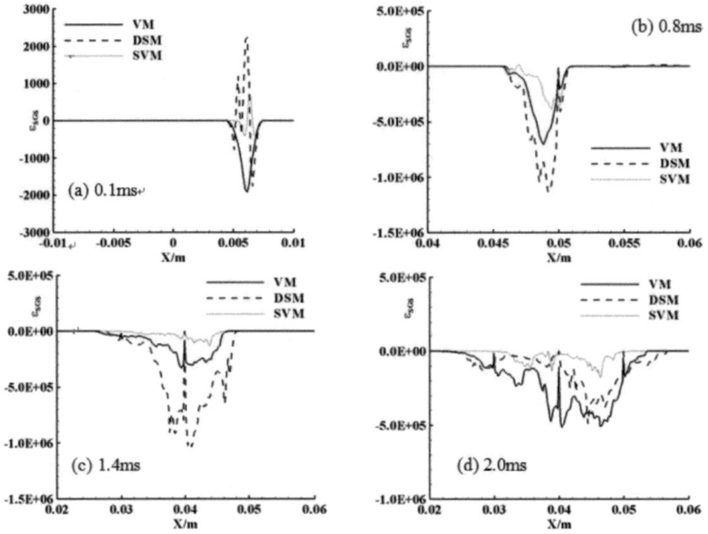

Figure 19. Distribution of the SGS turbulent dissipation in a streamwise direction for the Vreman, dynamic Smagorinsky and stretched-vortex models at four times.

larger and its range is wider. After reshock, the turbulent fluctuations are stronger extremely the turbulent dissipation also increases. The dynamic model's dissipation is the highest, then followed by the Vreman model, and the stretched-vortex model's dissipation is the lowest. At the late time, the SGS dissipation of the dynamic and Vreman models is the same basically, and the dynamic model is also to be absolutely dissipative, yet the stretched-vortex model is still able to predict the local phenomenon of energy backscatter in a small range. So, the dynamic model is poor in representing the energy backscatter. The Vreman and stretched-vortex models are all robust, but the former is absolutely dissipative.

4.3. Effects of the initial conditions on the growth of RMI and the turbulent mixing

4.3.1. Effects of the initial conditions on the growth of single-mode RMI

First, we consider the effects of initial conditions of perturbation amplitude and wavelength on the growth of single-mode Richtmyer-Meshkov instability without reshock [18]. The incident shock wave with Mach number 1.2 impacts the air/SF_6 interface. The initial perturbation amplitude and wavelength are listed in **Table 1**. The calculations are carried out in two dimensions, and the RMI does not develop into turbulent mixing completely. **Figure 20** shows the effects of the initial perturbation amplitude on the growth of single-mode RMI for the fixed initial perturbation wavelength $\lambda_0 = 60$ mm. The perturbation amplitude (a) and growth rate (b) increase gradually with the increasing of initial amplitude. And when the initial amplitude is much larger, the growth of amplitude enters a linear stage earlier at the late times, and the growth rate remains a constant. The growth rate increases fast and reaches the highest peak at the early times. After the peak, the effect of compressibility is reduced and the flow nonlinearity starts to play a dominant role and causes the growth rate to decay with time. **Figure 21** shows the effects of the initial perturbation wavelength on the growth of single-mode RMI for the fixed initial

http://dx.doi.org/10.5772/67750

Wavelength (mm)										60	30	40	50	70	80
Amplitude (mm)	1	2	3	4	5	6	8	10	12	14	3				

Table 1. Initial conditions of perturbation for single-mode RMI.

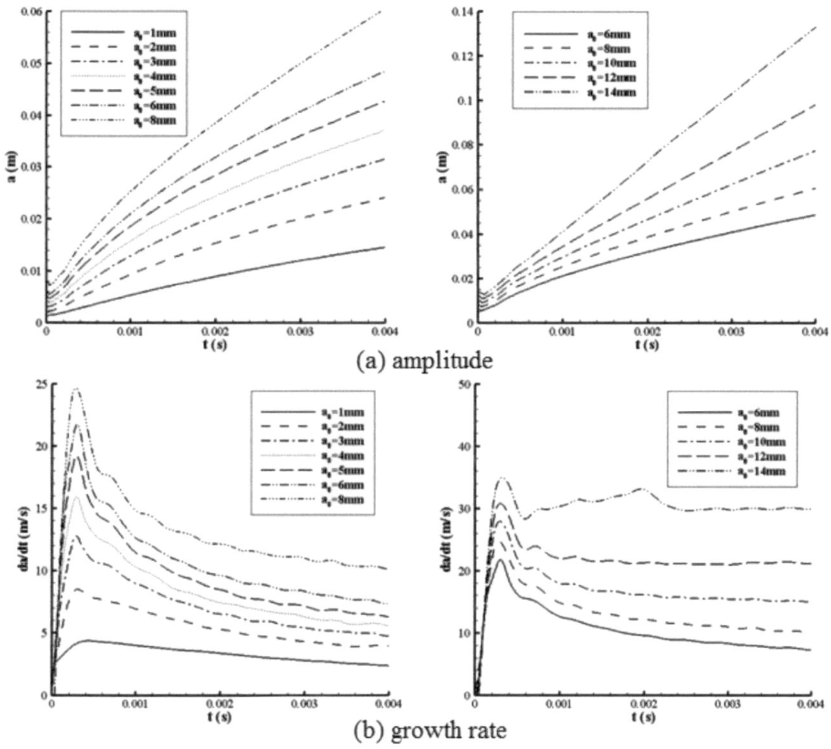

Figure 20. Effects of the initial perturbation amplitude on the growth of single-mode RMI for the fixed initial perturbation wavelength $\lambda_0 = 60$ mm.

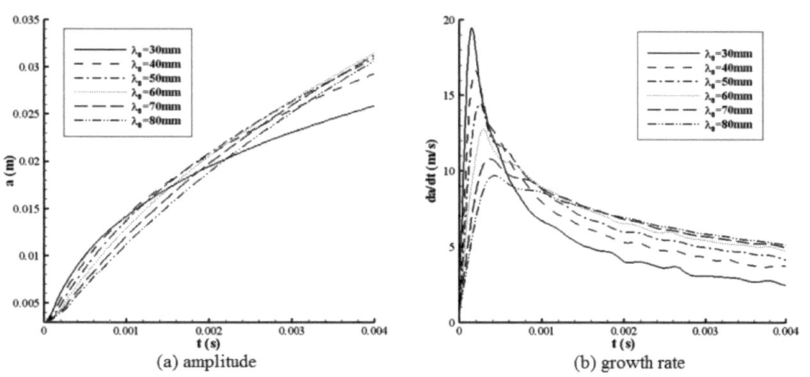

Figure 21. Effects of the initial perturbation wavelength on the growth of single-mode RMI for the fixed initial perturbation amplitude $a_0 = 3$ mm.

perturbation amplitude $a_0 = 3$ mm. The perturbation amplitude (a) and growth rate (b) reduce gradually at the early times and increase gradually at the late times with the increasing of initial wavelength. And when the initial perturbation strength (ration of initial amplitude to wavelength) is much smaller, the perturbation growth is mainly dependent on the initial perturbation amplitude and slightly dependent on the initial perturbation wavelength at the late times of RMI. The same conclusion can be obtained from three-dimensional calculations.

4.3.2. *Effects of the initial conditions on the growth of multi-mode RMI and the induced turbulent mixing*

For the effects of three-dimensional initial multi-mode conditions, the case of Richtmyer-Meshkov instability is same as the above Section 4.1. **Table 2** lists the initial condition of perturbations, the perturbation strength (*PS*) is defined as the ratio of initial amplitude to wavelength. Turbulent kinetic energy *K*, dissipation rate ε, and enstrophy Ω are defined as follows [31]:

$$K = \frac{\langle \rho u''_i u''_i \rangle}{2\langle \rho \rangle} + \frac{\langle \tau_{ii} \rangle}{2\langle \rho \rangle} \tag{46}$$

$$\varepsilon = \frac{\langle \sigma_{ij} \partial u''_i / \partial x_j \rangle}{\langle \rho \rangle} - \frac{\langle \tau_{ij} \partial u''_i / \partial x_j \rangle}{\langle \rho \rangle} \tag{47}$$

$$\Omega = \frac{1}{2} |\omega|^2 \tag{48}$$

where u''_i is the velocity fluctuation and $\langle\ \rangle$ denotes the transverse plane-average. For large-eddy simulations, the turbulent kinetic energy and dissipation rate both include two parts named as the resolved-scales (the first term on the right side of Eqs. (46) and (47)) and the subgrid-scales (the second term on the right side of Eqs. (46) and (47)).

Figure 22 shows the growth history of TMZ width. **Figures 23** and **24** show the time histories of the perk values of turbulent kinetic energy and enstrophy, respectively. For the larger initial perturbation strength, the TMZ grows faster, the turbulent kinetic energy is also larger or the turbulence strength is also stronger, the deposited vorticity is larger too. The development of turbulent mixing has a strong dependence on the initial conditions between the initial shock and the impingement of the first reflected rarefaction wave, after that the evolution of the turbulent mixing has lost the memory of the initial conditions.

	CASE1	CASE2	CASE3	CASE4	CASE5
η_0 (mm)	0.07	0.14	0.28	0.56	1.12
PS	0.035	.0.7	0.14	0.28	0.56

Table 2. Model parameters.

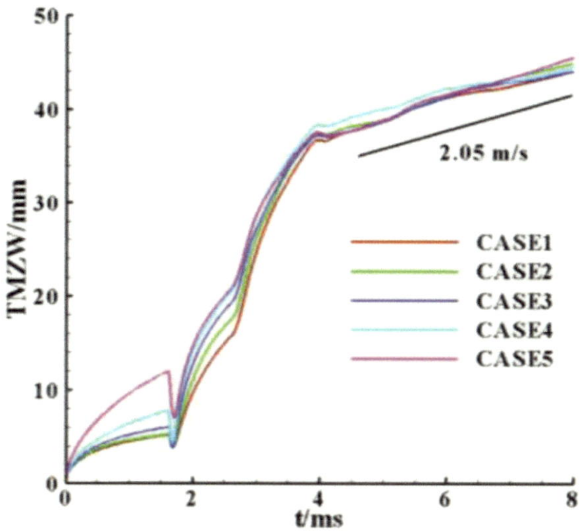

Figure 22. Growth history of the TMZ width for different models.

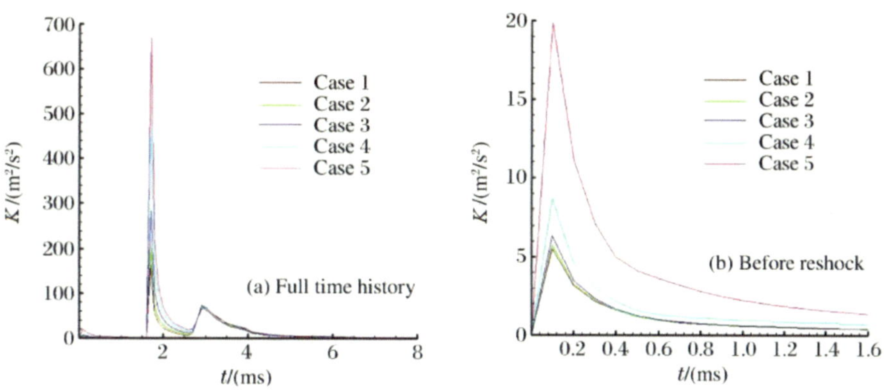

Figure 23. History of turbulent kinetic energy for different models.

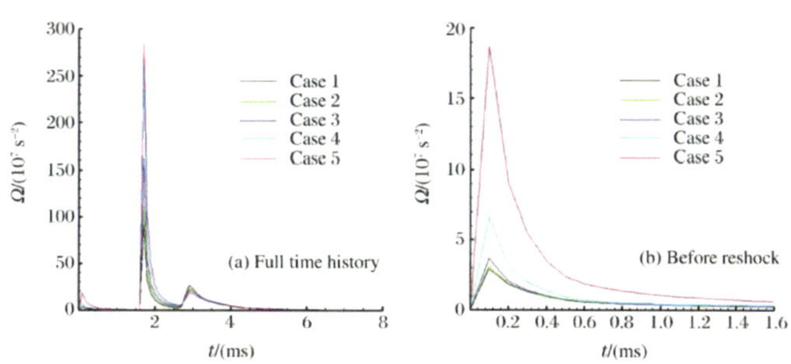

Figure 24. History of enstrophy for different models.

4.4. Dynamic characters of multi-mode RM instability and induced turbulent mixing

Figures 25–27 show the spatial profiles of the turbulent kinetic energy, dissipation rate and enstrophy along the motion direction of shock wave at different times before and after reshock individually. They all have a spatial profile similar to Gaussian distribution. The strongest turbulent intensity is located in the center of TMZ. **Figures 28–30** show the temporal evolutions of the peak values of the turbulent kinetic energy, dissipation rate, and enstrophy in the spatial profiles, along with their fitted results, respectively. The turbulent kinetic energy, dissipation rate, and enstrophy decay gradually because of dissipation and diffusion. After the initial shock and before the reshock, the turbulent kinetic energy and enstrophy decay with time as a power law ~t^{θ} except the dissipation rate which decays with time as an exponential law ~$e^{-t/t^{*}}$. One reason is that the TMZ is not fully developed turbulence before reshock and the other reason is that the flow in TMZ is stronger anisotropic between the transverse direction and the axis direction. After reshock and the first reflected rarefaction wave, they all decay with time as the negative exponential law with the closed decay factors at the same

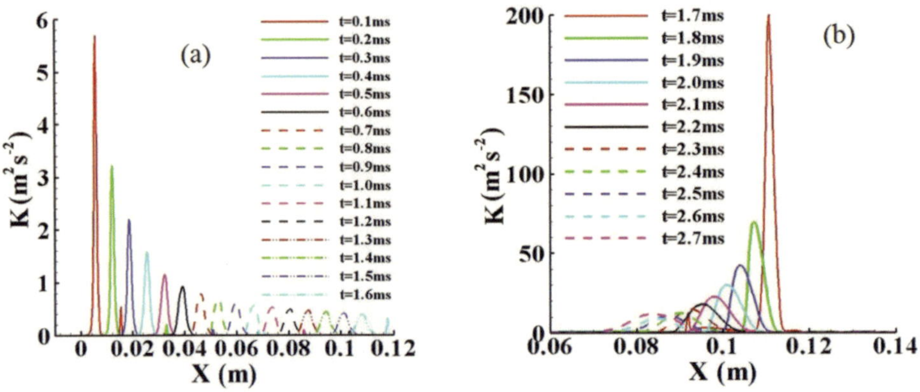

Figure 25. Spatial profiles of the turbulent kinetic energy at different times. (a) Before reshock and (b) after reshock.

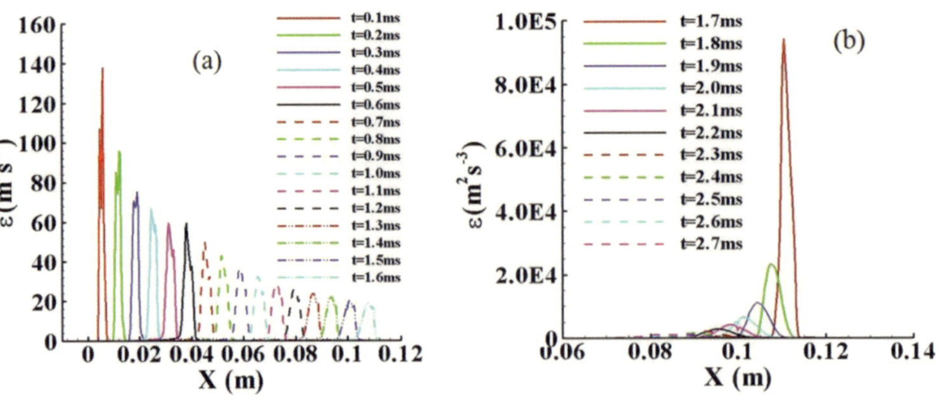

Figure 26. Spatial profiles of the turbulent kinetic energy dissipation rate at different times. (a) Before reshock and (b) after reshock.

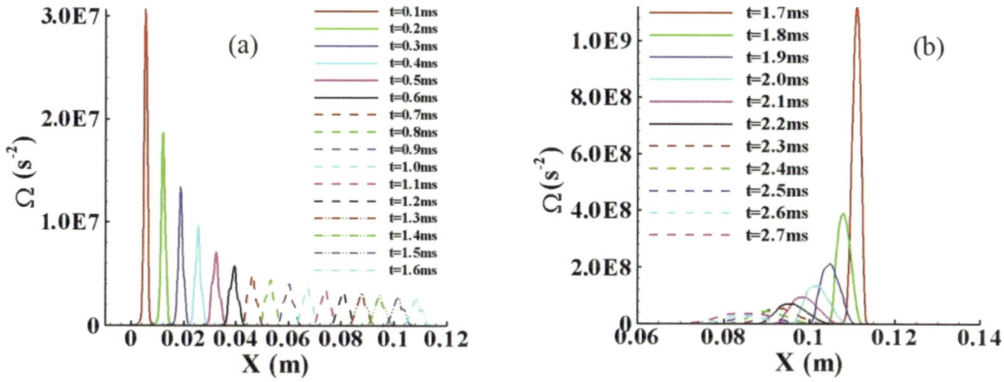

Figure 27. Spatial profiles of the enstrophy at different times. (a) Before reshock and (b) after reshock.

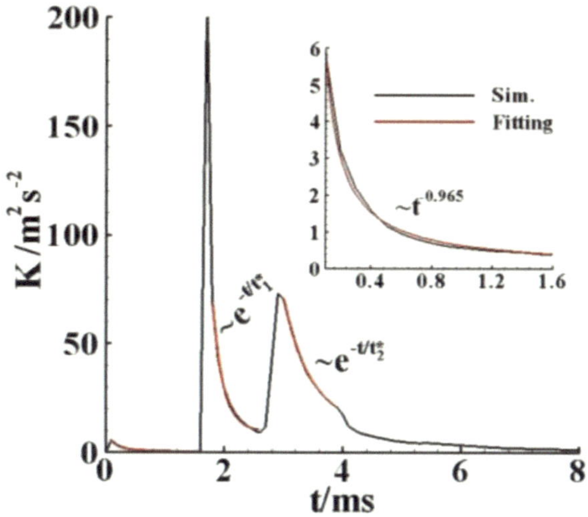

Figure 28. Time history of the peak value of turbulent kinetic energy.

stage, and be similar to the growth of TMZ width. And then, they all decay asymptotically due to no remarkable energy deposition. Therefore, the turbulent mixing behaves in a statistical self-similar pattern. **Figure 31** shows the one-dimensional global energy spectra of three velocity components on a log-log scale at three times. The energy spectra of two transverse components of velocity are too close, and there is a difference between transverse and axis components. The turbulent mixing flow is continuous anisotropic yet the anisotropy weakens gradually. That is to say, the development of the turbulent mixing presents a trend of isotropy.

4.5. Numerical study of the elliptic gas cylinder instability

Our group first performed the experimental and numerical investigations of the elliptic gas cylinder instability. As we know that the initial density distribution of the gas cylinder is hard

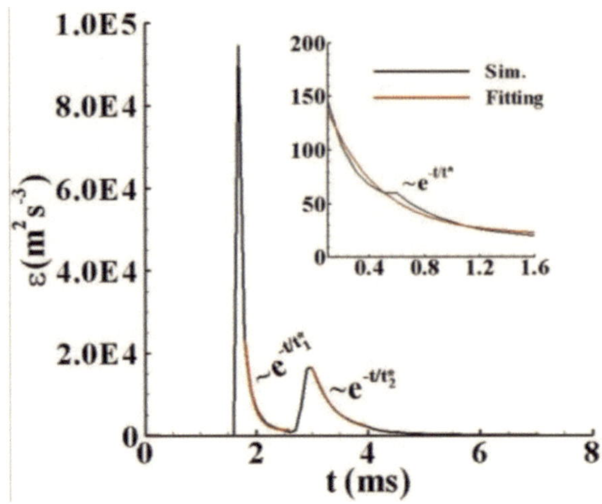

Figure 29. Time history of the peak value of turbulent kinetic energy dissipation rate.

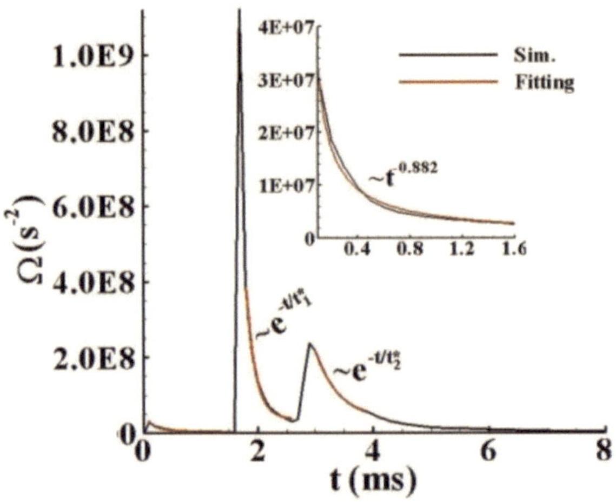

Figure 30. Time history of the peak value of enstrophy.

to determine in the experiment, which can only give the one-dimensional radial concentration distribution for circular gas cylinder as an approximate Gaussian function. Drawing to the experience on the one-dimensional diffusive interfacial transition layer with finite thickness for circular gas cylinder, we constructed a two-dimensional diffusive interfacial transition layer with finite thickness for elliptic gas cylinder through numerical simulation [32],

$$\rho(\alpha,\beta) = \chi_0 \rho_{SF_6} e^{-[(\alpha-\alpha_{\min})^2+(\beta-\beta_{\min})^2]/\delta^2} \tag{49}$$

$$\frac{(x-x_0)}{\alpha^2} + \frac{(y-y_0)}{\beta^2} = 1 \tag{50}$$

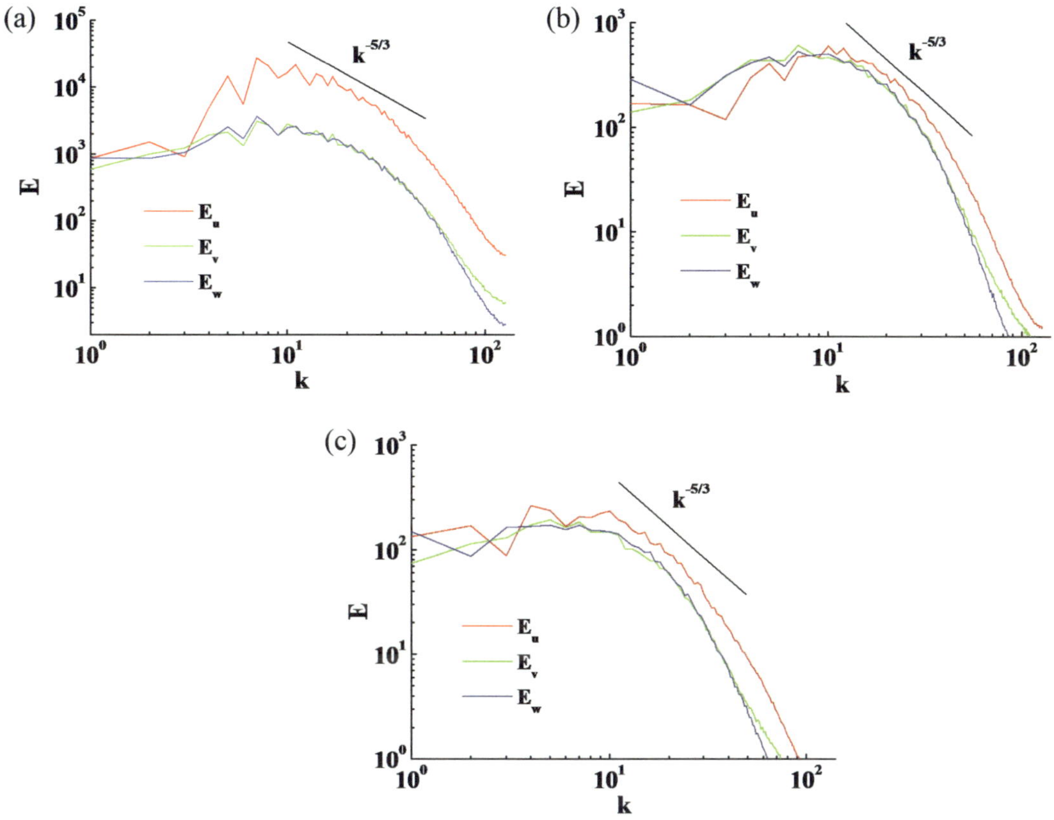

Figure 31. One-dimensional global energy spectra of three components of velocity on a log-log scale.

where $x_0 = y_0 = 0$, $\alpha \in [\alpha_{min}, \alpha_{max}]$, $\beta \in [\beta_{min}, \beta_{max}]$, $\alpha_{min} = \beta_{min} = 1.0 \times 10^{-5}$ mm, $\alpha_{max} = 6.30$ mm, $\beta_{max} = 2.30$ mm, and $\delta = 6.16$ mm. $\chi_0 = 0.71$ is the concentration of the elliptic center. The density distribution is plotted in **Figure 32**. The Mach number of air shock wave is 1.25. The simulations (see **Figure 33**) reproduce the elliptic gas cylinder instability experiment very well, they achieve to a good agreement qualitatively and quantitatively, some salient features of the vortex pairs are obtained clearly.

Figure 34 shows the vorticity at the center of the core and the distance between the two vortex cores of these two simulations. When the incident shock accelerates the elliptic gas cylinder along the major axis, the absolute vorticity in the vortex core $|\omega_{core}|$ is larger; but the distance between two vortex cores is larger when the incident shock accelerates the cylinder along the minor axis. So, the effect of convergence is stronger when the incident shock accelerates the elliptic gas cylinder along the major axis, for which the gas jet appears at the symmetry center.

4.6. Foundation of new RM instability and mechanism

At present, the investigation for the Richtmyer-Meshkov instability is performed in a uniform flow field. We first study the Richtmyer-Meshkov instability and turbulent mixing in a

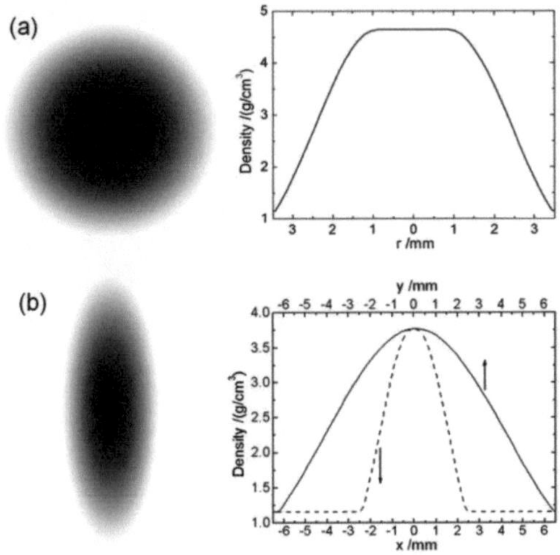

Figure 32. Density images and distributions of the (a) circular and (b) elliptic SF_6 gas cylinder initially constructed.

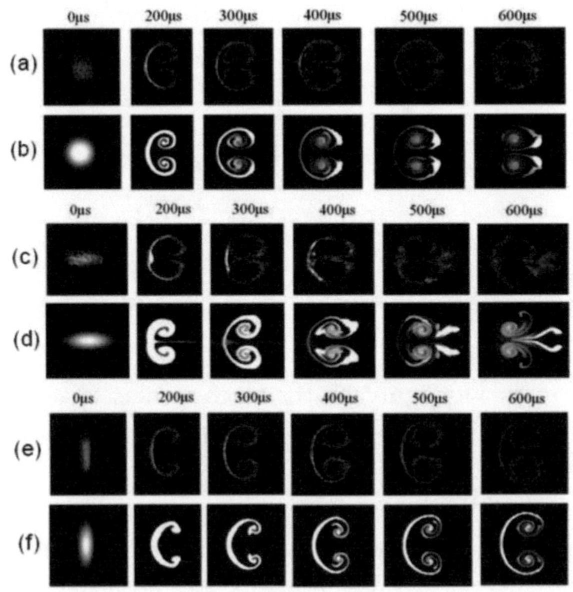

Figure 33. Experimental evolution images and numerical simulation results by MVFT at $t = 200, 300, 400, 500$, and 600 μs, the experiments corresponding (a), (c), and (e), and the simulations corresponding (b), (d), and (f).

nonuniform flow field by experiment and numerical simulations and find some new phenomena. **Figure 35** shows the test section schematic of shock tube. The air shock wave with Mach number 1.27 accelerates the two-mode sinusoidal air/SF_6 interface (amplitude $A_{01} = 5$ m, $A_{02} =$ 7.5 mm). The experimental Schlieren images (gray images in **Figure 38**) show the transmitted shock wave and the interface all incline which is different from the familiar RMI. We think this

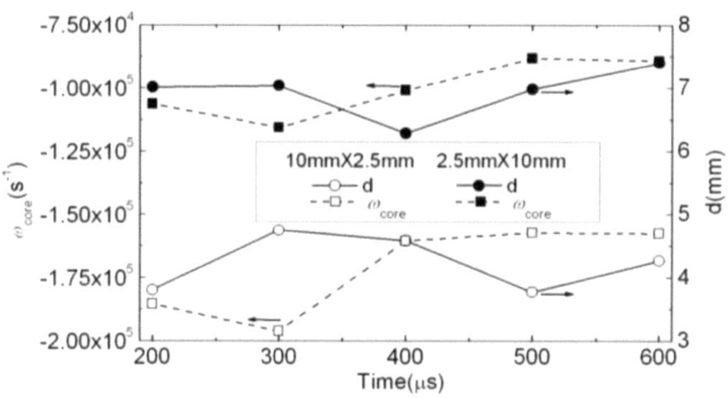

Figure 34. Vorticity at the center of the cores and the distance between the two vortex cores from simulations.

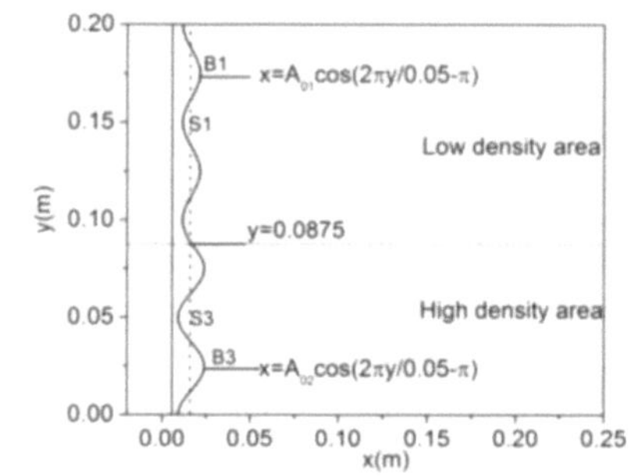

Figure 35. Initial structure diagram in the shock tube.

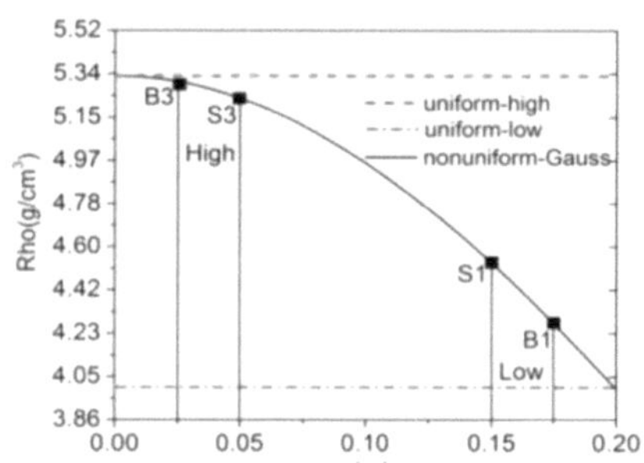

Figure 36. Density profiles of nonuniform flows with Gaussian function and uniform flows in a vertical direction.

may be owing to the nonuniformity of flow field. Also, drawing to the experience on the diffusive interfacial transition layer of circular gas cylinder, we construct a nonuniform flow field with a Gaussian distribution of density along the direction perpendicular to the shock motion direction (as shown in **Figure 36**). The numerical result (shown in **Figure 37**) confirms this idea [12, 33, 34].

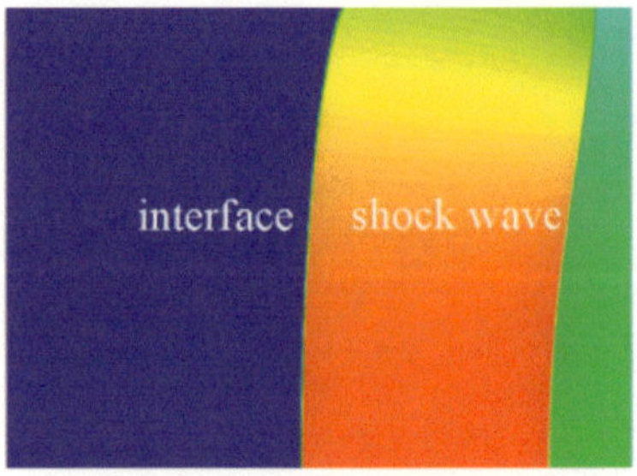

Figure 37. Simulated density distribution in a nonuniform flow field by using Gaussian function.

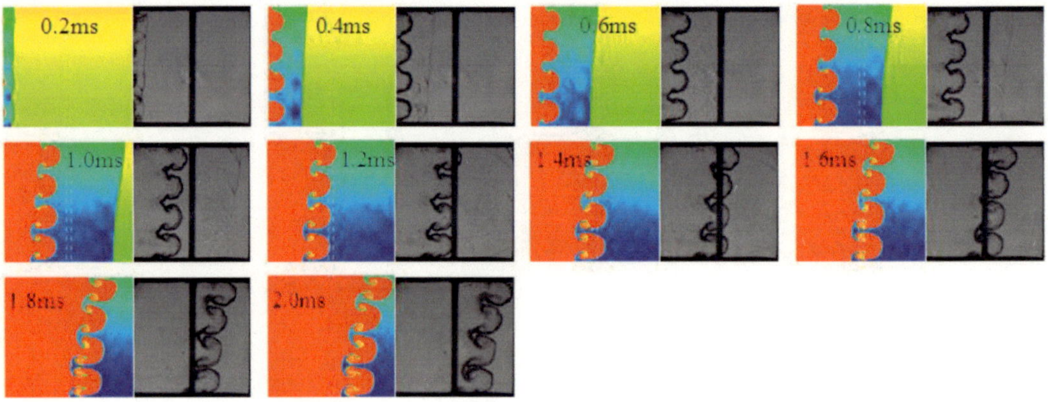

Figure 38. Experimental schlieren photography images and numerical simulation results by MVFT2D at a certain time series (the sizes of the pictures are ones of the test window [0.038, 0.25 m] × [0.0, 0.2 m]).

Figure 38 shows the evolution of the interface and the propagation of the transmitted shock wave, the calculated results agree well with the experiment. Due to the nonuniform flow field of SF_6 gas, the propagating velocity of transmitted shock wave in the upper part of shock tube is faster than in the bottom of shock tube, and it forms an oblique shock wave front and the interface. **Figure 39** shows the shock front locations between experiment and numerical simulations, **Figure 40** shows the perturbation amplitude of experiment, numerical simulation, and theories, they are in good agreement. **Figure 41** shows the calculated perturbation amplitude of RM instability for different modes in the initial uniform and nonuniform flow fields. As we can see that, at late times, the growth of small perturbation in a low-density zone catches up and exceeds the large perturbation in a high-density zone, which is opposite to the case of uniform flow field. **Figure 42** shows the perturbation amplitudes of four different initial amplitudes 2.5, 5.0, 7.5, and 10 mm in low- and high-density zones of nonuniform flow fields. It shows that the effect of initial amplitude on the growth of RM instability in the nonuniform flow field is different from the case of uniform one, which can be explained by the baroclinic vorticity shown in **Figures 43** and **44**, the baroclinic vorticity produced in the low-density zone is larger than that in the high-density zone.

http://dx.doi.org/10.5772/67750

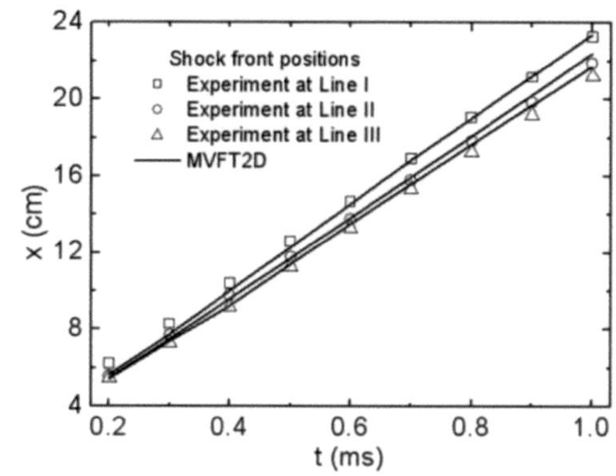

Figure 39. Shock front locations of the experiment and calculated results on the three test lines.

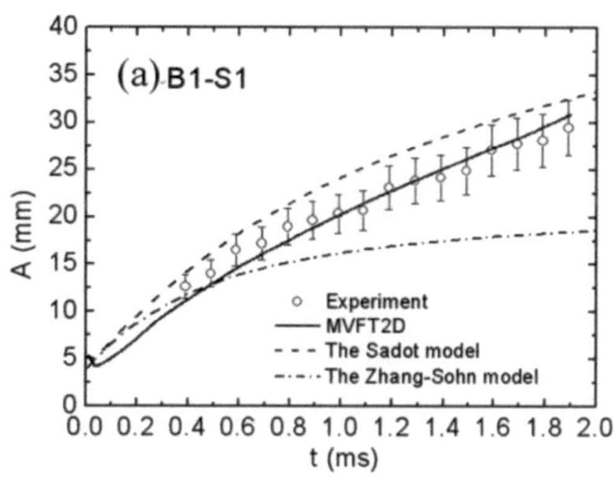

Figure 40. Perturbation amplitudes of the experiment, simulation, and comparison with theories.

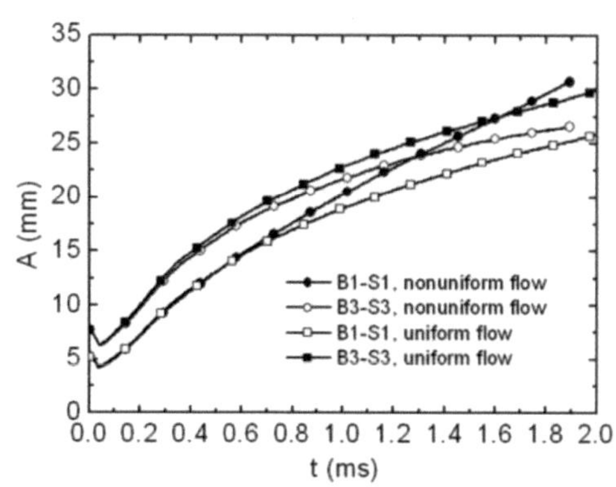

Figure 41. Calculated perturbation amplitude of RM instability for different modes in the initial uniform and nonuniform flow fields.

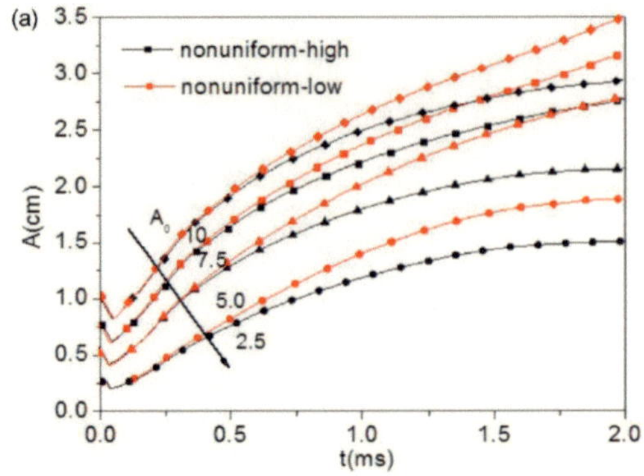

Figure 42. Perturbation amplitudes of four different initial amplitude 2.5, 5.0, 7.5 and 10 mm in the low and high density zones of nonuniform flow field.

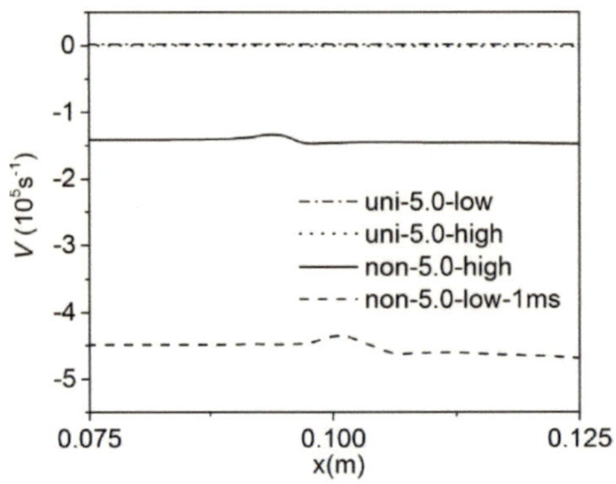

Figure 43. Average vorticity when initial amplitude is 5.0 mm which is in the low and high density zones of nonuniform flow field at 1 ms.

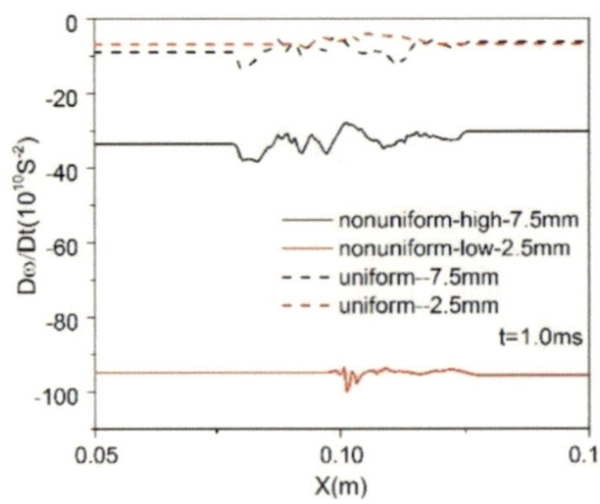

Figure 44. Baroclinic vorticity in the uniform and nonuniform flow fields with the initial amplitude group ($A_{01} = 2.5$ mm, $A_{02} = 7.5$ mm) at 1.0 ms.

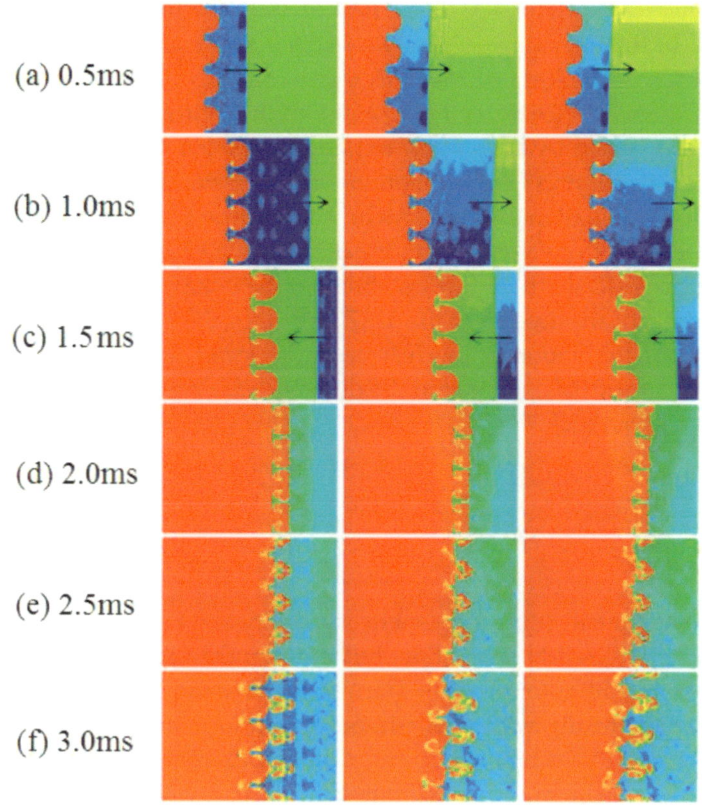

Figure 45. Density contour of the numerical simulated by MVFT at a time series. The left column with a uniform initial condition, the middle column with a δ_1 nonuniform Gaussian function, and the right column with a δ_2 nonuniform Gaussian function. The small arrow denotes the direction of propagation of the shock wave fronts before reshock the interface.

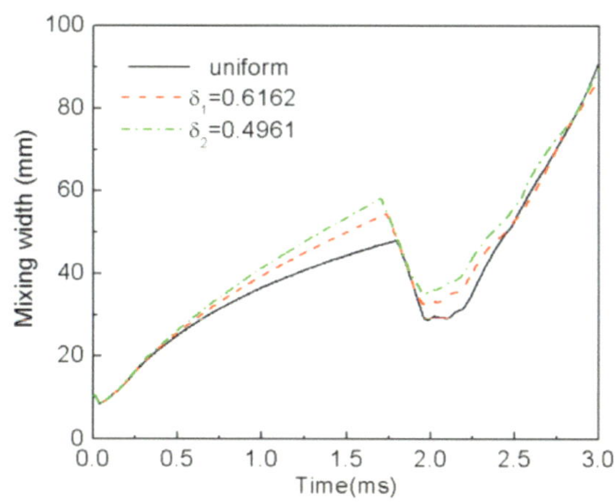

Figure 46. TMZ width of RM instability in the initial uniform and nonuniform flow fields.

For the above investigations of RM instability in the nonuniform flow field, they are all without reshock. In the following section, we will study the effect of nonuniformity of flow field and the reshock on the RM instability. Here, we consider two kinds of nonuniformity: the nonuniform coefficient $\delta_1 = 0.6162$ m and $\delta_2 = 0.4961$ m. **Figure 45** shows the density contour of the numerical simulated by MVFT at a time series, the left column with a uniform initial conditions, the middle column with a δ_1 nonuniform Gaussian function, and the right column with a δ_2 nonuniform Gaussian function. There is a significant difference between the uniform and nonuniform flows before reshock, but the difference decreases in evidence after reshock. **Figure 46** shows the TMZ width of RM instability in the initial uniform and nonuniform flow fields. It points out that the growth of the TMZ width for the initial nonuniform flow field is greater than that for the uniform flow field, and the lesser the nonuniform coefficient, the higher the growth rate of TMZ width, but the difference for the three different flow configurations diminishes after reshock.

5. Prospect

Now, we investigated the Richtmyer-Meshkov instability and the induced turbulent mixing in fluid flow by using large-eddy simulation, but the turbulent mixing is a complicated three-dimensional problem with multiple time-space scales, and the more engineering applications involve the materials mixing with strength. So, the future work will be carried out in the following aspects: (a) the direct numerical simulations with high precision and high resolution on the platform of supercomputer, (b) the interface instability and turbulent materials mixing with strength, this may involve the material properties such as the deformation, fracture, melt, phase transition, material microstructures, etc., and it may make this problem to be more difficult.

Acknowledgements

Professor Jingsong Bai and Associate professor Tao Wang from the Institute of Fluid Physics, China Academy of Engineering Physics, People's Republic of China and some other from the same institute should be thanked for the contributions and helps to these works, such as Prof. Ping Li, Dr. Liyong Zou, Kun Liu, Mr. Jinhong Liu, Min Zhong and Bing Wang, and Jiaxin Xiao.

The authors would like to thank the support from "Science Challenge Project" (no. TZ2016001), the National Natural Science Foundation of China (nos. 11372294, 11202195, 11072228, and 11532012), the Science Foundation of China Academy of Engineering Physics (nos. 2011B020 2005, 2011A0201002, and 2008B0202011), and the Foundation of National Key Laboratory of Shock Wave and Detonation Physics (no. 9140C670301150C67290).

http://dx.doi.org/10.5772/67750

Author details

Jingsong Bai[1,2] and Tao Wang[1*]

*Address all correspondence to: wtao_mg@163.com

1 Institute of Fluids Physics, China Academy of Engineering Physics, Mianyang, Sichuan, People's Republic of China

2 National Key Laboratory of Shock Wave and Detonation Physics (LSD), Institute of Fluid Physics, China Academy of Engineering Physics, Mianyang, Sichuan, People's Republic of China

References

[1] Richtmyer RD. Taylor instability in a shock acceleration of compressible of fluids. Communications on Pure and Applied Mathematics. 1960;**13**(2):297–319

[2] Meshkov EE. Instability of the interface of two gases accelerated by a shock wave. Izv Akad Nauk SSSR Mekh Zhidk Gaza. 1969;**4**(5):151–157

[3] Zoldi CA. Simulation of a shock-accelerated gas cylinder and comparison with experiment images and velocity fields. In: 8th International Workshop on the Physics of Compressible Turbulent Mixing; 9–14 December; Pasadena, California, USA, 2001. p. 12

[4] Amendt P, Colvin JD, Tipton RE, et al. Indirect-drive noncryogenic double-shell ignition targets for the National Ignition Facility: Design and analysis. Physics of Plasmas. 2002;**9**:2221–2233

[5] Khokhlov AM, Oran ES, Thomas GO. Numerical simulation of deflagration-to-detonation transition: The role of shock-flame interaction in turbulence flames. Combustion and Flames. 1999;**117**:323–339

[6] Waitz IA, Marble FE, Zukoski EE. Investigation of a contoured wall injector for hypervelocity mixing augmentation. AIAA Journal. 1993;**31**:1014–1021

[7] Maran SP, Sonneborn G, Pun CSJ, et al. Physical conditions in circumstellar gas surrounding SN 1987A 12 years after outburst. Journal of Astrophysical. 2000;**545**:390–398

[8] Arnett WD, Bahcall JN, Kirshner RP, Woosley SE. Supernova 1987 A. Annual Review of Astronomy Astrophysics. 1989;**27**:629

[9] Bai JS, Li P, Tan DW. Simulations of the instabilities experiment s in St ratified cylindrical shells. Chinese Physics Letter. 2006;**23**(7):1850–1852

[10] Bai JS, Wang T, Li P, Zou LY, Liu CL. Numerical simulation of the hydrodynamic instability experiments and flow mixing. Science in China Series G: Physics, Mechanics & Astronomy. 2009;**52**(12):2027–2040

[11] Wang T, Bai JS, Li P, Liu K. Large-eddy simulations of the Richtmyer-Meshkov instability of rectangular interface accelerated by shock waves. Science in China Series G: Physics, Mechanics & Astronomy. 2010;**53**(5):905–914

[12] Bai JS, Liu JH, Wang T, Zou LY, Li P, Tan DW. Investigation of the Richtmyer-Meshkov instability with double perturbation interface in nonuniform flows. Physical Review E, Statistical Physics, Plasmas, Fluids, and Related Interdisciplinary Topics. 2010;**81**(2): 056302

[13] Smagorinsky J. General circulation experiments with the primitive equations. Monthly Weather Review. 1963;**91**:99–164

[14] Vreman AW. An eddy-viscosity subgrid-scale model for turbulent shear flow: Algebraic theory and applications. Physics of Fluids. 2004;**16**(10):3670–3681

[15] Moin P, Squires K, Cabot W, Lee S. A dynamic subgrid-scale model for compressible turbulence and scalar transport. Physics of Fluids A (Fluid Dynamics). 1991;**3**(11): 2746–2757

[16] Hill DJ, Pantano C, Pullin DI. Large-eddy simulation and multiscale modelling of a Richtmyer-Meshkov instability with reshock. Journal of Fluid Mechanics. 2006;**557**:29–61

[17] Bai JS. High resolution numerical methods and adaptive mesh refinement algorithm for compressible multi-fluid dynamics [PhD thesis]. China Academy of Engineering Physics; 2003. (in Chinese)

[18] Wang T, Bai JS, Li P, et al. Two and three dimensional numerical investigations of the single-mode Richtmyer-Meshkov instability. Chinese Journal of High Pressure Physics. 2013;**27**(2):277–286

[19] Wang T, Bai JS, Li P, Zhong M. The numerical study of shock-induced hydrodynamic instability and mixing. Chinese Physics B. 2009;**18**(3):1127–1135

[20] Malamud G, Elabz Y, Leinov E, et al. Two- and three-dimensional numerical simulations of the interaction of a Richtmyer-Meshkov instability induced turbulent mixing zone with several re-shocks. In: 10th IWPCTM; 17–21 July 2006, Paris, France. 2006.

[21] Bai JS, Wang T, Liu K, et al. Large-Eddy simulation of the Three-dimensional experiment on Richtmyer-Meshkov instability induced turbulence. International Journal of Astronomy and Astrophysics. 2012;**2**:28–36

[22] Poggi F, Thorembey MH, Rodriguez G. Velocity measurements in turbulent gaseous mixtures induced by Richtmyer–Meshkov instability. Physics of Fluids. 1998;**10**:2698–2712

[23] Wang T, Li P, Bai JS, Tao G, Wang B, Zou LY. Large-eddy simulation of the Richtmyer-Meshkov instability. Canadian Journal of Physics. 2015;**93**(10):1124–1130

http://dx.doi.org/10.5772/67750

[24] Bates KR, Nikiforakis N, Holder D. Richtmyer-Meshkov instability induced by the interaction of a shock wave with a rectangular block of SF_6. Physics of Fluids. 2007;**19**:036101

[25] Holder DA, Smith AV, Barton CJ, Youngs DL. Shock-tube experiments on Richtmyer-Meshkov instability growth using an enlarged double-bump perturbation. Laser and Particle Beams. 2003;**21**:411–418

[26] Li P, Bai JS, Wang T, et al. Large eddy simulation of a shocked gas cylinder instability induced turbulence. Science China Physics, Mechanics & Astronomy. 2010;**53**(2):262–268

[27] Mikaelian KO. Rayleigh-Taylor and Richtmyer-Meshkov instabilities and mixing in stratified spherical shells. Physical Review A (Atomic, Molecular, and Optical Physics). 1990;**42**(16):3400–3420

[28] Bai JS, Wang T, Zou LY, et al. Numerical simulation for shock tube and jelly interface instability experiments. Chinese Journal of Applied Mechanics. 2009;**26**(3):418–425

[29] Erez L, Sadot O, Oron D, et al. Study of the membrane effect on turbulent mixing measurements in shock tubes. Shock Waves. 2000;**10**:241–251

[30] Wang T, Tao G, Bai JS, et al. Numerical comparative analysis of Richtmyer-Meshkov instability simulated by different SGS models. Canadian Journal of Physics. 2015;**93** (5):519–525

[31] Wang T, Tao G, Bai JS, et al. Analysis of dependence of multi-mode Richtmyer-Meshkov instability on initial conditions. Chinese Journal of High Pressure Physics. 2016;**30** (5):380–386

[32] Bai JS, Zou LY, Wang T, et al. Experimental and numerical study of the shock-accelerated elliptic heavy gas cylinders. Physical Review E, Statistical Physics, Plasmas, Fluids, and Related Interdisciplinary Topics. 2010;**82**(5):056318

[33] Bai JS, Wang B, Wang T, Liu K. Numerical simulation of the Richtmyer-Meshkov instability in initially nonuniform flows and mixing with reshock. Physical Review E, Statistical Physics, Plasmas, Fluids, and Related Interdisciplinary Topics. 2012;**86**(6):066319

[34] Xiao JX, Bai JS, Wang T. Numerical study of initial perturbation effects on Richtmyer-Meshkov instability in nonuniform flows. Physical Review E, Statistical Physics, Plasmas, Fluids, and Related Interdisciplinary Topics. 2016;**94**(1):013112

Chapter 4

Statistical Modeling for the Energy-Containing Structure of Turbulent Flows

Yuriy Nuzhnov

Additional information is available at the end of the chapter

http://dx.doi.org/10.5772/67844

Abstract

The development of statistical theory for the energy-containing structure of turbulent flows, taking the phenomenon of internal intermittency into account, is proposed, and new differential equations for conditional means of turbulent and nonturbulent fluid flow are established. Based on this fact, a new principle of constructing mathematical models is formulated as the method of autonomous statistical modeling of turbulent flows, *ASMTurb* method. Testing of the method is attained on the example of constructing a mathematical model for the conditional means of turbulent fluid flow in a turbulent mixing layer of co-current streams. Test results showed excellent agreements between the predictions of the *ASMTurb* model and known experimental data.

Keywords: turbulence, statistical modeling, intermittency, *ASMTurb* method

1. Introduction

The Reynolds-averaged Navier-Stokes equations (*RANS*) method does not take the intermittency of turbulent and nonturbulent fluid into consideration. As a result, this method allows us to model only the unconditional averages of a turbulent flow and does not provide a description of the conditional averages for each of the intermittent region, taking place in a turbulent stream. At the same time, the intermittency is an inherent property of such flows and that is why the conditional average modeling is necessary, for example see [1–4]. The phenomenon of intermittency (hydrodynamic intermittency) represents an interleaving process of the space-time domains of the flow, hydrodynamic structures of which are different. As is known, such domains contain so-called "turbulent" and "nonturbulent" fluid [1]. In this connection, the turbulent fluid contains a hierarchy of all possible scales and amplitudes of the fluctuations (pulsations) of hydrodynamic values, i.e., the whole spectrum of wavenumbers, while the nonturbulent fluid may contain only the large-scale fluctuations or absolutely does not contain any ones (when the nonturbulent fluid is far away from the mixing layer). The main purpose of this chapter is to justify a new method of statistical modeling of turbulent flows as the *ASMTurb* method, which enables to construct mathematical models of such flows with a high efficiency. The presented *ASMTurb* method, declared in [5], fundamentally differs from the previously proposed (for example, see Refs. [6–8]) in that it is based on the conditional statistical averaging of the Navier-Stokes equations, as applied to each of the intermittent region of turbulent flow, while the generating process of the turbulent fluid begins in a thin superlayer between turbulent and nonturbulent fluid and finishes in separate small areas, involved inside the turbulent flow. The first attempts to substantiate such an approach [5] have been presented previously [9, 10]. However, the deficiency of the mathematical body of statistical hydrodynamics under the intermittency conditions makes such an approach vulnerable. In this regard, we need primarily to develop a mathematical body for statistical modeling of turbulent flows.

2. Development of the statistical modeling theory

A spectacular example of the intermittent turbulent flow is the flow in the mixing layer of co-current streams, **Figure 1**. With that at point $x = x_0$, with the course of time, will be observed an interleaving of the turbulent and nonturbulent fluid. The behavior of the instantaneous longitudinal velocity $u(x, t)$ in the flow range with strong intermittency at the point x_0 is shown in **Figure 2**. As it seen, the structure of the turbulent fluid flow is fundamentally different from the structure of the nonturbulent fluid flow (the nonturbulent fluid involvement is shown with the arrows in **Figure 1**). It is evident that the behavior of any other hydrodynamic variable $f(x, t)$ will be the same. It is important to note that the conditional averaging of variable $f(x, t)$ is interpreted as the result of the averaging operation, referring only to as the turbulent ($r = t$) or nonturbulent ($r = n$) fluid, i.e., for the conditional time averaging

$$\overline{f(x,t)}^r = \lim_{\tau_0 \to \infty} \frac{1}{\tau_r} \int_0^{\tau_r} f(x, t; \tau_0) dt, \quad r = t,\ n \tag{1}$$

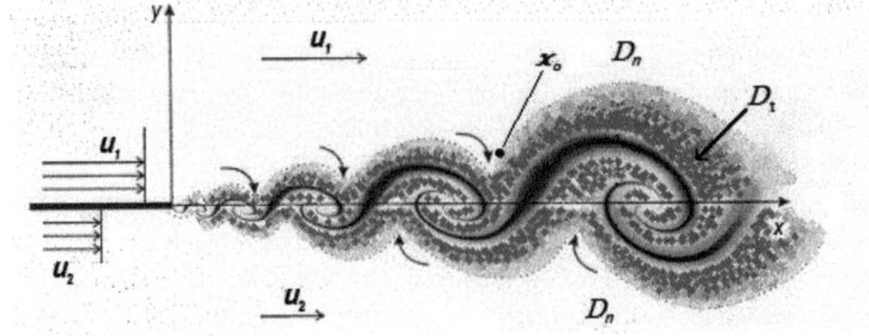

Figure 1. A sketch of the turbulent and nonturbulent fluid in the mixing layer of co-current streams. Here D_t is the region with the turbulent fluid and D_n– is the region with the nonturbulent fluid.

http://dx.doi.org/10.5772/67844

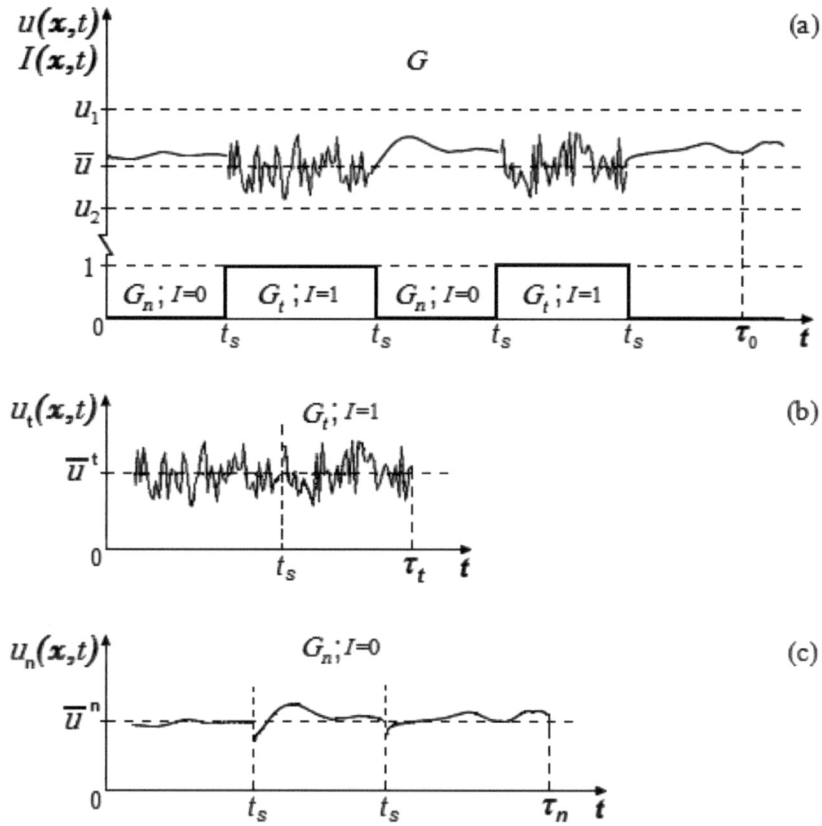

Figure 2. Behavior of the instantaneous longitudinal velocity in different regions, interleaving at the preset point $x = x_0$ in **Figure 1**: (a) unconditional velocity $u = u(x, t)$, $(x, t) \in G$; (b) "cross-linking" of the velocity over the turbulent fluid domain, $u_t = u_t(x, t)|_{I(x,t)=1}$, $(x, t) \in G_t$; (c) "cross-linking" of the velocity over the nonturbulent fluid domain, $u_n = u_n(x, t)|_{I(x,t)=0}$, $(x, t) \in G_n$. Here, $G = G_t + G_n$, $\overline{u}$ is the total time average, $\overline{u}^t$ and $\overline{u}^n$ is the conditional time means, t_s is the time point of observing the interfacial joint between the turbulent and nonturbulent flow domain in which cross-linking is carried out.

where $\overline{f(x,t)}^r \equiv \overline{f_r(x,t)}^r$, $f_t(x,t) = f(x,t)|_{I=1}$, $f_n(x,t) = f(x,t)|_{I=0}$, $I = I(x,t)$ is the intermittency function and $\tau_0 = \tau_t + \tau_n$. At that the total average is

$$\overline{f(x,t)} = \gamma(x)\overline{f(x,t)}^t + \left(1 - \gamma(x)\right)\overline{f(x,t)}^n \tag{2}$$

and $\gamma(x) = \overline{I(x,t)}$ is an intermittency factor. At the same time in the theory of statistical modeling are used the statistical characteristics, i.e., instead of the averaging operation of Eq. (1) is required the operation of statistical averaging.

To construct the mathematical model, first of all, it is necessary to determine what kind of statistical characteristics are the most suitable for modeling. In the classical *RANS* method, such characteristics are the unconditional means. In the methods, taking the intermittency into consideration, such characteristics are the conditional means of each intermittent region of turbulent flow. But in this case, it requires the development of a theory of statistical hydrodynamics under the conditions of intermittency.

The mathematical body of this theory may be developed from both the theoretical-probabilistic approach, based on the N-th repetition of the turbulent flow experiment [4, 11], and the theoretical-set approach [12, 13], which elementary events can be represented as a some set in the generalized space of the specifically considered turbulent flow. The advantage of the theoretical-set approach is that it can be implemented in the experimental research.

2.1. The mathematical body of statistical hydrodynamics

Getting started to the development of the mathematical body of statistical hydrodynamics in terms of intermittency, first of all we need to create a probability space $(\Omega, \mathcal{F}, P)$ of a random field of any hydrodynamic value as a random process in the generalized physical space of turbulent flow, where Ω – is the sample space, $\mathcal{F}$ – *is* σ-algebra of subsets, P– is the probability measure in $\mathcal{F}$.

2.1.1. The introduction of the sample space into the body of statistics

For the introduction of the sample space Ω, we consider the behavior of the value of $f(\mathbf{x}, t)$, measured by the sensor at the point $\mathbf{x} = \mathbf{x}_0$ of statistically stationary turbulent flow with strong intermittency, i.e., when $\gamma(\mathbf{x}) \cong 0.5$, see **Figure 3**. According to **Figure 2**, function $f(\mathbf{x}, t)$ forms a random continuous field in the space $G = D \times [0, \tau_0]$. Hence it follows that at the point $\mathbf{x}_0 \in D$ we have a continuous random varying function of time $f(\mathbf{x}_0, t)$. Let the measurements of $f = f(\mathbf{x}_0, t)$ were carried out in a fairly wide range (order to the averaged statistical value of this function was stable) of the observation time $t = [0, \tau_0]$. It allows us to form an ensemble of values Ω as the one-point countable set of elementary events f, if we split the range of values of function $f(\mathbf{x}_0, t)$ at

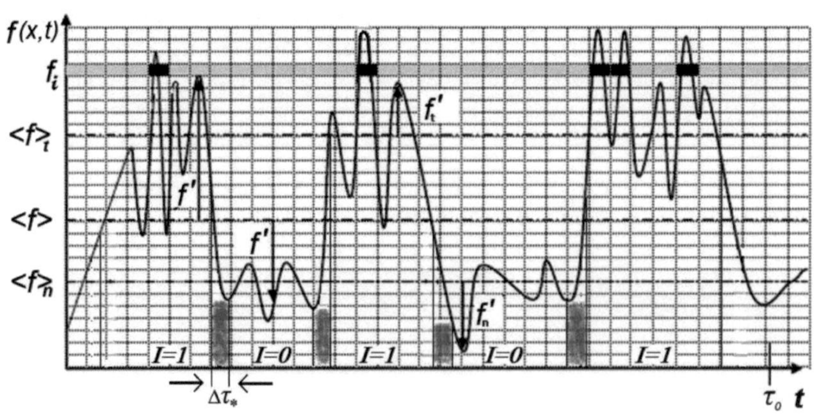

Figure 3. Illustration of statistical averaging of the instantaneous hydrodynamic variable. Here $f = \{f(\mathbf{x}, t)\}$ is the range of function $f(\mathbf{x}, t)$ at the point $\mathbf{x} = \mathbf{x}_0$; $\langle f \rangle$ is the total statistical average; $\langle f \rangle_t$ and $\langle f \rangle_n$ is the conditional statistical mean in each of the intermittent media of the turbulent flow; $f' = f - \langle f \rangle$, $f'_t = f_t - \langle f \rangle_t$, and $f'_n = f_n - \langle f \rangle_n$ are fluctuations (pulsations), measured from its own statistical means; τ_0 is the period of averaging time, sufficient to ensure sustainable statistical mean of values f; $\Delta\tau_*$ is the characteristic time of the superlayer observation, $I = I(\mathbf{x}, t)$ is the intermittency function of the turbulent fluid domain.

http://dx.doi.org/10.5772/67844

sufficiently small intervals Δf, and the range $[0, \tau_0]$ at sufficiently small intervals Δt, **Figure 2**. Having fixed a certain value of function f_i in each of the selected intervals Δf we come to the Lebesgue integral in terms of the set theory, formed in the physical space. Indeed, in the *i*-th layer there are N_i sampled values f_i in the form of shaded elementary cells $\Delta f\,\Delta t$, **Figure 2** (the selection of one particular value f_i from these cells plays no special role due to their small value). The total number of cells $N = \tau_0 / \Delta t$ and represented as the ensemble of values f, and also in the limit $\Delta f \to 0$ and $\Delta t \to 0$ this set will be dense, and the numerical value of f will be an element of this set, i.e., an elementary event.

In other words, for every fixed point $\boldsymbol{x} = \boldsymbol{x}_0$, the total number of all sample values f_i forms a random continuous field of values $f \in \Omega$ in the physical space $G = D \times [0, \tau_0]$. As a result, we come to a random process in the Borel space, in which a random variable $f(\boldsymbol{x}, t)$ takes all values of $f = \{f(\boldsymbol{x}, t)\}$, which are the elements of continuous set

$$\Omega = \{f : f_{min} \leq f(\boldsymbol{x}, t) < f_{max},\ N|_{\boldsymbol{x}=\boldsymbol{x}_0} = \lim_{\tau_0 \to \infty} \tau_0/\Delta t,\ (\boldsymbol{x}, t) \in G\} \tag{3}$$

So, from the physical space G with the hydrodynamic quantity $f(\boldsymbol{x}, t)$ we went to sample space Ω, elements of which are a set of values of $f = \{f(\boldsymbol{x}, t)\}$, i.e.

$$f(\boldsymbol{x}, t) \to f = \{f(\boldsymbol{x}, t)\};\ f(\boldsymbol{x}, t)|_{I=1} \to f_t = \{f(\boldsymbol{x}, t)|_{I=1}\};\ f(\boldsymbol{x}, t)|_{I=0} \to f_n = \{f(\boldsymbol{x}, t)|_{I=0}\} \tag{4}$$

$$\overline{f(\boldsymbol{x}, t)} \to \langle f \rangle;\ \overline{f(\boldsymbol{x}, t)}^{t} \to \langle f \rangle_t,\ \overline{f(\boldsymbol{x}, t)}^{n} \to \langle f \rangle_n \tag{5}$$

$$f'(\boldsymbol{x}, t) \to f' = \{f(\boldsymbol{x}, t) - \overline{f(\boldsymbol{x}, t)}\};\ f_r'(\boldsymbol{x}, t) \to f_r' = \{f_r(\boldsymbol{x}, t) - \overline{f(\boldsymbol{x}, t)}^{r}\} \tag{6}$$

Now we need develop the apparatus of statistics together with the operations of statistical averaging of the hydrodynamic quantities. For this we represent the apparatus of statistics based on a formal using of the probability density function (one-point PDF) of some hydrodynamic quantity $f = \{f(\boldsymbol{x}, t)\}$. At that the intermittency function $I = I(\boldsymbol{x}, t)$ will be used to obtain conditional one-point statistics.

2.1.2. The introduction of the algebra of events and PDFs

Let us introduce a one-point probability density function $p(f) = p(f; \boldsymbol{x}, t)$ into the body of statistics. According to the Kolmogorov axioms [12], it can be carried out via the Lebesgue-Stieltjes integral:

$$\boldsymbol{P}(\Omega) - \int_{\Omega} p(f)df \tag{7}$$

where $p(f) = \lim_{N \to \infty} p(f;\ f \in \Omega,\ N)$ and $\int_{\Omega} p(f)df = 1$.

For introduction of the algebra of events, we suppose that the space Ω, defined by Eq. (3), contains two independent subspaces (subsets)

$$\Omega_1 = \Omega|_{\mathcal{J}_1=1}, \quad \Omega_2 = \Omega|_{\mathcal{J}_2=1} \tag{8}$$

i.e., the generalized set $\Omega = \Omega_1 + \Omega_2$, and we have $\mathcal{F}$ as σ-algebra of the subsets. The indicators of these subsets are the characteristic functions

$$\mathcal{J}_1 = \begin{cases} 1 \; if \; f \in \Omega_1 \\ 0 \; otherwise \end{cases}, \quad \mathcal{J}_2 = \begin{cases} 1 \; if \; f \in \Omega_2 \\ 0 \; otherwise \end{cases} \tag{9}$$

at that the set of values f, belonging to the super-layer, is excluded. In the results, we have the Borel algebra subsets of the set Ω with the Kolmogorov axioms, which according to the total probability formula gives

$$\boldsymbol{P}(\Omega) = \sum_{k=1}^{2} \boldsymbol{P}\{\Omega|_{\mathcal{J}_k=1}\}\boldsymbol{P}\{\mathcal{J}_k = 1\} \tag{10}$$

where $\boldsymbol{P}\{\mathcal{J}_k = 0\} = 0$ as an impossible event.

2.2. Statistical averaging of hydrodynamic quantities

2.2.1. *Applied to the intermittent turbulent flows*

For the intermittent turbulent flows, the sample sets, which we designated as $\Omega_t = \Omega|_{\mathcal{J}_1=1}$ and $\Omega_n = \Omega|_{\mathcal{J}_2=1}$, are the set of values of hydrodynamic variable, belonging to the turbulent and nonturbulent fluid of turbulent flow. Indicators of these sets are the functions $\mathcal{J}_1 = \mathcal{J}_t$ and $\mathcal{J}_2 = \mathcal{J}_n$, while $\boldsymbol{P}\{\mathcal{J}_t = 1\} = \gamma_t$ and $\boldsymbol{P}\{\mathcal{J}_n = 1\} = \gamma_n$ are the measures of these sets with the condition $\gamma_t + \gamma_n = 1$, and represent the intermittency factors as the probability of observing the turbulent and nonturbulent fluid at the point $\boldsymbol{x}$ of turbulent flow, i.e., $\gamma_t = \gamma_t(\boldsymbol{x})$ and $\gamma_n = \gamma_n(\boldsymbol{x})$. Now, according to Eq. (10),

$$\boldsymbol{P}(\Omega) = \gamma_t \boldsymbol{P}(\Omega_t) + \gamma_n \boldsymbol{P}(\Omega_n) \tag{11}$$

where $\boldsymbol{P}(\Omega_t)$ и $\boldsymbol{P}(\Omega_n)$ – conditional random set of value f, belonging to the turbulent $\Omega_t = \{f_t\}$ and nonturbulent $\Omega_n = \{f_n\}$ fluid at the point $\boldsymbol{x}$; the values $\gamma_t = \gamma_t(\boldsymbol{x})$ and $\gamma_n = \gamma_n(\boldsymbol{x})$, while the one-point PDF

$$p(f) = \gamma_t p_t(f) + \gamma_n p_n(f) \tag{12}$$

where $p_t(f) = p_t(f; \boldsymbol{x}, t)$, $(\boldsymbol{x}, t) \in G_t$ and $p_n(f) = p_n(f; \boldsymbol{x}, t)$, $(\boldsymbol{x}, t) \in G_n$ represent the conditional one-point PDFs. As it turns out, a PDF may have or not to have an explicit dependence on $\boldsymbol{x}$. In actual fact, if the flow is intermittent, it has a dual structure [1] and in the generalized set we have $\Omega = \Omega_t + \Omega_n$ so that the measures of sample sets γ_t and γ_n are depend on $\boldsymbol{x}$; if the flow is not intermittent (when the phenomenon of intermittency is not considered) it occurs in a "single" space as a set of elementary events $\Omega = \Omega_R$, the measure of which does not depend on $\boldsymbol{x}$. In the case of the explicit dependence, we denote the PDF $p(f)$ in Eq. (7) as

http://dx.doi.org/10.5772/67844

$$P(f) = \lim_{N\to\infty} p\Big(f;\ f \in (\Omega_t + \Omega_n),\ N,\ \boldsymbol{x}\Big) \tag{13}$$

and we call this function as the "*total*" PDF, and the flow—flow of the "*intermittent*" continuous media with turbulent and nonturbulent fluid. In the absence of such dependence, we denote it as

$$p_R(f) = \lim_{N\to\infty} p(f;\ f \in \Omega_R,\ N) \tag{14}$$

and called the "*unconditional*" PDF $p_R(f)$, and the flow—flow of the "*nonintermittent*" continuous medium, which is modeled by the *RANS* method. The explicit dependence of the PDF $P(f)$ Eq. (13) on the coordinates creates certain difficulties in its using in the statistical modeling and also leads to the necessity of introducing in the theory of statistical hydrodynamics the conditional PDF for the hydrodynamic characteristics of turbulent and nonturbulent media.

So, to perform the conditional averaging of the instantaneous characteristics of the flow, we introduce into statistical body the conditional PDF, i.e., the CPDFs:

$$p_t(f) = p(f|_{I=1}),\ p_n(f) = p(f|_{I=0}) \tag{15}$$

with the indicator (characteristic) function of the turbulent fluid

$$I = \begin{cases} 1\ if\ f \in \Omega_t \\ 0\ if\ f \in \Omega_n \end{cases} \tag{16}$$

represents a probability of observing the turbulent flow at the given point $\boldsymbol{x}$, i.e., it is the intermittency factor $\gamma = \gamma(\boldsymbol{x})$. Now the expression for the "*total*" PDF in Eq. (12), by virtue of the fact that $\gamma_n = 1 - \gamma$, is transformed into

$$P(f) = \gamma p_t(f) + (1-\gamma)p_n(f) \tag{17}$$

with the explicit dependence on $\boldsymbol{x}$, while the CPDF $p_t(f)$ and $p_n(f)$ obviously do not depend on $\boldsymbol{x}$.

Now conduct the operations of statistical averaging of hydrodynamic quantities. These operations we will conduct with the help of a formal using of the PDFs, i.e., when a particular form of this function does not necessarily need to know.

2.2.2. *Operations of statistical averaging of the hydrodynamic quantity*

The statistical averaging of the hydrodynamic quantity $f(\boldsymbol{x}, t)$ can be performed by a formal using of the PDF. The results of statistical averaging operation are the *conditional* statistical means when $r = t$ for turbulent and $r = n$ for nonturbulent fluid

$$\langle f \rangle_r = \int_{\Omega_r} f p_r(f) df, \qquad r = t,\ n \tag{18}$$

and also the *total* statistical average

$$\langle f\rangle = \int_{\Omega} fP(f)df \tag{19}$$

which by virtue of the expression in Eq. (17) gives

$$\langle f\rangle = \gamma\langle f\rangle_t + (1-\gamma)\langle f\rangle_n \tag{20}$$

At that by definition the value $f_t = f|_{I=1}$ and $f_n = f|_{I=0}$ and for the "pulsations" we have

$$f' = f - \langle f\rangle\,;\ f'_r = f_r - \langle f_r\rangle_r;\ f'|_{I=1} = f'_t + \langle f\rangle_t - \langle f\rangle\,;\ f'|_{I=0} = f'_n + \langle f\rangle_n - \langle f\rangle \tag{21}$$

whence it follows that

$$\langle f_r\rangle_r = \langle f\rangle_r;\ \langle f'_r g'_r\rangle_r \neq \langle f'g'\rangle_r \tag{22}$$

As is evident that the total average in Eq. (20) represents the statistical characteristic of a rather complex structure, while the unconditional mean, when in Eq. (18) we have $r = R$, is a characteristic of the "simplified" flow without considering effects of intermittency. At that $\langle f\rangle_R \cong \langle f\rangle$ because the total average $\langle f\rangle$ does not contain the values of f belonging to the superlayer [16].

2.2.3. *Operations of statistical averaging of derivative of the hydrodynamic quantity*

The statistical averaging of the derivative of hydrodynamic quantity $\xi = \partial f/\partial x$ gives the following. In point of fact, on the one side using the joint CPDF $p_r(f,\xi)$ we have

$$\iint \frac{\partial f}{\partial x} p_r(f,\xi)df d\xi = \frac{\partial}{\partial x}\int \langle f|_\xi\rangle_r p_r(\xi)d\xi = \frac{\partial\langle f\rangle_r}{\partial x} \tag{23}$$

because that in accordance with [14]

$$p_r(f,\xi) = p_r(f|_\xi)p_r(\xi)$$

and

$$\iint \left(\partial f p_r(f,\xi)/\partial x - f\partial p_r(f,\xi)/\partial x\right)df d\xi = \int \left(\partial\left(\int f p_r(f|_\xi)df\right)/\partial x\right)p_r(\xi)d\xi \tag{24}$$

where $\langle f|_\xi\rangle_r$ is the conditional mean values of f in turbulent ($r = t$) or nonturbulent ($r = n$) medium for all possible fixed values of ξ. At that $\partial p_r(f,\xi)/\partial x = 0$ due to the fact that the function (f, ξ) does not depend obviously on the coordinate x. On the flip side, we have

$$\iint \frac{\partial f}{\partial x} p_r(f,\xi)df d\xi = \left\langle \frac{\partial f}{\partial x}\right\rangle_r \tag{25}$$

because that

$$\iint \xi p_r(\xi|_f) p_r(f) d\xi df = \int \langle \xi|_f \rangle_r p_r(f) df \tag{26}$$

where $\langle \xi|_f \rangle_r$ is the conditional mean of the gradient $\xi = \partial f / \partial x$ in turbulent or nonturbulent medium, given for all possible fixed values of f. As a result, we have

$$\left\langle \frac{\partial f}{\partial x} \right\rangle_r = \frac{\partial \langle f \rangle_r}{\partial x}, \quad r = t, n \tag{27}$$

Thus, the operation of conditional statistical averaging of derivatives is permutational. So, we have proved that the permutation of conditional averaging operation has a strict mathematical justification.

It is appropriate to note that in the classical method of *RANS*, the operation of unconditional statistical averaging of derivatives gives the same result. Actually, the *unconditional* joint PDF $p_R(f, \xi)$ of Eq. (14) does not depend on the coordinates obviously and therefore it is correctly Eqs. (23)–(27) with index $r = R$, i.e.,

$$\left\langle \frac{\partial f}{\partial x} \right\rangle_R = \frac{\partial \langle f \rangle_R}{\partial x} \tag{28}$$

that proves the rule of permutation of the operation of unconditional averaging of derivatives in the method of *RANS*.

About the permutation of the operation of derivatives total averaging I must say the following. The operation of total statistical averaging of partial derivatives of type $\xi = \partial f / \partial x$ by Eq. (19) for intermittent continuous media with turbulent and nonturbulent fluid cannot be a permutational. This operation is carried out similarly in Eqs. (23)–(27). Here, however, must keep in mind that now the total PDF $P(f)$ in Eq. (17) obviously depend on the coordinates due to $\gamma = \gamma(\boldsymbol{x})$. The legitimacy of such a permutation of the operation is easy to establish if we attract Eq. (20) as applied to the partial derivatives. In this case

$$\left\langle \frac{\partial f}{\partial x} \right\rangle = \gamma \left\langle \frac{\partial f}{\partial x} \right\rangle_t + (1 - \gamma) \left\langle \frac{\partial f}{\partial x} \right\rangle_n \tag{29}$$

and

$$\frac{\partial \langle f \rangle}{\partial x} = \gamma \frac{\partial \langle f \rangle_t}{\partial x} + (1 - \gamma) \frac{\partial \langle f \rangle_n}{\partial x} + (\langle f \rangle_t - \langle f \rangle_n) \frac{\partial \gamma}{\partial x} \tag{30}$$

after comparing of which, with regard to Eq. (27) we get

$$\left\langle \frac{\partial f}{\partial x} \right\rangle = \frac{\partial \langle f \rangle}{\partial x} - (\langle f \rangle_t - \langle f \rangle_n) \frac{\partial \gamma}{\partial x} \tag{31}$$

It follows that the permutation of the operation of total statistical averaging of derivatives is not legitimate, i.e.,

$$\left\langle \frac{\partial f}{\partial x} \right\rangle \neq \frac{\partial \langle f \rangle}{\partial x} \tag{32}$$

With regard to the total statistical averaging of time derivatives, instead of expression (31) we have $\langle \partial f/\partial t \rangle = \partial \langle f \rangle/\partial t$ because of $\partial \gamma/\partial t = 0$, i.e., for statistically stationary turbulent flows such a permutation is possible. The same applies to the conditional averaging of derivatives. So, we showed that the statistical modeling of turbulent flows, in the case of taking into account the effects of intermittency, should be based on Eqs. (20), (27), and (32).

2.2.4. *The statistical averaging of hydrodynamics equations*

The Navier-Stokes equations for an incompressible fluid together with the continuity equation are accepted as the basis of the hydrodynamic equations system [11]. When the external forces are absent, this system has the following form:

$$\begin{cases} \dfrac{\partial u_i}{\partial t} + \dfrac{\partial}{\partial x_k}\left(u_i u_k + \dfrac{p\delta_{ik}}{\rho} - \sigma_{ik} \right) = 0, \; i = 1, 2, 3 \\ \dfrac{\partial u_k}{\partial x_k} = 0 \end{cases} \tag{33}$$

where $\sigma_{ik} = \mu(\partial u_i/\partial x_k + \partial u_k/\partial x_i)$ is the tensor of viscous stress, μ is the dynamic factor of viscosity, p is the pressure, and ρ is the density. Our primary goal is to conduct an operation of statistical averaging of the SE (33) so as to obtain a system of equations for the conditional mean $\langle u_i \rangle_t$. At the beginning, we will conduct an operation of conditional statistical averaging of the continuity equation in SE (33). For this, we introduce the joint CPDF

$$p_r(u_i, \xi_i) = p_r(u_1, u_2, u_3, \xi_1, \xi_2, \xi_3) \tag{34}$$

with index $r = t$ for turbulent and $r = n$ for nonturbulent fluid, $\xi_1 = \partial u_1/\partial x_1$, $\xi_2 = \partial u_2/\partial x_2$, $\xi_3 = \partial u_3/\partial x_3$, and use the procedure of conditional averaging (23). As a result, we reach the following averaging procedure:

$$\int..\int \frac{\partial u_k}{\partial x_k} p_r(u_i, \xi_i) du_1, \ldots, d\xi_3 = \int..\int \frac{\partial u_k p_r(u_i, \xi_i)}{\partial x_k} du_1, \ldots, d\xi_3 = \frac{\partial}{\partial x_k}\int..\int u_k p_r(u_i, \xi_i) du_1 \ldots, d\xi_3 = 0 \tag{35}$$

because that the function $p_r(u_i, \xi)$ does not depend on x_k, i.e., $\partial p_r(u_i, \xi_i)/\partial x_k = 0$ and $p_r(u_i, \xi_i)\partial u_k/\partial x_k = \partial u_k p_r(u_i, \xi_i)/\partial x_k$. From here toward $k = 1$ in Eq. (35), we deduce

$$\frac{\partial}{\partial x_1}\int u_1 p_r(u_1) du_1 = \frac{\partial \langle u_1 \rangle_r}{\partial x_1} \tag{36}$$

as $p_r(u_i) = \int \ldots \int p_r(u_1, u_2, u_3, \xi_1, \xi_2, \xi_3) du_{i-1}, du_{i+1}..d\xi_3$, $du_0 = 1$, for example, $p_r(u_1) = \int \ldots \int$ $p_r(u_1, u_2, u_3, \xi_1, \xi_2, \xi_3) du_2 \ldots d\xi_3$. The same operation is carried out for $k = 2, 3$ using $p_r(u_2)$ and

$p_r(u_3)$. Now then, the conditionally averaged continuity equation for each of the intermittent media of turbulent flow has the form

$$\frac{\partial\langle u_k\rangle_r}{\partial x_k} = 0, \;\; r = t, n \tag{37}$$

To conduct the operation of conditional statistical averaging of the Navier-Stokes SE (33), we use the CPDF $p_r = p_r(\xi_1, \xi_2)$, where $\xi_1 = \frac{\partial u_i}{\partial t}$, $\xi_2 = \frac{\partial}{\partial x_k}\left(u_i u_k + \frac{p\delta_{ik}}{\rho} - \sigma_{ik}\right)$ with the summation over k = 1, 2, 3. Then, according to Eq. (25) for the momentum equation in SE (33) we obtain $\iint \xi_1 p_r(\xi_1, \xi_2) d\xi_1 d\xi_2 = \langle \partial u_i/\partial t\rangle_r$. Similarly, it conducted the averaging operation of the value ξ_2:

$$\iint \xi_2 p_r(\xi_1, \xi_2) d\xi_1 d\xi_2 = \left\langle \frac{\partial}{\partial x_k}\left(u_i u_k + \frac{p\delta_{ik}}{\rho} - \sigma_{ik}\right)\right\rangle_r \tag{38}$$

Applying the rule of permutation (27) and using the Reynolds development $f_r = \langle f\rangle_r + f_r'$, we deduce $\langle u_i u_k\rangle_r = \langle u_i\rangle_r\langle u_k\rangle_r + \langle u'_{ir}u'_{kr}\rangle_r$ in view of Eq. (22). As a result of the above-performed operation of statistical averaging of SE (33) now for the statistically stationary turbulent flow, we have the system of equations with two autonomous subsystems for the flow's conditional means of each of the intermittent media with turbulent and nonturbulent fluid:

$$\begin{cases} \dfrac{\partial\langle u_i\rangle_r}{\partial t} + \dfrac{\partial\langle u_i\rangle_r\langle u_k\rangle_r}{\partial x_k} + \dfrac{\langle u'_{ir}u'_{kr}\rangle_r}{\partial x_k} + \dfrac{\partial\langle (p\delta_{ik} - \sigma_{ik})/\rho\rangle_r}{\partial x_k} = 0 \\ \dfrac{\partial\langle u_k\rangle_r}{\partial x_k} = 0, \;\; r = t, \; n \end{cases} \tag{39}$$

where the fluctuating velocity of the turbulent or nonturbulent fluid flow $u'_{ir} = u_{ir} - \langle u_{ir}\rangle_r$ and $\langle u_i\rangle_r \equiv \langle u_{ir}\rangle_r$, but the one-point covariances $\langle u'_{ir}u'_{kr}\rangle_r \neq \langle u'_i u'_k\rangle_r$ according to Eq. (22). Besides, $\partial\langle u_i\rangle_r/\partial t = 0$ for statistically stationary turbulent flows. Each SS (39) with index $r = t$ or $r = n$ is statistically independent and is determined by the fact that the one-point correlation of the hydrodynamic quantities of turbulent and nonturbulent media is absent, i.e., $\langle f_t f_n\rangle = 0$. These subsystems allow the conditional means modeling of each of the intermittent media with turbulent and nonturbulent fluid independently from the each other.

The derivation of the turbulent kinetic energy budget equation by the *RANS* method is well known [1, 4]. The procedure of the budget equations derivation for conditional means of kinetic energy fluctuations in each of the intermittent medium of the turbulent flow will be the same. In the approximation of a free boundary layer, these equations have the following form:

$$\frac{\partial\langle E_r\rangle_r}{\partial t} + \langle u_k\rangle_r \frac{\partial\langle E_r\rangle_r}{\partial x_k} + \frac{\partial\langle (E_r + p'_r/\rho)v'_r\rangle_r}{\partial x_k} + \langle u'_{ir}u'_{kr}\rangle_r \frac{\partial\langle u_i\rangle_r}{\partial x_k} + \langle \varepsilon_r\rangle_r = 0 \tag{40}$$

Hereinafter $E_r = 0.5(u_r'^2 + v_r'^2 + w_r'^2)$ and $\langle E_r\rangle_r = 0.5(\langle u_r'^2\rangle_r + \langle v_r'^2\rangle_r + \langle w_r'^2\rangle_r)$.

3. The *ASMTurb* method

The new principle of constructing mathematical models of the energy-containing structure of turbulent flows (the large-scale turbulent motion) is as follows: (1) as the main statistical characteristics of modeling are chosen the conditional averages of hydrodynamic quantities of the turbulent and nonturbulent fluid; (2) to describe the conditional means of hydrodynamic quantities are used two statistically independent (autonomous) systems of differential equations; (3) each of the autonomous systems for the conditional averages is closed by its own closure hypothesis; and (4) the total average of hydrodynamic quantities is obtained by the algebraic relations of statistical hydrodynamics, which bind the total and conditional means through the mediation of the intermittency factor. To realize this principle, the mechanism of the turbulent fluid formation in a turbulent flow is proposed. This is achieved by the introduction of the "superlayer" between turbulent and nonturbulent fluid, where shear rate and pressure fluctuations in the turbulent fluid generate the pressure fluctuations in the nonturbulent fluid. This process leads to the so-called "nonlocal" transfer of the impulse and initiates the occurrence of velocity fluctuations (for particulars see in [15, 16]). The formulated principle of constructing mathematical models is called the *ASMTurb* method [5].

3.1. Mathematical tools of the *ASMTurb* method

3.1.1. General system of equations for conditional means

According to the *ASMTurb* method, we have two autonomous subsystems (SS) of the difference equations corresponding to Eqs. (39) and (40) in the form of

$$\begin{cases} \dfrac{\partial\langle u_i\rangle_t}{\partial t}+\dfrac{\partial\langle u_i\rangle_t\langle u_k\rangle_t}{\partial x_k}+\dfrac{\partial\langle u'_{it}u'_{kt}\rangle_t}{\partial x_k}+\dfrac{\partial\langle (p_t\delta_{ik}-\sigma_{tik})/\rho\rangle_t}{\partial x_k}=0 \\ \dfrac{\partial\langle u_k\rangle_t}{\partial x_k}=0 \\ \dfrac{\partial\langle E_t\rangle_t}{\partial t}+\langle u_k\rangle_t\dfrac{\partial\langle E_t\rangle_t}{\partial x_k}+\dfrac{\partial\langle (E_t+p'_t/\rho)v'_t\rangle_t}{\partial x_k}+\langle u'_{it}u'_{kt}\rangle_t\dfrac{\partial\langle u_i\rangle_t}{\partial x_k}+\langle\varepsilon_t\rangle_t=0 \end{cases} \tag{41}$$

and

$$\begin{cases} \dfrac{\partial\langle u_i\rangle_n}{\partial t}+\dfrac{\partial\langle u_i\rangle_n\langle u_k\rangle_n}{\partial x_k}+\dfrac{\partial\langle u'_{in}u'_{kn}\rangle_n}{\partial x_k}+\dfrac{\partial\langle (p_n\delta_{ik}-\sigma_{nik})/\rho\rangle_n}{\partial x_k}=0 \\ \dfrac{\partial\langle u_k\rangle_n}{\partial x_k}=0 \\ \dfrac{\partial\langle E_n\rangle_n}{\partial t}+\langle u_k\rangle_n\dfrac{\partial\langle E_n\rangle_n}{\partial x_k}+\dfrac{\partial\langle (E_n+p'_n/\rho)v'_n\rangle_n}{\partial x_k}+\langle u'_{in}u'_{kn}\rangle_n\dfrac{\partial\langle u_i\rangle_n}{\partial x_k}+\langle\varepsilon_n\rangle_n=0 \end{cases} \tag{42}$$

that describe the conditional mean flow characteristics of each of the intermittent media with the turbulent ($r = t$) and nonturbulent ($r = n$) fluid. Let us note that each of the SS (41) and SS (42) is statistically independent, in terms of the one-point correlations $\langle f_t f_n\rangle=0$, so after the

http://dx.doi.org/10.5772/67844

completion of these subsystems using the corresponding expressions for $\langle u'_{ir}u'_{kr}\rangle_r$, $\langle(E_r + p'_r/\rho)v'_r\rangle_r$ and $\langle\varepsilon_r\rangle_r$ as the closure hypothesis we obtain mathematical models for the flow of the turbulent and nonturbulent fluid.

3.1.2. The closure hypothesis

The closure hypothesis for SS (41) and SS (42) we will choose in the form of a simple expression gradient relation [16]:

$$-\langle u'_r v'_r\rangle_r = \nu_r \frac{\partial\langle u\rangle_r}{\partial y} \tag{43}$$

$$\left\langle\left(E_r + \frac{p'_r}{\rho}\right)v'_r\right\rangle_r = -\nu_r \frac{\partial\langle E_r\rangle_r}{\partial y}, \langle\varepsilon_r\rangle_r = c_* \frac{\nu_r\langle E_r\rangle_r}{L_r^2} \tag{44}$$

where ν_r is the coefficient of turbulent viscosity, expressed by the "second" Prandtl formula

$$\nu_r = k_r(u_1 - u_2)x \tag{45}$$

It is clear that the use of Eq. (45) allows us to solve our "dynamic task" (i.e., the continuity and momentum equations in SS (41) and (42)) without distinction of "fluctuating task" (i.e., turbulent-kinetic-energy budget equations in SS (41) and (42)). This approach greatly simplifies the modeling process.

3.1.3. Modeling of the total averages

To calculate the total statistical averages, we will use the statistical ratio (20). For example, for the velocity components

$$\langle u_i\rangle = \gamma\langle u_i\rangle_t + (1-\gamma)\langle u_i\rangle_n \tag{46}$$

To determine the total averages for correlations of velocity pulsations (the covariances), we will use the ratios of the type

$$\langle u'v'\rangle = \gamma\langle u'_t v'_t\rangle_t + (1-\gamma)\langle u'_n v'_n\rangle_n + \gamma(1-\gamma)(\langle u\rangle_t - \langle u\rangle_n)(\langle v\rangle_t - \langle v\rangle_n) \tag{47}$$

This equation can be obtained according to our theory. In actual fact, for the velocity pulsations we have $u_i'|_{I=1} = \langle u_{i_t}\rangle_t + u'_{i_t} - \langle u_i\rangle$ and $u_i'|_{I=0} = \langle u_{i_n}\rangle_n + u'_{i_n} - \langle u_i\rangle$ according to Eqs. (21) and (22) whence it follows from Eq. (47), since $\langle u'v'\rangle = \gamma\langle u'v'\rangle_t + (1-\gamma)\langle u'v'\rangle_n$ and $\langle u'_r\langle v_r\rangle_r\rangle_r = 0$, $\langle\langle u\rangle\rangle_t = \langle u\rangle$ and so on. Eq. (47) aligns with the expression in [4, 17].

The fluctuating structure modeling is determined by the separate terms of equations for kinetic energy of the velocity fluctuations in each of the intermittent media, i.e., the turbulent kinetic energy budget equations in SS (41) and (42). In addition, the expression for the total average of turbulent energy is the same as Eq. (47), viz.,

$$\langle E\rangle = \gamma\langle E_t\rangle_t + (1-\gamma)\langle E_n\rangle_n + Ed \tag{48}$$

where

$$Ed = 0.5\gamma(1-\gamma)[(\langle u\rangle_t - \langle u\rangle_n)^2 + (\langle v\rangle_t - \langle v\rangle_n)^2 + (\langle w\rangle_t - \langle w\rangle_n)^2] \tag{49}$$

Let us note that Eq. (48) can also be obtained as Eq. (47). According to Eq. (47), wherein $u' = v'$, we have

$$\langle u'^2\rangle = \gamma\langle u_t'^2\rangle_t + (1-\gamma)\langle u_n'^2\rangle_n + u_d,\ u_d = \gamma(1-\gamma)(\langle u\rangle_t - \langle u\rangle_n)^2 \tag{50}$$

To calculate the total averages, as is evident from the foregoing, distribution of the intermittency factor γ is required. To model the intermittency factor γ we will use the expression in [16]:

$$\gamma \cong \langle \varepsilon_t\rangle_t / \langle \varepsilon\rangle \tag{51}$$

4. Testing of the *ASMTurb* method

The *ASMTurb* method has been tested in [15, 16] on the example of constructing the mathematical models for self-similar turbulent shear flows such as: I, the two-stream plane mixing layer; II, the outer region of the boundary layer on the wall; III, the far wake behind a cross-streamlined cylinder; and IV, the axisymmetric submerged jet. Test results were presented in the form calculating the main conditional and total statistical averages applied to a self-similar region of turbulent flows. A comparison was performed between the predictions and known experimental data for the energy-containing structure of turbulent flow, and excellent agreements were noted. By this means, it was shown that these *ASMTurb* models are more accurate and more detailed than the RANS models.

In view of the fact that construction of each mathematical model requires a significant volume, here we will present without details only testing results the *ASMTurb* method on the example of constructing a mathematical model for the turbulent fluid flow in a self-similar mixing layer. It is doing because all turbulence processes existing only into turbulent fluid. Calculations of the main "dynamic" and "fluctuating" characteristics we will compare with known experimental data. More detailed of this model see in [16].

4.1. Construction of the model for two-stream plane mixing layer

The mathematical *ASMTurb* model for two-stream mixing layer (see [18, 19], etc.), formed as a result of turbulent mixing of two co-current streams with identical fluid and ρ = Const, moving with different velocities $u_1 = u_{\max}$ and $u_1 = u_{\min}$, includes two subsystems SS (41) and SS (42) for conditional means of each of the intermittent media of the turbulent and nonturbulent fluid. In this case, first of all, we use the SS (41) that was written in approximation of a free boundary layer and reduced to a nondimensional form after the introduction of nondimensional variables. The task of modeling only the velocity field of turbulent flow has been called as "dynamic task."

http://dx.doi.org/10.5772/67844

4.1.1. Modeling of the turbulent fluid flow

So, the dynamic task for modeling the velocity field of the turbulent fluid is reduced to solving the following system of equations

$$\begin{cases} \langle u\rangle_t \dfrac{\partial\langle u\rangle_t}{\partial x} + \langle v\rangle_t \dfrac{\partial\langle u\rangle_t}{\partial y} + \dfrac{\partial\langle u_t' v_t'\rangle_t}{\partial y} = 0 \\ \dfrac{\partial\langle u\rangle_t}{\partial x} + \dfrac{\partial\langle v\rangle_t}{\partial y} = 0 \end{cases} \quad (52)$$

with boundary conditions, which initially assuming as asymptotical, namely

$$\langle u\rangle_t = \begin{cases} u_1, \ y \to \infty \\ u_2, \ y \to -\infty \end{cases}, \ (\boldsymbol{x}, t) \in G_t \quad (53)$$

At that the closure hypothesis in Eqs. (43) and (45) take the form of

$$-\langle u_t' v_t'\rangle_t = \nu_t \frac{\partial\langle u\rangle_t}{\partial y}, \ \nu_t = k_t(u_1 - u_2)x \quad (54)$$

where ν_t is the coefficient of turbulent viscosity, $k_t = k_t(m)$, $m = u_2/u_1$). For transformation of the SE (52) to the self-similar mode in order to deduce the self-similar solution of our task, let us introduce dimensionless variables

$$\frac{\langle u\rangle_t}{u_1} = F_t'(\eta), \ \eta = \frac{y}{x} \quad (55)$$

where $F_t' = \partial F_t/\partial\eta$ with transformation $\partial/\partial x = -\eta/xd/d\eta$, $\partial/\partial y = 1/xd/d\eta$. The boundary conditions (53) take the form of

$$F_t'(\eta) \cong \begin{cases} 1, \ \eta \to \infty \\ m, \ \eta \to -\infty \end{cases} \quad (56)$$

The nondimensional transverse velocity is defined after integrating the continuity equation in SE (52):

$$\frac{\langle v\rangle_t}{u_1} = \eta F_t' - F_t \quad (57)$$

the while correlation in Eq. (52) is

$$-\frac{\langle u_t' v_t'\rangle_t}{u_1^2} = k_t(1-m)F_t'' \quad (58)$$

As a consequence, the momentum equation in SE (52) takes the form of ordinary differential equation

$$F_t''' + 2\sigma_t{}^2 F_t F_t'' = 0 \tag{59}$$

where σ_t is a first empirical parameter of the model, the value of which is determined by the condition of the best agreement of calculated and measurements of the longitudinal velocity. We now represent a function $F_t(\eta)$ as a power series in the small parameter $(m-1)$:

$$F_t = \sum_{i=0}^{\infty}(m-1)^i F_{it} = F_{0t} + (m-1)F_{1t} + (m-1)^2 F_{2t} + \ldots \tag{60}$$

If we substitute this expression into Eq. (59) and compare the components at the same powers of parametric value $(m-1)^i$, we obtain a system of sequentially interconnected ordinary differential equations (here we confine ourselves to the second approximation of our task):

$$\begin{cases} F_{0t}''' + 2\sigma_t{}^2 F_{0t}F_{0t}'' = 0, \ i = 0 \\ F_{1t}''' + 2\sigma_t{}^2(F_{0t}F_{1t}'' + F_{1t}F_{0t}'') = 0, \ i = 1 \\ F_{2t}''' + 2\sigma_t{}^2(F_{0t}F_{2t}'' + F_{1t}F_{1t}'' + F_{2t}F_{0t}'') = 0, \ i = 2 \end{cases} \tag{61}$$

From the boundary conditions (56) it follows that

$$\begin{cases} F_{0t}' = 1\,, \ F_{1t}' = 0, \ F_{2t}' = 0 \ as \ \eta \to \infty \\ F_{0t}' = 1\,, \ F_{1t}' = 1, \ F_{2t}' = 0 \ as \ \eta \to -\infty \end{cases} \tag{62}$$

where we get after integration

$$F_{0t}' = 1, \ F_{0t} = \eta - \eta_{0t} \tag{63}$$

where $\eta_{0t} = \eta_{0t}(m)$ is the displacement of the symmetry plane of the mixing layer $\eta = 0$. Now the SE (61) takes the form of

$$\begin{cases} \tilde{F}_{1t}''' + 2\varphi\tilde{F}_{1t}'' = 0 \\ \tilde{F}_{2t}''' + 2\varphi\tilde{F}_{2t}'' = -2\tilde{F}_{1t}\tilde{F}_{1t}'' \end{cases} \tag{64}$$

where $\tilde{F}_t' = \partial\tilde{F}/\partial\varphi$, $\tilde{F}_t'' = \partial^2\tilde{F}/\partial\varphi^2$, etc. and

$$\tilde{F}_t(\varphi) = \sigma_t F_t(\eta), \ \varphi = \sigma_t(\eta - \eta_{0t}) \tag{65}$$

To determine the value η_{0t} will use the known Karman's condition, namely

$$\langle v(\infty)\rangle_t + m\langle v(-\infty)\rangle_t = 0 \tag{66}$$

The boundary conditions (56) because of $\langle u\rangle_t/u_1 = F_t'(\eta) = \tilde{F}_t{}'(\varphi)$ are converted in accordance with the boundary conditions (62) to the form

$$\begin{cases} \tilde{F}'_{0t} = 1, \; \tilde{F}'_{1t} = 0, \; \tilde{F}'_{2t} = 0 \; as \; \varphi \to \infty \\ \tilde{F}'_{0t} = 1, \; \tilde{F}'_{1t} = 1, \; \tilde{F}'_{2t} = 0 \; as \; \varphi \to -\infty \end{cases} \tag{67}$$

The solution of the dynamic task in the first approximation is easy to obtain in an analytical form [16]. At that according to momentum equation in SE (64) we have

$$\frac{\langle u \rangle_t}{u_1} = \tilde{F}'_t = 1 + \frac{m-1}{2}(1 - erf\varphi) \tag{68}$$

while the transverse velocity in Eq. (57) and correlation in Eq. (58) take the following form

$$\frac{\langle v \rangle_t}{u_1} = \frac{(\varphi + \sigma\eta_{0t})\tilde{F}'_t - \tilde{F}_t}{\sigma_t} \tag{69}$$

$$-\frac{\langle u'_t v'_t \rangle_t}{\Delta U^2} = \frac{\tilde{F}''_t}{2\sigma_t(1-m)^2} \tag{70}$$

where $\Delta U = u_1 - u_2$ and the flow function (65) is

$$\tilde{F}_t = \varphi + \frac{m-1}{2}\left(\varphi - \varphi erf\varphi - \frac{1}{\sqrt{\pi}}e^{-\varphi^2} + 2c_{1t}\right) \tag{71}$$

Now, we can calculate both the longitudinal velocity profile by Eq. (68) and the correlation profile by Eq. (70) to evaluate the accuracy of our model in the first approximation. These calculations showed that the velocity profile in Eq. (68) coincides with the known experimental data at σ_t = 18.0, while the correlation profile of fluctuating velocities in Eq. (70) greatly overestimated (see **Figure 4b** where according to (47) we have to have $\langle u'v' \rangle = \langle u'_t v'_t \rangle_t$ as γ = 1). Therefore, for specification of our model, we must consider the second approximation of our task.

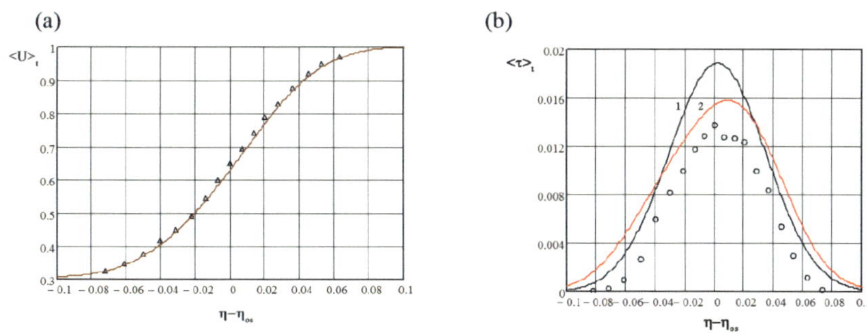

Figure 4. (a) The self-similar profile of the normalized conditional mean longitudinal velocity $\langle U \rangle_t = \langle u \rangle_t / u_1$ over the turbulent fluid. (b) Profiles of the normalized conditional mean shear stress $\langle \tau \rangle_t = -\langle u'_t v'_t \rangle_t / \Delta U^2$: 1, calculation $\langle \tau \rangle_t$ corresponding to the solution of the dynamic task in the first approximation, $\sigma_{t1} = 18.0$,; and 2, calculation $\langle \tau \rangle_t$ corresponding to the solution of the dynamic task in the second approximation, $\sigma_t = 21.5$. Symbol o is the measurements of the total average $\langle \tau \rangle = -\langle u'v' \rangle / \Delta U^2$ (measurements $\langle \tau \rangle_t$ of [20] are absent). From now on the curves—our calculations, symbols—experimental data [20] in the self-similar mixing layer at the parameter m = 0.305.

The solution of the dynamic task in the second approximation was found in such a manner. The solution of the second equation in SE (64)

$$\tilde{F}_{2t}''' = e^{-\varphi^2}(c_0 - \int \tilde{F}_{1t}\tilde{F}_{2t}'' e^{-\varphi^2} d\varphi) \tag{72}$$

was found by numerical calculation. At that according to Eq. (60), function $\tilde{F}_t$ in the second approximation contains two constants of integration c_{1t} и c_{2t}. To determine these constants, the integral relation was involved (for example, see [16]):

$$\lim_{\varphi_1, \varphi_2 \to \infty} \int_{-\varphi_2}^{\varphi_1} (\tilde{F}_t''' + 2\tilde{F}_t \tilde{F}_t'') d\varphi = 0 \tag{73}$$

Hence, the values of the constants are defined with the help of numerical calculation, $c_{1t} \cong 0.4$, $c_{2t} \cong -0.1$. To determine the value $\eta_{0t} = \eta_{0t}(m)$ in the expression of the dimensionless coordinates $\varphi = \sigma_t(\eta - \eta_{0t})$ we will use Eqs. (69) and (66).

The results of calculations of conditional means of this dynamic task for the mixing layer with the parameter $m = 0.305$ in comparison with the experimental data of [20] are shown in **Figure 4**. In this case, according to our model the calculated value $\eta_{0t} = -0.0181$ when $\sigma_t = 21.5$ (in [20] empirical value $\eta_{0S} = -0.02$, i.e., we have a good accuracy for $\eta_{0t} \cong \eta_{0S}$). Hereinafter curves – our calculations, symbols —measurements are mentioned [20].

Solution of the "fluctuating" task was found in such a manner. The equation of kinetic energy of the velocity fluctuations in SS (41) for the statistically stationary flow of the turbulent fluid, now in the approximation of a free boundary layer, has the following form:

$$\underbrace{\langle u\rangle_t \frac{\partial \langle E_t\rangle_t}{\partial x} + \langle v\rangle_t \frac{\partial \langle E_t\rangle_t}{\partial y}}_{Conv_t} + \underbrace{\frac{\partial \langle (E_t + p_t'/\rho) v_t'\rangle_t}{\partial y}}_{Turb\ D_t} + \underbrace{\langle u_{it}' u_{kt}'\rangle_t \frac{\partial \langle u\rangle_t}{\partial y}}_{Prod_t} + \underbrace{\langle \varepsilon_t\rangle_t}_{Diss_t} = 0 \tag{74}$$

For completion of Eq. (74) was used the known expressions (43)–(45) with index $r = t$. Transformation of Eq. (74) taking into account to an automodel form gives

$$\frac{d^2\langle E_t\rangle_{*t}}{d\varphi^2} + 2\tilde{F}_t \frac{d\langle E_t\rangle_{*t}}{d\varphi} - 2\nu_{Et}\langle E_t\rangle_{*t} = -\frac{\tilde{F}_t''^2}{(1-m)^2} \tag{75}$$

Here $\langle E_t\rangle_{*t} \equiv \langle E_t\rangle_t/\Delta U^2$, $\Delta U = u_1 - u_2$, $L_t = a_{0t}x$; the second empirical parameter of the model is

$$\nu_{Et} = \frac{c_*}{2(a_{0t}\sigma_t)^2} \tag{76}$$

and is determined by the condition of the best agreement of calculated and experimental data of turbulent kinetic energy. The separate components in Eq. (75) correspond to Eq. (74) and have a definite physical meaning:

http://dx.doi.org/10.5772/67844

$$Conv_t = \tilde{F}_t d\langle E_t\rangle_{*t}/d\varphi \rightarrow \langle u\rangle_t \partial\langle E_t\rangle_t/\partial x + \langle v\rangle_t \partial\langle E_t\rangle_t/\partial y - \text{convective transfer} \quad (77)$$

$$TurbD_t = 0.5d^2\langle E_t\rangle_{*t}/d\varphi^2 \rightarrow \partial\langle (E_t + p'_t/\rho)v'_t\rangle_t/\partial y - \textit{diffusion through the velocity fluctuations} \quad (78)$$

$$Prod_t = \tilde{F}_t''^2/2(1-m)^2 \rightarrow \langle u'_{it}u'_{kt}\rangle_t \partial\langle u\rangle_t/\partial y - \textit{production of the energy fluctuations} \quad (79)$$

$$Diss_t = -\nu_{Et}\langle E_t\rangle_{*t} \rightarrow \langle \varepsilon_t\rangle_{*t} - \textit{dissipation rate of the energy fluctuations} \quad (80)$$

Eq. (75) was solved with boundary conditions in the form

$$\frac{d\langle E\rangle_{*t}}{d\varphi} = 0\ ,\ \varphi = \begin{cases} 1.65 \\ -1.65 \end{cases} \quad (81)$$

To calculate separate components of intensity (variance) of fluctuating velocity, we will use approximate ratios:

$$\langle u_t'^2\rangle_t \cong \langle E_t\rangle_t,\ \langle v_t'^2\rangle_t \cong \langle w_t'^2\rangle_t \quad (82)$$

Eq. (75) was solved by the numerical method (mathematical package *MathCad* was used). Interestingly, the solution of Eq. (75) with using asymptotic boundary conditions $\langle E_t\rangle_{*t} \rightarrow 0$ *as* $\varphi \rightarrow \pm\infty$ gives the bad calculation data. In this regard for the flow of the turbulent fluid have been used the hard boundary conditions in the form (for $m = 0.305$)

$$\frac{\langle u\rangle_t}{u_1} = \begin{cases} 0.99,\ \varphi_1 = 1.65 \\ 0.32,\ \varphi_2 = -1.65 \end{cases} \quad (83)$$

and

$$\frac{d\langle E\rangle_{*t}}{d\varphi} = \begin{cases} 0,\ \varphi_1 = 1.65 \\ 0,\ \varphi_2 = -1.65 \end{cases} \quad (84)$$

The results of our calculations of conditional means of this *"fluctuating" task* are presented in **Figure 5**. **Figure 5a** shows the calculation $\langle u_t'^2\rangle_t/\Delta U^2$ corresponding to Eqs. (75) and (82). **Figure 5b** shows the turbulent kinetic energy budget according to Eqs. (77)–(80). At that value of the parameter $\nu_{Et} = 2$. It is worth noting that only Eq. (75) gives the hard edges $-0.075 \leq \eta - \eta_{0S} \leq 0.079$ (the same (84)) for the flow of the turbulent fluid due to the fact that the solution of Eq. (75) loses its physical sense outside these boundaries (see **Figure 5a**). So, we got the hard edges only to the flow of the turbulent fluid.

4.1.2. Modeling of the nonturbulent fluid flow

Solution of the dynamic task for the flow of a nonturbulent fluid was defined in such a manner. Modeling of this flow was carried out according to the SS (42) and was related to modeling of the flow of a turbulent fluid by means of statistical ratios in the central field of the mixing layer. It appeared that division of this subsystem into two with high velocity and low velocity

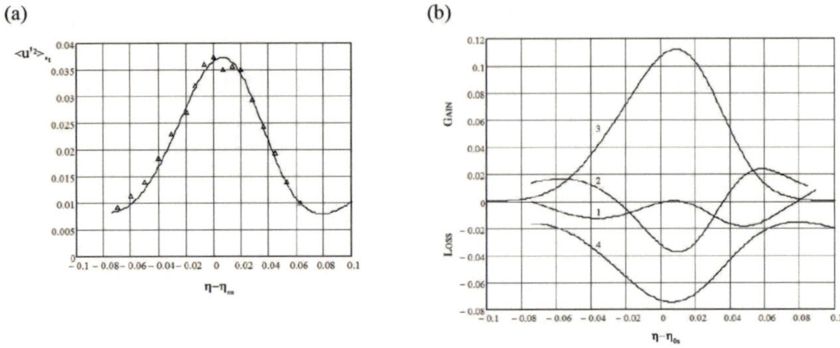

Figure 5. (a) The profile of the normalized conditional mean intensity of longitudinal velocity fluctuations $\langle u_t'^2\rangle_{*t} = \langle u_t'^2\rangle_t/\Delta U^2$. (b) The turbulent kinetic energy budget over the turbulent fluid: 1, Conv$_t$; 2, TurbD_t, 3, Prod$_t$; 4, Diss$_t$. The calculated parameter $\nu_{Et} = 2$. Measurements in [20] are absent.

regions $G_{n1} + G_{n2} = G_n$ gives more precise results of modeling. The flow of a nonturbulent fluid in one of these regions has not only its own parameters ($\sigma_{n1} = 51.44$, $\eta_{0n1} = -0.015$; $\sigma_{n2} = 36.4$, $\eta_{0n2} = -0.016$) but also boundary conditions: asymptotic ones in external regions and hard ones inside the mixing layer. A butting of the obtained solutions was carried out on the line $\eta - \eta_{0s} \cong 0.009$ where the condition $\langle u\rangle_t \cong \langle u\rangle_{n1} \cong \langle u\rangle_{n2}$ is satisfied. Solution of the fluctuating task for the flow of a nonturbulent fluid was defined in such a manner. In this case, the solution of the fluctuating kinetic-energy budget equation in SS (42) was found the same as task for the flow of turbulent fluid. Here, however, boundary conditions were given as asymptotic ones. The results of the modeling are presented in **Figures 7** and **8a**.

4.1.3. *Modeling of the total averages*

The total averages calculation is required a distribution of the intermittency factor γ. Modeling of this factor can be performed with the help of the statistical ratio (51) in view of the dissipation rate $\langle \varepsilon\rangle = \gamma\langle \varepsilon_t\rangle_t + (1-\gamma)\langle \varepsilon_n\rangle_n$ and $\langle \varepsilon\rangle_n \cong 0$. In this case, the value $\langle \varepsilon_t\rangle_t$ is in the process of modeling of turbulent fluid by Eq. (74). To calculate the total average of the dissipation rate $\langle \varepsilon\rangle$, we propose to use the assumption on its equality to the unconditional mean, which is found from the *RANS* model constructed for the mixing layer. At that the empirical constants $\sigma_R = 29.0$ and $\eta_{0R} = -0.0134$ are chosen only from the condition of agreement of the intermittency factor γ calculation with the experimental data. **Figure 6** presents the

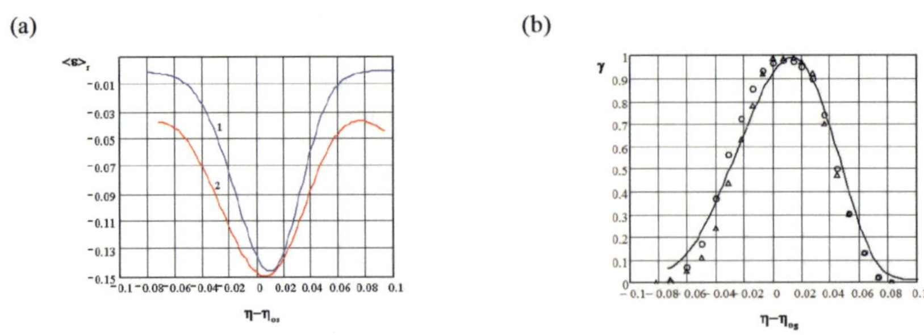

Figure 6. (a) Profiles of the dissipation rate of the energy fluctuations $\varepsilon_r = \text{Diss}_r$: 1, Diss$_R = -\nu_{ER}\langle E_R\rangle_{*R}$ and 2, Diss$_t = -\nu_{Et}\langle E_t\rangle_{*t}$ at the calculated parameter $\nu_{ER} = \nu_{Et} = 2$. (b) The profile of the intermittency factor γ of the turbulent fluid.

http://dx.doi.org/10.5772/67844

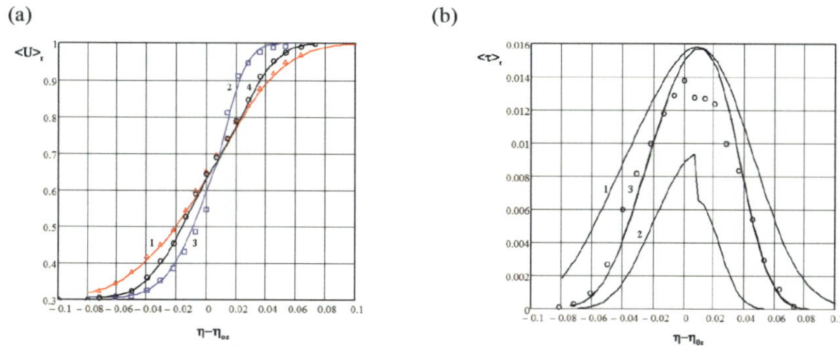

Figure 7. (a) Profiles of normalized conditional and total average longitudinal velocity $\langle U\rangle_r = \langle u\rangle_r/u_1$: 1-Δ-$\langle U\rangle_t$; 2-□-$\langle U\rangle_{n1}$; 3-□-$\langle U\rangle_{n2}$; 4-o-$\langle U\rangle$. (b) Profiles of normalized conditional and total average shear stress $\langle\tau\rangle_r = -\langle u_r' v_r'\rangle_r/\Delta U^2$: 1–$\langle\tau\rangle_t$; 2–$\langle\tau\rangle_n$; 3–o–$\langle\tau\rangle$ (measurements of $\langle\tau\rangle_t$ and $\langle\tau\rangle_n$ in [20] are absent).

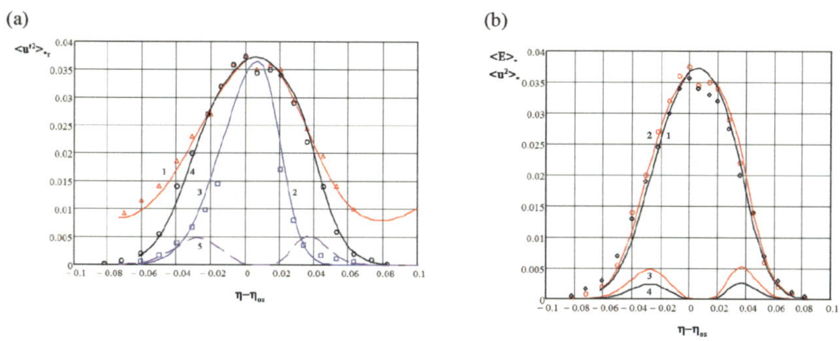

Figure 8. (a) Profiles of normalized conditional and the total average intensity of longitudinal velocity fluctuations $\langle u'^2\rangle_{*r} = \langle u_r'^2\rangle_r/\Delta U^2$: 1-Δ-$\langle u'^2\rangle_{*t}$; 2-□-$\langle u'^2\rangle_{*n1}$, 3-□-$\langle u'^2\rangle_{*n2}$; 4-o-$\langle u'^2\rangle_*$; 5 – $u_d/(1-m)^2$. (b) Profiles of normalized total average turbulent-kinetic-energy $\langle E\rangle_* = \langle E\rangle/\Delta U^2$ and intensity of longitudinal velocity fluctuations $\langle u'^2\rangle_* = \langle u'^2\rangle/\Delta U^2$: 1-o-$\langle E\rangle_*$; 2–Δ–$\langle u'^2\rangle_*$; 3 – $u_d/(1-m)^2$. 4 – $E_d/(1-m)^2$.

calculation. As *RANS* models give a good result only in the regions with insignificant intermittency, such a method for determination of the intermittency factor should be considered only as an approximate one. The results of the modeling of the total averages are presented in **Figures 7** and **8**. The some results of the unconditional means, obtained by the *RANS* model, are presented in **Figure 9**. As it seen that the *RANS* model does not give good results.

5. Conclusion

The new differential equations for the conditional means of turbulent flow are the theory result of this chapter. On the basis of these equations, the method of autonomous statistical modeling *ASMTurb* of such flow was justified. The main feature of this method is that it allows us to construct the mathematical models for the conditional means of each of the intermittent media taking place into a turbulent stream autonomously, i.e., independently. The main advantage of this method is that the system of differential equations for the conditional means does not contain the source terms. According to this method, the process of transformation the nonturbulent fluid in the turbulent fluid (as a generator of the turbulent fluid) occurs in the superlayer. Even more, the *ASMTurb* method allows us to construct the model only for the turbulent fluid flow, without considering the nonturbulent fluid flow. As far as all the mixing turbulent processes (and, as consequence, the processing modeling of turbulent heat and mass transfer) take place only into the turbulent fluid, this peculiarity essentially simplifies the modeling of such processes. Especially it refers to the turbulent combustion processes, in which modeling is attended by difficulties. It is important to note that *ASMTurb* SS (41) and (42) for conditional means of the turbulent and nonturbulent fluid differ from the known ones (for example [7]). It should be emphasized that the presented model contains only two empirical parameters σ_t and ν_{Et}. With regard to these parameters, it must be said that their appearance is due to the fact that we do not know neither the expansion rate of the turbulent fluid downstream nor the maximum value of the turbulent energy generated by the shear rate.

We now make several important remarks.

On the operation of conditional statistical averaging. Sometimes the value $\langle Y_1|Y_2\rangle$ is also called "conditional" mean that makes some confusion in comparison with the conditional means $\langle Y_1|Y_2\rangle_r$, $r = t$ or $r = n$. Indeed, variable $\langle Y_1|Y_2\rangle = \gamma\langle Y_1|Y_2\rangle_t + (1-\gamma)\langle Y_1|Y_2\rangle_n$ where $\langle Y_1|Y_2\rangle_t$ and $\langle Y_1|Y_2\rangle_n$ are the conditional means of the characteristics for the turbulent and nonturbulent fluid, respectively. So, the value of $\langle Y_1|Y_2\rangle$ actually is the *total* average of the random variable Y_1, obtained under the condition of the variable Y_2.

On the source terms. The known equations for conditional means contain the source terms, which are intended to describe the increase in volume of the turbulent fluid downstream. Here, it is interesting to discover the reasons of such source terms appearance. For this, we consider the procedure of statistical "unconditional" averaging of the continuous equation, premultiplied by the intermittency function

$$\frac{\partial u_k}{\partial x_k} = 0 \rightarrow I\frac{\partial u_k}{\partial x_k} \rightarrow \left\langle I\frac{\partial u_k}{\partial x_k}\right\rangle = \gamma\left\langle\frac{\partial u_k}{\partial x_k}\right\rangle_t = 0 \tag{85}$$

However, the permutation of averaging and differentiation operations, used in the approach, gives

$$\left\langle I\frac{\partial u_k}{\partial x_k}\right\rangle = \left\langle\frac{\partial I u_k}{\partial x_k} - u_k\frac{\partial I}{\partial x_k}\right\rangle \rightarrow \frac{\partial\langle I u_k\rangle}{\partial x_k} - \left\langle u_k\frac{\partial I}{\partial x_k}\right\rangle \rightarrow \frac{\partial\gamma\langle u_k\rangle_t}{\partial x_k} - \left\langle u_k\frac{\partial I}{\partial x_k}\right\rangle = 0 \tag{86}$$

i.e., gives rise to the appearance of the source terms of a singular type. It stands to reason that the appearance of such source term is only due to the accepted commutation of the averaging operation of the partial derivates and has no physical justification.

On the mathematical model for the turbulent fluid flow. The *ASMTurb* method allows us to construct a model for the turbulent fluid flow without considering the nonturbulent fluid flow. As far as all mixing turbulent processes take place only in the turbulent fluid, this peculiarity essentially simplifies the modeling. Even more, this approach allows us to take into account the source term, using one of the semi-empirical parameters of the mathematical model. To solve the "pulsation" task we use the turbulent-kinetic-energy budget equation. To distribute the intensity of the longitudinal velocity pulsations we use the ratio $\langle u_r'^2\rangle_r \cong \langle E_r\rangle_r$.

http://dx.doi.org/10.5772/67844

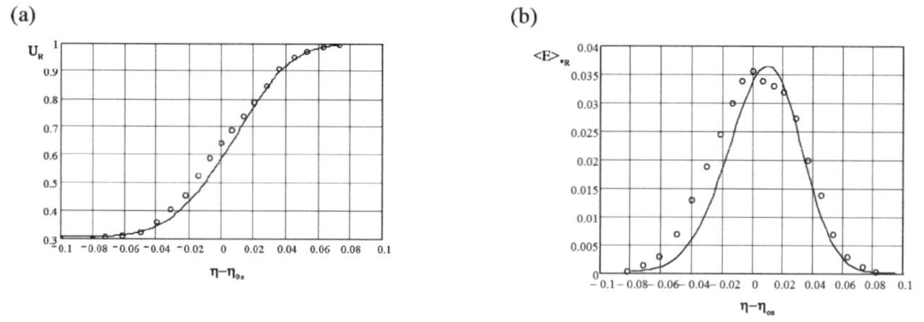

Figure 9. (a) Unconditional mean longitudinal velocity *RANS* $\langle U\rangle_R = \langle u\rangle_R/u_1$. (b) Profiles of the normalized unconditional mean *RANS* turbulent kinetic energy $\langle E\rangle_{*R} = \langle E\rangle_R/\Delta U^2$.

What gives the ASMTurb method. The results of testing *the ASMTurb method* showed a "surprising" precision for the turbulent flows modeling—calculations of the conditional and total averages of statistical characteristics practically completely agreed with the known measurements [20] (see **Figures 7** and **8** where curves—our calculation, symbols—experimental data are mentioned[20]).

What gives the RANS method. The some results of the unconditional means, obtained by the *RANS* model, are presented in **Figure 9**. As can be seen, the *RANS* model does not gives good results.

So, the *ASMTurb* differential equations for the conditional averaged characteristics of the turbulent and nonturbulent fluid flows coincide with each other in external view. Moreover, the *RANS* differential equations have the same external view. However, the boundary conditions and closure hypothesis for the turbulent and nonturbulent fluid flows in the *ASMTurb* models may be different. It is this circumstance allows us to construct highly efficient *ASMTurb* models of turbulent flows. The *RANS* method does not have this property and thus a searching for the "satisfactory" closure hypotheses for the *RANS* models will not give good results.

Author details

Yuriy Nuzhnov

Address all correspondence to: ynuzhnov@gmail.com

Institute of Combustion Problems, Almaty, Kazakhstan

References

[1] Townsend AA. The Structure of Turbulent Shear Flow. Cambridge University Press; 1956

[2] Libby PA, Williams FA. Turbulent Reacting Flows. Springer-Verlag; 1980

[3] Kuznetsov VR, Sabel'nikov VA. Turbulence and Combustion. Hemisphere Publishing Corporation; 1990

[4] Pope SB. Turbulent Flows. Cambridge University Press; 2000

[5] Nuzhnov YV. Method of autonomous modeling of turbulent flows (ASMTurb). IP 0010816, Bulletin No. 1392, issued on 21 October 2013. (in Russian)

[6] Libby P. On the prediction of intermittent turbulent flows. Journal of Fluid Mechanics. 1975;**68**(2):273–295

[7] Dopazo C. On conditioned averages for intermittent turbulent flows. Journal of Fluid Mechanics. 1977;**81**(3):433–438

[8] Dopazo C, O'Brien EE. Intermittency in free turbulent shear flows. In: Durst F, Launder BW, Schmidt FW, Whitelaw JH, editors. Turbulent Shear Flows. 1st ed. New York: Springer-Verlag; 1979. pp. 6–23

[9] Nuzhnov YV. Conditional averaging of the Navier-Stokes equations and a new approach to modeling intermittent turbulent flows. Journal of Fluid Dynamics. 1997;**32**:489–494

[10] Nuzhnov YV. On the theory of turbulent heat and mass transfer with allowance for intermittence effects. Journal of Engineering Physics and Thermophysics. 2011;**84**:160–170

[11] Monin AS, Yaglom AM. Statistical Fluid Mechanics: The Mechanics of Turbulence. MIT Press; 1971

[12] Kolmogorov AN. Foundations of the Theory of Probability. Chelsea Publishing Company; 1956

[13] Feller W. Introduction to Probability Theory and its Applications. Wiley; 1957

[14] Rozanov YA. Probability Theory: A Concise Course. Dover Publications Incorporated; 1977

[15] Nuzhnov YV. The Method of Autonomous Statistical Modeling ASMTurb and its Testing on the Example of Classical Turbulent Flows. In: ASME Congress (IMECE), Volume 7: Fluids Engineering Systems and Technologies. Quebec, Montreal; 2014.

[16] Nuzhnov YV. Statistical modeling of the intermittent turbulent flows (Қазақ университеті, Алматы, 2015 (in Russian)).

[17] Pope SB. Calculations of a plane turbulent jet. AIAA Journal. 1984;**22**(7):896–904

[18] Wygnanski G, Fiedler HE. The two-dimensional mixing region. Journal of Fluid Mechanics. 1970;**41**:327-361

[19] Dimotakis PE. Two-dimensional shear-layer entrainment. AIAA Journal. 1986;**11**:1791–1796

[20] Spenser BW, Jones BG. Statistical investigation of pressure and velocity fields in the turbulence two-stream mixing layer. AIAA Journal. 1971. Paper No. 71–613

Chapter 5

Turbulence Transport in Rotor-Stator and Stator-Rotor Stages of Axial Flow Fans

Jesús Manuel Fernández Oro,
Andrés Meana Fernández and Bruno Pereiras García

Additional information is available at the end of the chapter

http://dx.doi.org/10.5772/67751

Abstract

Turbulence analysis in turbomachines is a complex task. The combination of different turbulence sources and transport mechanisms poses the question of determining adequate metrics to quantify turbulence levels and provide insight into the flow structure and its evolution. Apart from experimental measurement techniques, numerical techniques arise as a useful tool to characterize this type of flow, especially hybrid LES techniques that allow a highly accurate description of the transport of turbulent structures, while turbulence generation at solid boundaries is addressed using RANS schemes. In this chapter, recent developments concerning numerical simulation of rotor-stator and stator-rotor interactions in low-speed axial fans using LES techniques are presented. A post-processing framework is introduced to segregate the deterministic and turbulent components of the unsteady flow, allowing an accurate description of both phenomena. Hence, turbulent transport over the different stage rows of the axial fan can be accurately addressed. Following, LES capacities to simulate turbulence transport mechanisms such as breaking-up of turbulent eddies, stretching and dissipation of vorticity or identification and convection of coherent vortices are discussed. The ability of LES computations to disclose flow turbulence in rotor-stator environments at off-design conditions is specially illustrated.

Keywords: turbulence metrics, turbulence transport, LES simulations, rotor-stator interaction

1. Introduction

Turbulence analysis in single-stage turbomachines, like axial fans, is a complex task. Firstly, there is turbulence generation in both fixed (stator) and moving (rotor) reference frames.

Then, these turbulent scales and patterns are transported across the frames, interacting between them and developing secondary turbulent sources. Furthermore, unless the stages are designed under free-vortex criteria, there is a further swirl of the motion that leads to radial migration of momentum and turbulence. Finally, there are additional types of interactions that contribute to generate further turbulence, like the interaction between wakes, the interaction between wakes and blades or vanes, or even interactions between wakes, tip vortices and endwalls.

This combination of different turbulence sources and transport mechanisms poses the question of determining adequate metrics to quantify turbulence levels and give insight into the flow structure and its evolution within this kind of machines.

Apart from experimental measurement techniques such as hot-wire anemometry (HWA) or particle-image velocimetry (PIV), able to capture the turbulent variations of the flow inside these axial fans, numerical techniques arise as a useful tool to characterize this type of flows. Nowadays, scale-resolving simulations (SRS) [1] are becoming the standard to simulate the turbulent flow in multistage environments, being large eddy simulations (LES) the cornerstone of the methodology. On the contrary, U-RANS modelling is being progressively abandoned due to the continuous improvements of the computing capabilities [2].

However, fully-LES techniques are still excessively time-consuming and, in most cases, unaffordable because of the extremely refined meshes required for the boundary layers of the solid walls.

In this context, wall-modelled LES (WMLES) [3] and embedded LES (ELES) are hybrid techniques that allow a highly accurate description of the transport of the turbulent structures, while the generation in solid boundaries is still addressed using RANS schemes. LES computations in the vicinity of the turbomachinery rows are able to capture the turbulent variations up to the frequency related to the LES filter, which depends on the grid size. With an accurate selection of this size, it is possible to capture most of the turbulent kinetic energy of the larger and mid-scale vortices, which content practically all the significant portion of turbulent motion.

In this chapter, recent developments concerning numerical simulation of rotor-stator (R-S) and stator-rotor (S-R) interactions in low-speed axial fans using LES techniques are presented. In particular, LES capacities to simulate transport mechanisms for turbulence such as breaking-up of turbulent eddies, stretching and dissipation of vorticity or identification and convection of coherent vortices are discussed. The ability of LES computations to deal with the flow turbulence in rotor-stator environments at off-design conditions is specially illustrated.

2. Turbulent scales and periodic unsteadiness in rotating fans

Turbulence generation associated to vortex shedding in rotor blades is perceived as a periodic impingement of turbulent wakes on stator flow structures in R-S stages. Alternatively, the rotor flow structures are also affected by the incoming non-uniformities from the upstream

http://dx.doi.org/10.5772/67751

vanes in S-R arrangements. Thus, turbulent structures are superimposed to the velocity deficit associated to the rotor blockage, travelling unsteadily across the fan stage [4]. Due to the high Reynolds numbers, turbulence in the shear layers is rolled-up at quite small scales, significantly smaller than typical displacement thicknesses of the wake deficit.

Figure 1 shows a typical representation of the amplitude of the LES-resolved velocity fluctuations in the vicinity of the rotor for a single stage axial fan. Typical scales (frequency ranges) corresponding to forced and unforced unsteadiness, as well as turbulence, have been indicated in **Figure 1** for nominal and offset working conditions. Turbulence is clearly associated to the smallest length scales (higher frequencies). The energy budget of turbulent phenomena, including free-stream, generation at endwall boundary layers and wake-induced turbulence, presents a classic broadband distribution. As expected, turbulent scales are more evident at off-design conditions (grey lines). Forced unsteadiness is perfectly identifiable at blade passing frequencies (BPF) and its harmonics (sharp peaks), while unforced contributors are related to other periodic features non-clocked with the BPF, like blade misalignments, tip vortex instabilities or unsteadiness in separation conditions [5]. This kind of unsteadiness is strictly associated to a particular time scale, so it contains very important flow patterns, but

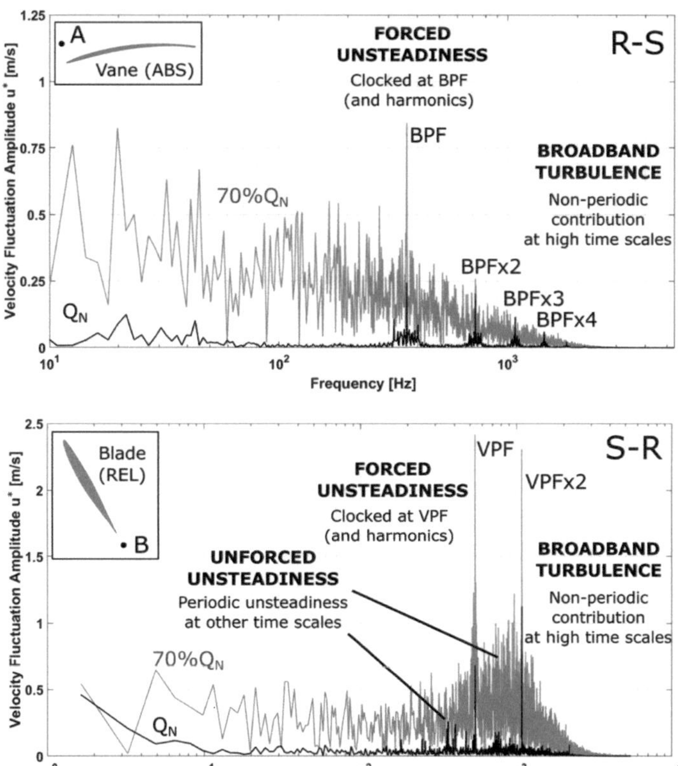

Figure 1. Typical scales of unsteadiness and turbulence for a single stage axial fan at different flow rates (top: R-S configuration, bottom: S-R configuration).

with a moderate energy budget. Note the particularity of the S-R configuration values, which are computed in the relative frame of reference, so the frequency peaks appear at an apparent 'vane-passing frequency' and its harmonics. Additionally, due to the flow complexity, interaction between the broadband turbulence and peaked frequencies (clocked at BPF/VPF or not) may arise, which obscures the analysis of the different contributions to the total unsteadiness (see the S-R configuration for the 70% Q_N case, which clearly presents a turbulent hump between VPF and VPFx2).

Consequently, to obtain a proper analysis of the turbulent characteristics, it is convenient to segregate between turbulent scales and unsteady events with special care. Otherwise, only a total unsteadiness is found in the analysis, including turbulent fluctuations as well as the non-stationary effects due to the periodical rotation of the machine blades. This procedure results in a superficial analysis of turbulence, especially for the turbulent transport.

The correct procedure requires that the instantaneous values of the measured variable are grouped in ensembles corresponding to the blade-passing frequency [6]. Then, mean values for every blade position are calculated, obtaining a signal that contains only the unsteadiness due to the rotation of the machine (periodical or deterministic unsteadiness). These ensemble-averaging values, combined with the time-averaging value of the complete signal, enable the segregation of the total unsteadiness of the measured variable into its deterministic and turbulent parts.

This segregation can be performed in both time and frequency domains [7]. Though strictly, performing ensemble averages at BPF in the time domain should be sufficient, the complex nature of the vortical motion and the interchange of energy in the viscous cascade within a sea of all-range eddies makes an exact segregation impossible. Sometimes, the use of a frequency decomposition is also required in order to decouple forced and unforced unsteadiness from turbulence [8].

3. Numerical methods: LES simulations in R-S and S-R configurations

A typical low-speed, single-stage axial fan will be considered for the computations. Because of its free-vortex rotor design, it will be assumed that the flow is quasi-planar in the mid-span sections, working in a cascade mode (**Figure 2**, left). The mesh density within the passages has been carefully selected in order to describe vortical structures in the range of the integral length scales when large separation arises (i.e. rotor wakes shedding at 70% Q_N). For both blades and vanes surfaces, a typical O-grid distribution with a [100 × 60] mesh density was considered sufficient to model the boundary layers. Additionally, the z-coordinate for the spanwise direction was covered with 15 cells, for a total height of the model of 0.12 blade chords, following the guidelines proposed by Davidson and Dahlström [9]. As a result, a total number of 3.4 and 3.6M cells were resolved for the R-S and S-R domains, respectively.

The sliding mesh technique has been employed in the computations with FLUENT® to simulate the relative movement of the rotor blades with respect to the stator vanes. In the present case, a time step of 9.26×10^{-5} s has been fixed, which implies 30 intermediate time steps per

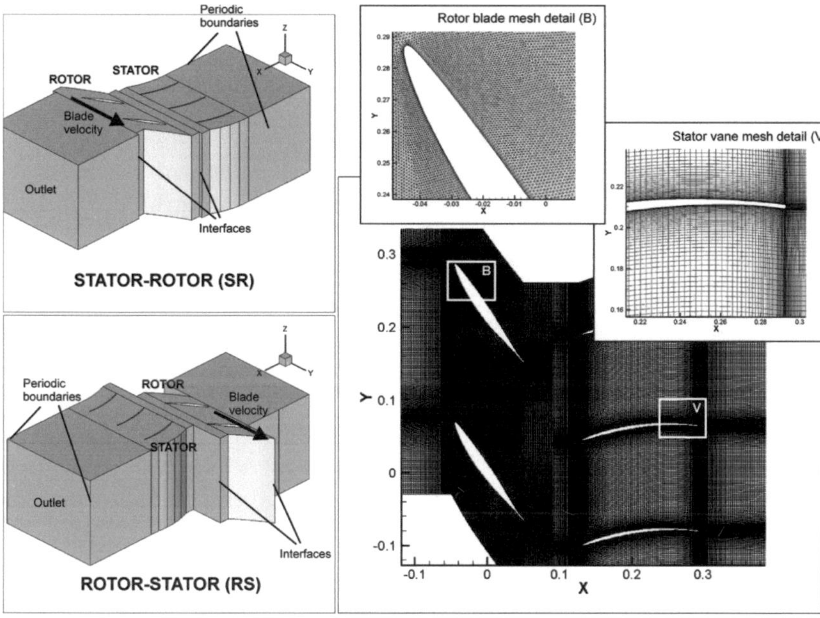

Figure 2. Stator-rotor and rotor-stator in linear cascade (left). Mesh densities in stator and rotor rows: details in the trailing edge of the vanes and leading edge of the blades (right).

blade event, being able to resolve the turn-out time of the resolved eddies in the LES scheme. All these assumptions, based on the extrusion of a two-dimension (2D) section of the fan with periodic conditions, have been already experimentally verified [10] for three different operating conditions: 100, 85 and 70% of the nominal flow rate. Furthermore, LES computations have been extended for a wide number of blade events, in order to obtain a representative statistical convergence of the averaged results [11].

All the computations have been completed on a home-made cluster, composed of four parallelized quad-core PCs at 2.67 GHz with 4 GB RAM in each node. The numerical routine adopted for both R-S and S-R configurations started with a steady simulation, followed by a transient unsteady simulation for 20 blade events (2 complete rotor turns) and a final statistically resolved stage for additional 200 blade events (an order of magnitude higher). As a consequence, 6000 time steps have been considered for the database for every simulated condition (7.2 GB of info per case). The CPU time required for each case reached up to 450 hours to obtain a converged statistical description of the unsteady flow patterns.

MATLAB® software was used for data processing of the instantaneous velocities and sub-grid turbulent viscosities stored for a planar window in the outlet domain of both configurations. The streamwise extension of these windows corresponded to five chord lengths of the blades, keeping the tangential width to three stator vane pitches (periodicity conditions). A number of home-made codes were developed for interpolation of grids, filtering, relocation in the relative frames and both phase and time averaging. The execution of all these routines also required a significant computational effort.

4. Assessment of turbulence transport: metrics and results

Different variables may be used to describe the level and the spatial distribution of the turbulent flow. The most convenient definition is the ensemble-averaging value of the random fluctuations to the square which, in turbomachinery environments, describes the unsteady turbulent field. In other words, it illustrates the transport of the turbulent structures in the unsteady, deterministic flow patterns. Moreover, the unsteady turbulence level can be defined comparing the RMS of that variable with the time-resolved velocity at each blade phase.

To obtain the ensemble-averaged values, a number of blade/vane passages are isolated and classified into a fixed number of phases. Then, averages are performed among the corresponding phases, so that chaotic flow variations may be eliminated, obtaining the unsteady deterministic field. Subsequently, the unsteady flow structures may be filtered out with a further time averaging, obtaining a steady flow pattern. Assuming that there is periodicity between consecutive blades/vanes, the phase- or ensemble-averaged value of a flow variable u at the phase ϕ (angular position) is obtained as :

$$\tilde{u}(\varphi)=\frac{1}{M}\sum_{m=1}^{M} u_m(t) \tag{1}$$

where, M is the total number of ensembles and $u_m(t)$ are the realizations made at times so that $\varphi = \frac{2\pi}{N_B}(m-1)+\omega t$. N_B is the number of blades/vanes and ω is the rotational speed. The tilde operator (~) will be used throughout the text to indicate the ensemble-averaging procedure.

Once the ensemble-averaged values are obtained, the different unsteady scales may be obtained. Turbulence scales are obtained from the subtraction of the ensemble-averaged values to the instantaneous values of the variable, according to $u'(t)=u(t)-\tilde{u}(\varphi)$. Finally, a second time-averaging operator is introduced to retrieve the mean-time value as follows:

$$\overline{u} = \frac{1}{N}\sum_{n=1}^{N} \tilde{u}(\varphi_n) \tag{2}$$

where N is the number of time steps between consecutive blade events (it may be also regarded as the temporal discretization of the ensemble-averaged signal). With the mean-time value, the deterministic unsteadiness for every phase is obtained as: $u''(\varphi)=\tilde{u}(\varphi)-\overline{u}$.

Once the different contributions to the total value of the signal have been segregated, the different metrics that describe the flow behaviour may be calculated.

4.1. Turbulence and deterministic unsteadiness intensity

The first variable of interest is the turbulence intensity, which represents the turbulent unsteadiness of the flow with respect to the mean values. It is defined as follows:

$$TU(\varphi)=\frac{\sqrt{\widetilde{u'^2}+\widetilde{v'^2}}}{\sqrt{\widetilde{u^2}+\widetilde{v^2}}} \tag{3}$$

Figure 3 shows the unsteady evolution of the turbulent structures in the axial fan stage in terms of the turbulence intensity (TU, in percentage). The flow patterns in both R-S and R-S configurations are compared at nominal (Q_N) and off-design conditions (70% Q_N). Five intermediate instants have been selected to track the time-resolved interaction of rotor and stator structures (t/T_R at 0.0, 0.2, 0.4, 0.6 and 0.8, being T_R the blade-passing time). In the S-R case, the relative frame of reference is represented for the analysis (t/T_S). Apart from the evident higher turbulence levels at off-design conditions, the impact of the turbulence of the rotor wakes onto the vanes in the R-S case is much more significant than the slight effect of the vane wakes onto the rotor (S-R). The turbulence level near the leading edge (LE) of the pressure side of the vanes is completely conditioned by the periodic passing of turbulent rotor wakes at 70% Q_N. In addition, the massive separation of the vanes suction side observed in the R-S is also a consequence of the poor underturning of the meanflow through the rotor passages when operating at low flow rates. Even at nominal conditions, the impact of the rotor wakes on the vanes is evident for the R-S configuration. In the S-R cases, the rotor wakes experience an abrupt thickening as a consequence of the flow rate reduction (from 100 to 70% Q_N). At nominal conditions, the rotor wakes do not seem to be significantly conditioned by the incoming stator wakes. On the contrary, rolling-up mechanisms of the larger vortices at 70% Q_N (see the vortex street on the right column) appear to be chopped by the stator wakes, generating hot spots of turbulence.

Likewise, deterministic fluctuations can be used to define a corresponding level of deterministic unsteadiness (DU). In this case, the deterministic fluctuation is further time-averaged, so a final mean-time footprint of the periodic variations in the stage is retrieved according to:

$$DU = \frac{\sqrt{\overline{u''^2} + \overline{v''^2}}}{\sqrt{\overline{u}^2 + \overline{v}^2}} \tag{4}$$

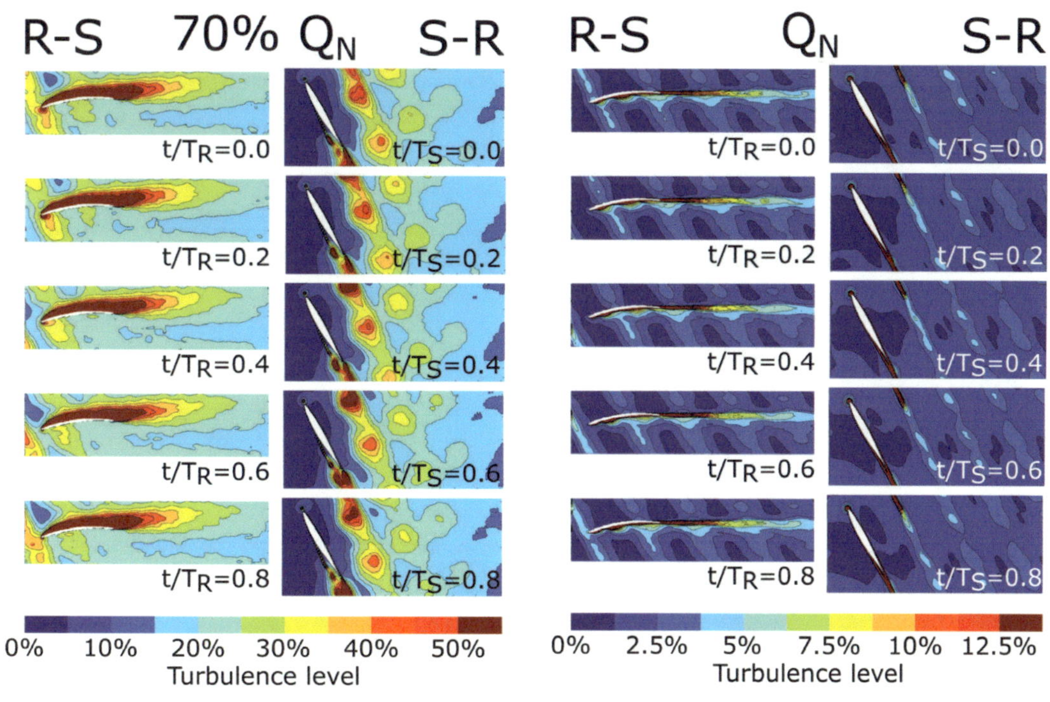

Figure 3. Unsteady convection of incoming wakes in stator (absolute in R-S) and rotor (relative in S-R) passages. Nominal flow rate (left) and off-design conditions (right).

Figure 4 (left) compares, for all the flow rates simulated, those regions in the stage where the intensity of the unsteadiness is prominent. In the case of the R-S configuration, major unsteadiness arises in the axial gap between the rows, where the periodic passing of blades generates a pulsating flow over the vanes and a significant interaction. Maximum levels around 12.5% are observed at off-design conditions, in the inter-row regions (observed as an 'averaged' tangential band due to the blade motion) and also on the trailing edge (TE) of the vanes suction side (precisely where turbulence generation appeared to be uncorrelated to deterministic fluctuations). On the contrary, the S-R configuration exhibits large values of unsteadiness at the core of the rotor wakes. This reveals that the intensity of the velocity deficit in the rotor wakes is modulated (up to a significant 15%) by the unmixed stator wakes convected out in the rotor passages. Note that the behaviour of the deterministic unsteadiness is completely opposite in both configurations, though in both cases it is more important at lower flow rates.

In addition, the levels of deterministic fluctuation and turbulence are also compared in **Figure 4**, after computing and representing the mean-time maps of turbulence on the right side. Turbulence levels are always higher than deterministic fluctuations, even at nominal conditions. Approximately, they are 3–5 times higher across the stage, depending on the working conditions. It is noticeable that both deterministic and turbulent sources of unsteadiness present similar distributions in the S-R configuration, while differing notably in the case of the R-S setup. In that case, turbulent mechanisms generated at the blades are not influenced by the periodic passing wakes, thus presenting a negligible wake-blade

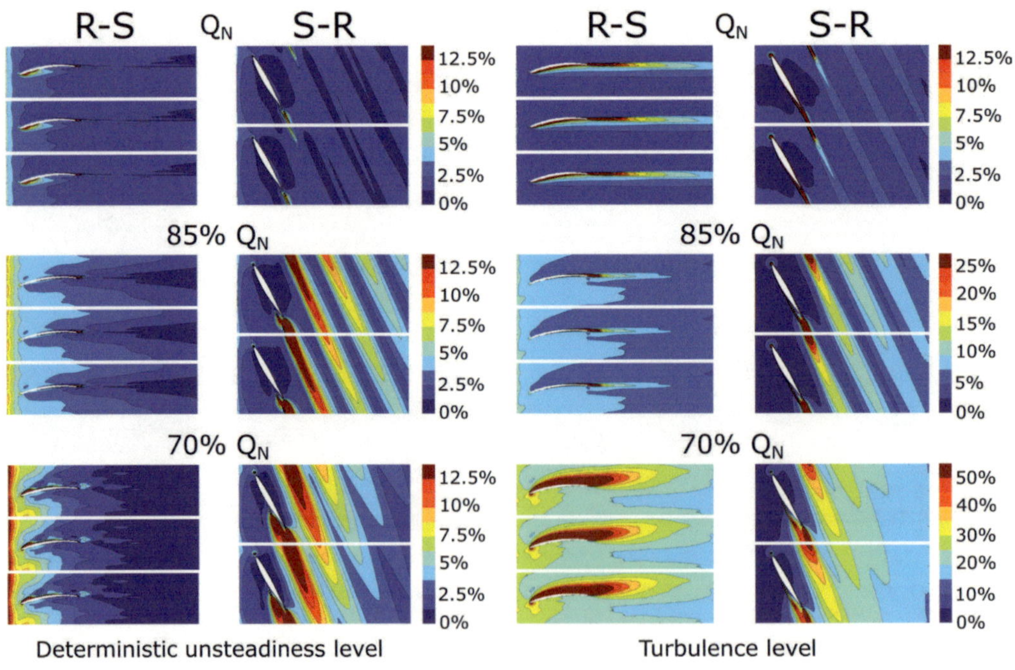

Figure 4. Comparison of deterministic unsteadiness (left) and mean time turbulence (right) for R-S and S-R configurations as a function of the flow rate.

interaction. On the other hand, the streamwise convection of the unmixed rotor wakes is clearly chopped by the passing vane wakes in the relative frame, resulting in an evident source of periodic fluctuation.

Previous maps in **Figure 4** can be further pitch-averaged to obtain the streamwise distribution of both deterministic and turbulence levels in the axial fan stage. This third (spatial) operator is essential to assess the impact of these unsteady sources for the modelling of total unsteadiness in the case of through-flow models, because the chordwise distribution of the velocity field obtained is independent of the reference frame (the tangential coordinate is swept out in the averaging). Since the fan stage presents bladed and no-bladed regions, a blockage factor (λ) [12] has to be introduced to take into account the solidity of the moving blades:

$$\overline{u}^{(\text{thru})} = \frac{1}{\lambda N_\theta} \sum_{\theta=1}^{\lambda N_\theta} \overline{u} \text{ where } \lambda = \frac{N_B(\theta_P - \theta_S)}{2\pi} \tag{5}$$

N_θ denotes the number of circumferential points per blade passage and $(\theta_P - \theta_S)$ is the tangential distance between the pressure and suction sides of the blades (function of the axial coordinate).

The pitch-averaged distributions of deterministic and turbulent sources are shown in **Figure 5**. In essence, these plots provide information about those axial positions where the generation of turbulence and deterministic unsteadiness arises. In the case of the R-S configuration (top plots), the deterministic unsteadiness decreases rapidly as it is convected throughout the vane passages, whereas turbulent contributions present a moderate decay. Some local peaks are found near the LE and TE of the vanes due to the influence of stagnant and separated conditions, respectively. The turbulence contours at 70% Q_N (red line in the right-top plot) reveal the massive separation occurring on the suction side of the vanes (also observed in **Figure 4**, bottom left) that leads to extremely high values of turbulence.

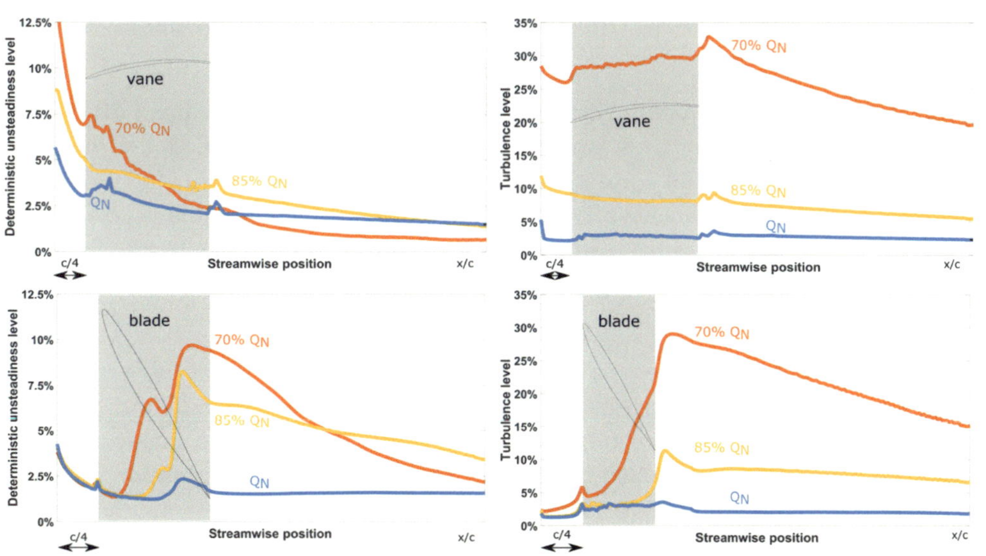

Figure 5. Streamwise distributions of pitch-averaged deterministic (left) and turbulent (right) levels for R-S and S-R configurations.

Concerning the S-R configuration, deterministic sources are progressively triggered at earlier chordwise positions as the flow rate is decreased. In addition, at partial flow rates (85 and 70% Q_N), a small calm region (local reduction of the unsteadiness) is developed where the turbulence separation is set off (see the corresponding location in the bottom right plot, where the turbulence level is abruptly increased from 10 to 30% at 70% Q_N). Once again, this suggests that regions with local turbulence generation are shielded from deterministic fluctuations, whereas further convection of decaying turbulence is more susceptible for intensity fluctuations due to deterministic wake-wake interactions.

4.2. Integral length scales

Another relevant variable is the integral length scale (ILS), which gives an idea of the spatial dimension of the largest eddies of the turbulent structure. Once all the periodic events have been removed from a velocity trace, the integral scale can be estimated from the area under the autocorrelation function—*ACF*—of the signal [13]. This procedure is valid if the average eddy size lies through the correlation of two velocity signals (Taylor's hypothesis of frozen turbulence). Note also that average eddies of a wave number larger than the grid filter of the LES modelling cannot be estimated. Therefore, for this numerical dataset, the ILS has been defined according to:

$$ILS = \overline{u}\int_0^{\infty} ACF(\tau)d\tau \text{ where } ACF(\tau) = \frac{\overline{u'(t)u'(t+\tau)}}{\overline{u^2}} \quad (6)$$

Again, the overbar denotes the time-averaged value, and τ is the time lag that is used to construct the autocorrelation. Due to the inherent randomness of turbulent fluctuations, it is typical to obtain significant scatter in the spatial distribution of the results. For that reason, only an estimated order of magnitude can be given concerning this variable. Moreover, pitchwise averaging has been also introduced to provide a more uniform representation of the results.

As largest vortices evolve across the fan stage, they are stretched and chopped inside the row passages. During convection, their kinematic characteristics are modified, and they can be also reinforced with additional vorticity shedding released from adjacent blades or vanes. In order to provide a comprehensive picture of these vortical dynamics, it is necessary to estimate the size of the largest vortices upstream of the last row in the stage. Besides, it is important to show the streamwise (averaged) evolution of these eddies and their final pitchwise distribution at the fan discharge.

Table 1 summarizes the characteristic size of the largest eddies within the inter-row region of the stage for every configuration. As expected, typical values around 15–20 mm (in average) are found in the case of the R-S configuration, in concordance with the characteristic thickness of the rotor wake. On the other hand, in the case of the S-R configuration, the value is slightly reduced to 8–10 mm, which also agrees with previous experimental measurements in the stator wakes (around 5 mm [14]). Generally, the ILS values are progressively reduced as the flow becomes turbulent (small scale turbulence in the shear layers of the wakes) with lower flow rates.

http://dx.doi.org/10.5772/67751

Flow rate	Rotor-stator (R-S) (mm)	Stator-Rotor (S-R) (mm)
100% Q_N	21.0	9.1
85% Q_N	15.3	8.3
70% Q_N	14.7	11.3

Table 1. Typical values of ILS within the inter-row region as a function of the flow rate.

Following, ILS distributions in the case of the R-S configuration for both streamwise and pitchwise positions at different flow rate conditions are shown in **Figure 6**. The progressive enlargement of the integral scale as the vortices are convected along a streamline (A-A′) is noticeable. This feature can be observed in the positive slope shared by the chordwise distributions of all the flow rates studied. ILS levels are significantly increased at the rotor exit (between 25 and 40 mm for all the flow rates) because of the dispersion and mixing-out process suffered by the rotating rotor wakes (the mesh filter is also enlarged, so smaller vortices are transferred to the sub-grid scales and the minimum size of resolved eddies is artificially increased). In the case of 70% Q_N, the massive vortex shedding (characterized by large recirculation vortices and large-scale instabilities) on the suction sides of the vanes causes a severe increment of the ILS chordwise evolution (bottom-left plot). In the pitchwise distribution (right plot), both 100 and 85% Q_N present a local value around 10 mm, corresponding to the stator wakes, which is progressively enlarged towards higher values (20–30 mm) in the mid-passage region. For 70% Q_N, the massive separation is observed as a double-peaked distribution with highest values up to 70 mm. The local minimum is also located in the stator wakes, but with slightly higher values than for the other flow rates, in the range of roughly 20 mm.

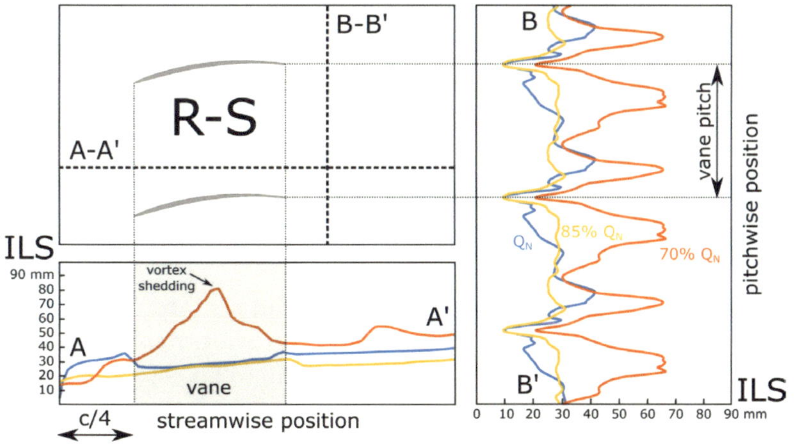

Figure 6. Distributions of ILS in the R-S configuration for different flow rates. Streamwise and pitchwise evolutions along lines A-A′ (bottom plot) and B-B′ (right plot), respectively.

Figure 7 analyses analogous results for the S-R configuration. The pitchwise distributions reveal again a jet-wake distribution, being clearer for the nominal flow rate. The local minimum is placed at the rotor TE, being similar for all the analysed flow rates, in the range of 4–6 mm. On the other hand, the plot comparing the streamwise evolution according to the flow rates is completely distorted by the excessive vortex shedding coming from the blade separation at 70% Q_N. This anomaly (ILS up to 70 mm) is rapidly vanished as the streamline reaches the blade TE, where values slightly higher than those found in the inter-row region (see **Table 1**) are recovered. Once again, the advection of large vortices in blade passages tends to increase the outlet values of the ILS. Moreover, a comparison with experimental results in the outlet positions reveals a reasonable agreement, where typical values around 10 mm were estimated with HWA for the nominal flow rate [14].

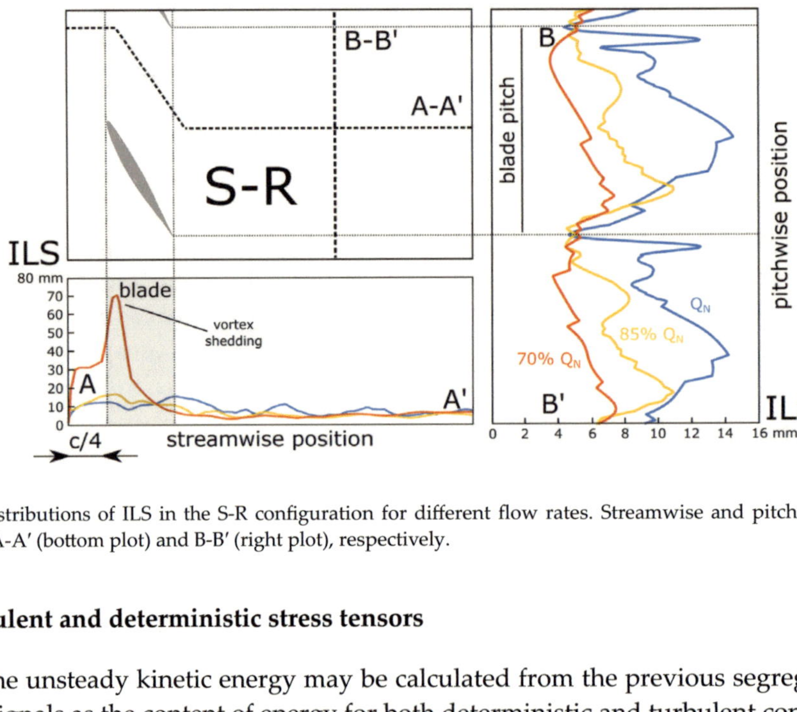

Figure 7. Distributions of ILS in the S-R configuration for different flow rates. Streamwise and pitchwise evolutions along lines A-A′ (bottom plot) and B-B′ (right plot), respectively.

4.3. Turbulent and deterministic stress tensors

Finally, the unsteady kinetic energy may be calculated from the previous segregation of the velocity signals as the content of energy for both deterministic and turbulent contributors. In addition, LES simulations reveal the residual kinetic energy modelled below the grid filter, using the sub-grid turbulent viscosity as its primary contributor.

The turbulent kinetic energy (TKE), defined from the addition of the components of the main diagonal of the Reynolds Stress Tensor at every time step, provides a collection of instantaneous values. For a simpler representation, it is more convenient to time-average the TKE distributions in order to provide a mean-time footprint of the turbulent scales in the flow structures:

$$\tau_{ij}^{(res)} = -\rho\overline{u_i^{'} u_j^{'}} \Rightarrow \overline{K_T} = \frac{1}{2}\tau_{ii}^{(res)} = \frac{1}{2}\rho\left(\overline{u^{'2}+v^{'2}}\right) \quad (7)$$

It is important to recall that this formulation gives only the resolved part of the whole turbulent energy budget. A complete estimation of the TKE requires the summation of the additional sub-grid TKE over the resolved part in the case of LES computations. However, the sub-grid component is quite marginal when a rough estimation of the resolved energy is around an 80%. In this case, it is acceptable to consider the resolved part as the major contributor, obviating the sub-grid scales. In Section 4.4, this assumption will be contrasted from the comparison of both resolved and sub-grid components.

In a similar fashion, the time-averaged value of the deterministic fluctuations leads to the definition of the deterministic stress tensor, which formulation resembles the turbulent one but in the case of the unsteady (periodic) fluctuations coming from the rotating blades. Mathematically speaking, this may be expressed as:

$$\tau_{ij}^{(\text{det})} = -\rho\overline{u_i'' u_j''} \Rightarrow \overline{K_D} = \frac{1}{2}\tau_{ii}^{(\text{det})} = \frac{1}{2}\rho\left(\overline{u''^2} + \overline{v''^2}\right) \tag{8}$$

where the deterministic kinetic energy (DKE) is computed from the addition of the components of the main diagonal of the tensor.

Figure 8 compares the mean-time maps of TKE and DKE for S-R and R-S configurations for both 100 and 70% Q_N working conditions. The values have been made dimensionless with the squared blade velocity for all the cases. Two significant features can be highlighted: firstly, concerning the configuration, it is evident that TKE is notably higher in the case of the R-S configuration with respect to the S-R case, independently of the operating conditions. Conversely, DKE is higher in the S-R arrangement, especially for the off-design flow rate. The massive separation in the vanes suction side for the R-S case and the flow detachment in the blades for the S-R case, both at 70% Q_N, are the responsible mechanisms for this inverted trend. Secondly, regarding the flow rate, it is evident the reinforcement of the different mechanisms for reduced flow conditions. In addition, in the R-S configuration, the distributions of TKE and DKE present local maxima that differ significantly from 100 to 70% Q_N. This characteristic is not replicated in the S-R case.

The complete Reynolds Stress Tensor is shown in **Figure 9**. As expected, in the R-S configuration, the pure axial components are clearly larger than the tangential components due to the chordwise alignment of the bulk flow. On the contrary, the relative flow in the S-R arrangement, with a blade outlet angle of 60°, makes both components equally important. Also, the

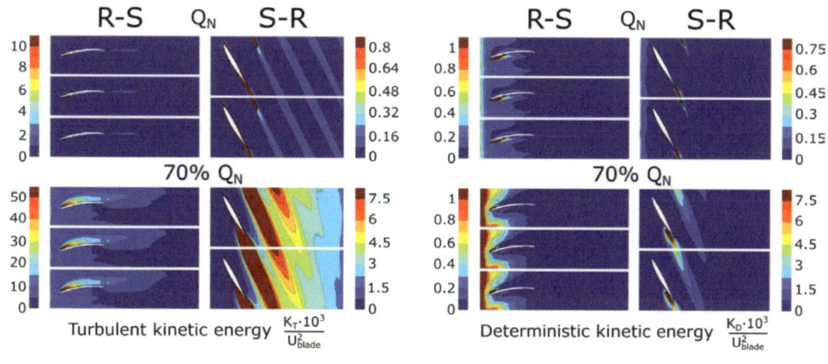

Figure 8. Time-averaged maps of Turbulent (left) and Deterministic (right) Kinetic Energy for R-S and S-R configurations. Comparison of nominal (top) and off-design (bottom) conditions.

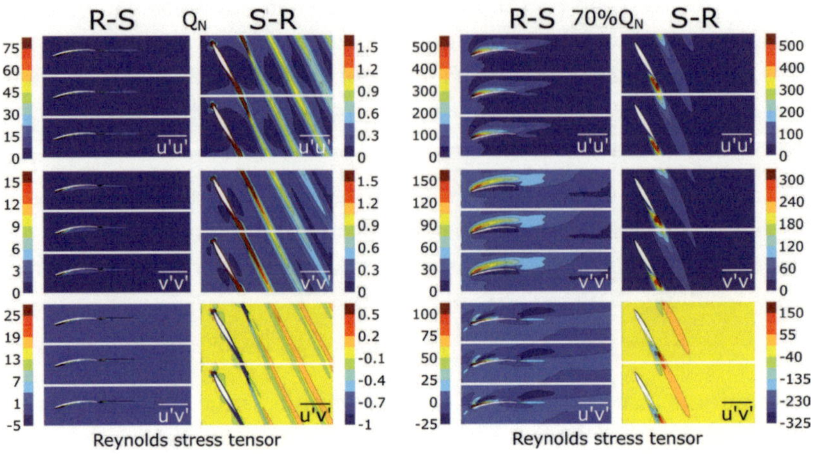

Figure 9. Components of the Reynolds Stress Tensor for both configurations. Comparison of nominal (left) and off-design (right) conditions.

crossed-component reveals the appearance of shear stress layers with opposite signs, which is a characteristic feature of wake or jet flows. At 70% Q_N, the turbulent crossed-component of the tensor shows significantly high levels, suggesting an appreciable correlation between axial and tangential randomness. This can be interpreted as an enhanced turbulent mixing in the rotor wakes.

This section is concluded presenting the deterministic stress tensor in **Figure 10**. The identification of the deterministic tensor is an essential tool to model with accuracy the effects of unsteadiness in the design of multistage turbomachinery using 1D through flow codes or conventional steady mixing-plane approaches [15, 16]. The proposal of transport models

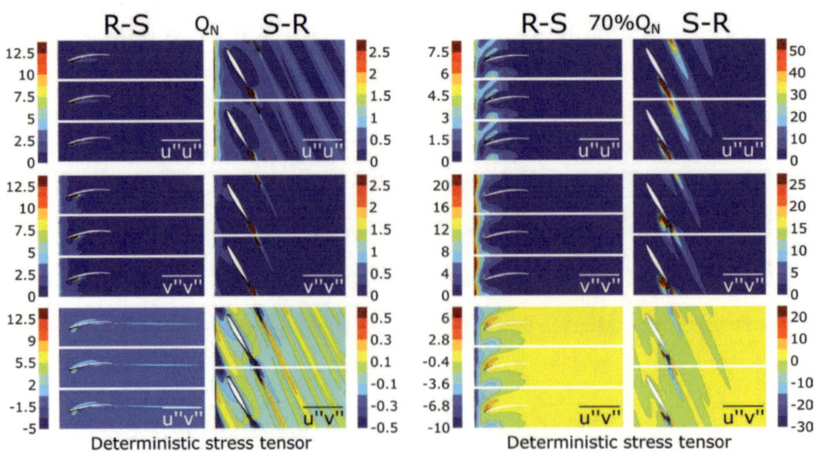

Figure 10. Components of the deterministic stress tensor for both configurations. Comparison of nominal (left) and off-design (right) conditions.

for deterministic stresses must be necessarily based on empirical correlations and, also, in a wide range of numerical experiments to validate the approaches [17]. In the S-R arrangement, modelling efforts must be driven towards the rotor wakes, where all the components are concentrated near the TE of the blades, being more intense at low flow rates. In the case of R-S configuration, the tensor components present a more complex behaviour, with local maxima near the LE of the vanes for all the components of the tensor at 100% Q_N, whereas the inter-row regions, with the predominance of the tangential correlation, are the major sources of unsteadiness at 70% Q_N.

4.4. Other metrics: degree of anisotropy and sub-grid TKE

Due to the importance of both axial and tangential components in the definition of the Stress Tensors, a final effort has been made to identify their relative importance. For that purpose, an additional indicator, the degree of anisotropy, has been defined to quantify the departures from isotropy of the flow structures associated to both turbulent and deterministic unsteadiness. For this investigation, the degree of anisotropy has been evaluated by comparing time-averaged values of turbulent and unsteadiness according to:

$$DA_T = \frac{\sqrt{\overline{u'^2}} - \sqrt{\overline{v'^2}}}{\sqrt{\overline{u'^2}} + \sqrt{\overline{v'^2}}} \text{ and } DA_D = \frac{\sqrt{\overline{u''^2}} - \sqrt{\overline{v''^2}}}{\sqrt{\overline{u''^2}} + \sqrt{\overline{v''^2}}} \tag{9}$$

With this definition (adapted from [18]), the degree of anisotropy is always comprised between −1 and 1. When this parameter is close to zero, the flow structure presents a clear isotropic characteristic. On the contrary, higher levels close to 1 indicate anisotropic patterns, being the axial component prominent, while lower values around −1 indicate also anisotropy but with the tangential component as the major contributor.

Turbulence has been found to be mainly isotropic for all the analysed cases. **Figure 11** (left) reveals low levels of anisotropy in the stator passages, mainly encapsulated around the vane surfaces. Only at nominal conditions, the stator wake is clearly anisotropic with the axial fluctuation being dominant. In the case of the S-R, the pattern is similar, with only a slightly higher anisotropy in the rotor wakes. The tangential components appear more important due to the background deviation of the flow. At 70% Q_N, the tangential anisotropy is also limited to the blade surfaces.

Alternatively, the deterministic flow exhibits a clear anisotropy behaviour for both configurations (**Figure 11**, right). Moreover, the anisotropy is more complex and intense in the case of low flow rates. For both R-S and S-R arrangements, the axial anisotropic is evident at nominal operating conditions. In the R-S, it is associated to the stator wakes, while in the S-R this feature is linked to the inner passages. However, at 70% Q_N, large regions of tangential anisotropy arise as a consequence of the enhanced mixing-out of the wakes (higher fluctuations). Even, for the R-S case, a hot spot is clearly established one chord downstream of the vanes; while for the S-R case a tangential band is formed instead.

In order to validate the previous assumptions, it is necessary to study the levels of sub-grid turbulent kinetic energy. Higher percentages of resolved TKE with respect to the total turbulent budget (at least, around an 80%) are required to make LES simulations representative of the turbulent motion [19]. For this dataset, the sub-grid TKE has been obtained from the post-processing

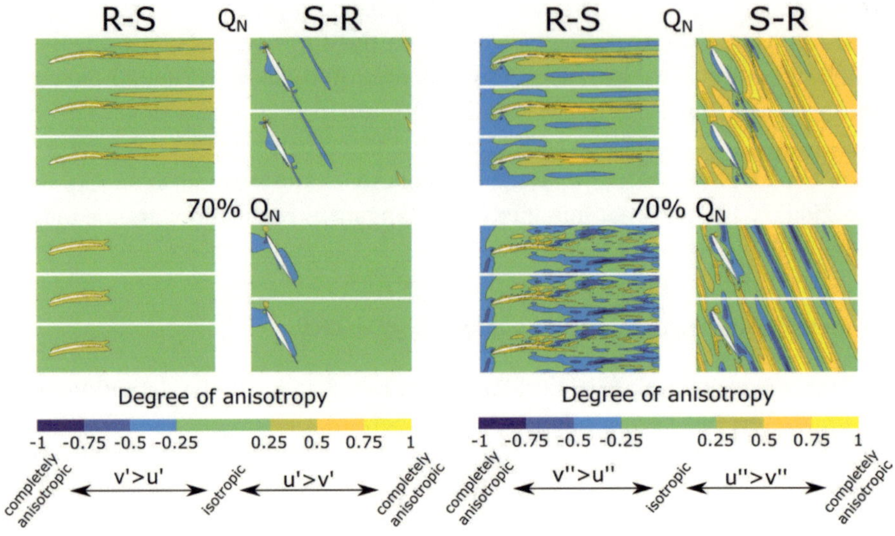

Figure 11. Degree of anisotropy for turbulent components (left) and deterministic unsteadiness (right) in both setups. Comparison of nominal (top) and off-design (bottom) conditions.

of the sub-grid turbulent viscosity maps given by the model equations. After some dimensional considerations [20], the sub-grid TKE can be calculated from:

$$K_{sgs} \simeq \mu_T^2/\rho^2 \left(C_S \Delta\right)^2 \tag{10}$$

where $\Delta = (\Delta x \Delta y \Delta z)^{1/3}$ stands for the averaged size of a computational cell and C_S is the Smagorinsky constant. This constant has been set to 0.1, as this value has been found to yield the best results for a wide range of flows.

Figure 12 shows instantaneous maps of the sub-grid turbulent viscosity ratio, μ_T/μ, for both R-S (top) and S-R (bottom) configurations for different flow rate conditions. These maps reveal the regions where sub-grid turbulence levels arise as consequence of an important turbulent generation at small-scales. As well, they highlight how the global level of disorder is progressively increased as the flow rate decreases. In the R-S case, there is high turbulence intensity in the stator passages, derived from the high levels of turbulence production in the rotor wakes, hence further convected downstream. In the relative frame of reference for the S-R configuration, the generation of turbulence in the rotor wakes is so large and evident that the stator turbulence is completely masked. Consequently, this second configuration is less turbulent and, despite of being a worse design in terms of noise generation, it is a moderate contributor for interaction with less viscous dissipation than the former case.

With these instantaneous maps, applying the previous equation for the estimation of the sub-grid TKE, it is possible to compare the sub-grid TKE values with the post-processed, time-resolved TKE and define the percentage of TKE that it is really resolved by the LES modelling. **Figure 13** summarizes, for both configurations, the typical levels of resolved TKE. It is determined that background levels around 90% provide an accurate framework for the present analysis of the turbulent transport in axial fan stages. These levels are sufficiently high to consider

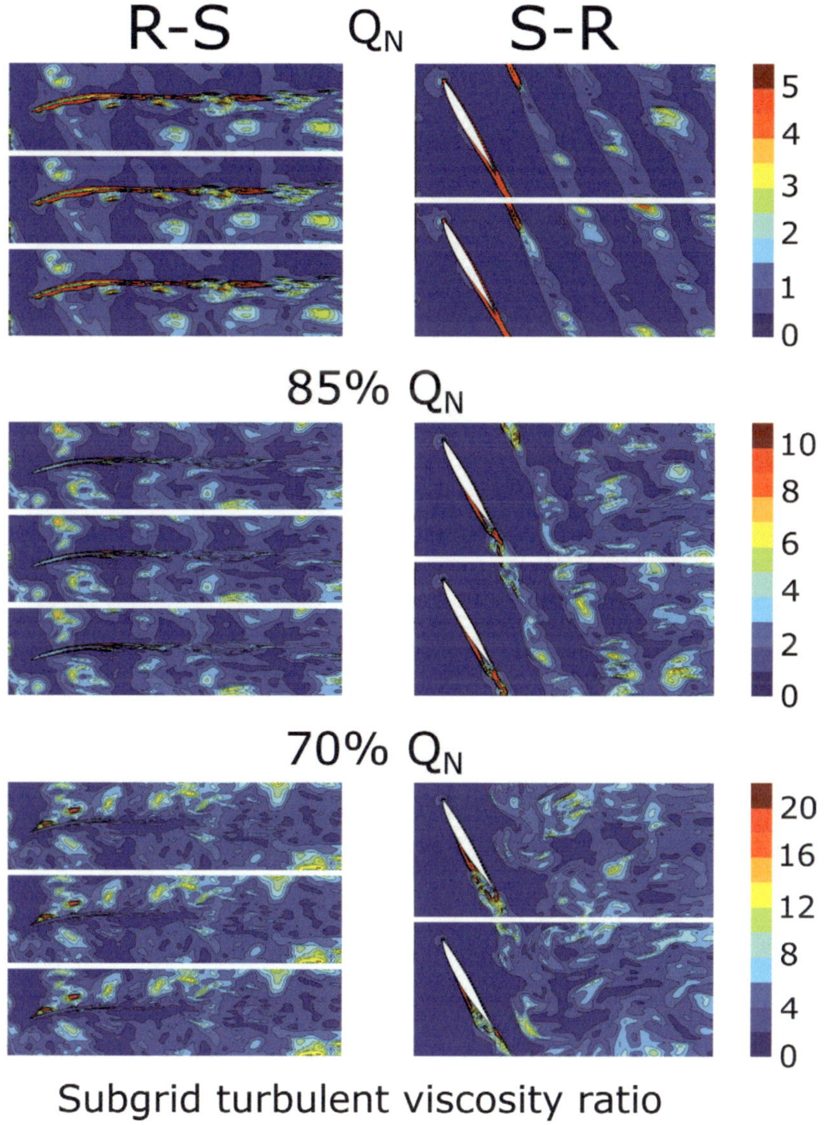

Figure 12. Generation of sub-grid turbulent viscosity at different flow rate conditions. Comparison of R-S (top) and S-R (bottom) configurations.

the present numerical database as a highly resolved LES simulation. Lower levels of resolved TKE are found in the viscous regions of the flow, especially associated to those regions of turbulence generation: on the vicinity of blade and vane surfaces and also in turbulent spots where superimposed viscous structures collide to reinforce vorticity sources (i.e. wake-wake interactions). Regarding the impact of the flow rate, it is evident that nominal conditions require a higher level of modelled turbulence, in both stator and rotor wakes, due to lower size of the vortices shed at high Reynolds numbers [21]. As the flow rate is progressively reduced, the

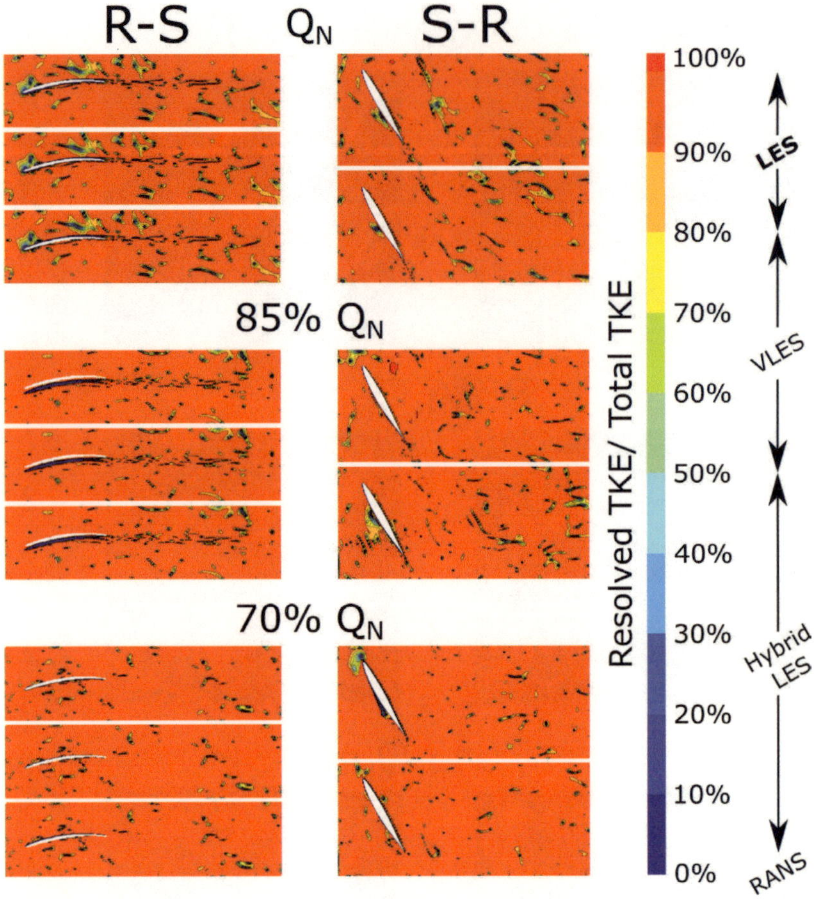

Figure 13. Percentage of resolved kinetic energy in the LES computations. Comparison of R-S and S-R configurations at nominal (top) and off-design (bottom) conditions.

larger size of the detached vortices reduces the percentage of modelled energy in the LES computations accordingly.

5. Conclusion

A comprehensive overview of unsteady deterministic scales, compared to turbulent ones, when both basic rotor-stator and stator-rotor configurations are analysed in the case of low-speed axial fans is presented. A 3D numerical simulation using LES algorithms has been employed to obtain the unsteady flow patterns within the single axial stages as a function of the operating conditions. Afterwards, an intensive post-processing has been performed to disclose deterministic maps as well as the turbulent structure of the flow, revealing its unsteady transport and the typical size of integral scales and their turbulence levels.

In summary, it has been shown that in the inter-row region of the R-S configuration, the periodic passing of rotor wakes is the main interaction mechanism, as it may be observed in the deterministic unsteadiness of the flow. On the other hand, for the S-R configuration, the major interaction mechanisms are viscous and are associated to wake-wake interactions between the unmixed lattice of rotor wakes and the stator shear layers.

This methodology may be applied to assess the unsteady effects on the dynamic performance and overall efficiency of prospective as well as commercial fan designs. Detrimental frequencies may be predicted, and the study of turbulent effects may lead to improvements in design to produce lower noise levels, as well as to achieve a better flow guidance that leads to less energy losses. Finally, the segregation and study of the deterministic part of the fluctuations leads to a better knowledge of the fan working principles. This represents a significant advantage for fan designers, as the deterministic part of the flow may be used to maximise energy exchange.

Author details

Jesús Manuel Fernández Oro*, Andrés Meana Fernández and Bruno Pereiras García

*Address all correspondence to: jesusfo@uniovi.es

Fluid Mechanics Area, Department of Energy, University of Oviedo, Asturias, Spain

References

[1] Menter, F.R., Schütze, J., Kurbatskii, K.A., Gritskevich, M., Garbaruk, A., (2011), "Scale–Resolving Simulation Techniques in Industrial CFD", Proceedings of the 6th AIAA Theoretical Fluid Mechanics Conference, AIAA2011-3474, June 27-30, Hawaii (USA).

[2] Tucker, P.G., (2013), "Trends in Turbomachinery Turbulence Treatments", Progress in Aerospace Sciences, 63, pp. 1-32.

[3] Shur, M.L., Spalart, P.R., Strelets, M.K. and Travin, A.K., (2008), "A Hybrid RANS–LES Approach with Delayed-DES and Wall-Modeled LES Capabilities", International Journal of Heat and Fluid Flow, 29, pp. 1638-1649.

[4] Adamczyk, J.J., (2000), "Aerodynamic Analysis of Multistage Turbomachinery Flow in Support of Aerodynamic Design", ASME Journal of Turbomachinery, 122, pp. 189-217.

[5] Goto, A., (1992), "Three–Dimensional Flow and Mixing in an Axial Flow Compressor with Different Rotor Tip Clearances", ASME Journal of Turbomachinery, 112, pp. 675-685.

[6] Lakshminarayana, B., (1981), "Techniques for Aerodynamic and Turbulence Measurements in Turbomachinery Rotors", ASME Journal of Engineering and Power, 103, pp. 374-392.

[7] Camp, T.R., Shin, H.-W., (1995), "Turbulence Intensity and Length Scale Measurements in Multistage Compressors", ASME Journal of Turbomachinery, 117, pp. 38-46.

[8] Fernández Oro, J.M., Argüelles Díaz, K.M., Blanco Marigorta, E., 2009, "Non-Deterministic Kinetic Energy within the Rotor Wakes and Boundary Layers of Low-Speed Axial Fans: Frequency-Based Decomposition of Unforced Unsteadiness and Turbulence", Journal of Turbulence, 10, N28.

[9] Davidson, L., Dahlström, S., (2005), "Hybrid LES–RANS: An Approach to Make LES Applicable at High Reynolds Numbers", International Journal of Computational Fluid Dynamics, 19 (6), pp. 415-427.

[10] Galdo Vega, M., Rodríguez Lastra, M., Argüelles Díaz, K.M., Fernández Oro, J.M., (2012), "Application of Deterministic Correlations in the Analysis of Rotor–Stator Interactions in Axial Flow Fans". Proceedings of the ASME 2012 Fluids Engineering Summer Meeting, FEDSM2012-72450, July 8-12, Puerto Rico (USA).

[11] Fernández Oro, J.M., Argüelles Díaz, K.M., Rodríguez Lastra, M., Galdo Vega, M., Pereiras García, B., (2014), "Converged Statistics for Time-Resolved Measurements in Low-Speed Axial Fans using High-Frequency Response Probes", Experimental Thermal and Fluid Science, 54, pp. 71-84.

[12] Persico, G., Rebay, S., (2012), "A Penalty Formulation for the Throughflow Modeling of Turbomachinery". Computers and Fluids, 60, pp. 86-98.

[13] Tropea, C., Yarin, A.L., Foss, J.F., (eds.), Handbook of Experimental Fluid Mechanics. Ch. 10. "Measurement of Turbulent Flows", pp. 745-855, Springer, New York, (2008).

[14] Fernández Oro, J.M., Argüelles Díaz, K.M., Santolaria Morros, C., Blanco Marigorta, E., (2007), "On the Structure of Turbulence in a Low–Speed Axial Fan with Inlet Guide Vanes", Experimental Thermal and Fluid Science, 32, pp. 316-331.

[15] Adamczyk, J.J., (1996), "Modelling the Effect of Unsteady Flows on the Time Average Flow Field of a Bladerow embedded in an Axial Flow Multistage Turbomachine", VKI Lecture Series, on Unsteady Flows in Turbomachines, 1996-05, Von Karmán Institute for Fluid Mechanics, Brussels (Belgium).

[16] Meneveau, C., Katz, J., (2002), "A Deterministic Stress Model for Rotor-Stator Interactions in Simulations of Passage-Averaged Flow", ASME Journal of Fluids Engineering, 124, pp. 550-554.

[17] Stollenwerk, S., Kügeler, E., (2013), "Deterministic Stress Modeling for Multistage Compressor Flowfields", Proceedings of the ASME Turbo Expo 2013, GT2013-84860, June 3-7, San Antonio (USA).

[18] Porreca, L., Hollestein, M., Kalfas, A.I., Abhari, R.S., (2005) "Turbulence Measurements and Analysis in a Multistage Axial Turbine", Proceedings of the 17th International Symposium on Air Breathing Engines, ISABE2005-1032, September 4-9, Munich (Germany).

[19] Pope, S.B., Turbulent Flows, Cambridge University Press, New York, (2000).

[20] Bailly, C., Comte-Bellot, G., Turbulence, Springer, Switzerland, (2015).

[21] Tennekes, H., Lumley, J.L., A First Course in Turbulence, MIT Press, Massachusetts, (1972).

Chapter 6

On Turbulence and its Effects on Aerodynamics of Flow through Turbine Stages

Galina Ilieva Ilieva

Additional information is available at the end of the chapter

http://dx.doi.org/10.5772/intechopen.68205

Abstract

In reality, the flows encountered in turbines are highly three-dimensional, viscous, turbulent, and often transonic. These complex flows will not yield to understanding or prediction of their behavior without the application of contemporary and strong modeling techniques, together with an adequate turbulence model, to reveal effects of turbulence phenomenon and its impact on flow past turbine blades. The discussion primarily targets the turbulence features and their impact on fluid dynamics; streaming of blades, and efficiency performance. Turbulence as a phenomenon, turbulence effects and the transition onset in turbine stages are discussed. Flow parameters distribution past turbine stages, approaches to turbulence modeling, and how turbulent effects change efficiency and require an innovative design, among others are presented. Furthermore, a comparison study regarding the application and availability of various turbulence models is fulfilled, showing that every aerodynamic effect, encountered of flow pass turbine blades can be predicted via different model. This work could be very helpful for researchers and engineers working on prediction of transition onset, turbulence effects, and their impact on the overall turbine performance.

Keywords: interaction effects, separation, transition, turbine blade, turbulence, vortices

1. Introduction

The phenomenon known as "turbulence" was already recognized as a distinct fluid behavior more than 500 years ago [1]. **Figure 1** shows a sketch of L. Da Vinci, related to observations of free-stream turbulence.

Turbulence is defined as flow regime, characterized by changes in pressure and velocity, boundary layer separation, creation of vortex structures, and flow disturbances, etc.

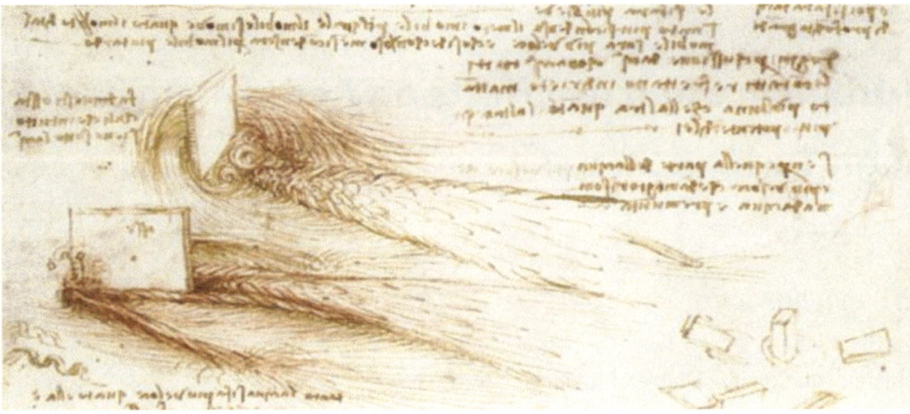

Figure 1. A sketch of turbulence by Leonardo Da Vinci, picture taken from [2].

Turbulence can be discovered in our everyday life and surrounding phenomena such as ocean waves, wind storms and smoke coming up from a chimney, among many others. Most flows observed in nature and existing in various machines and aggregates, such as pumps, fans, turbines, dryers, cyclones, stirred vessels, propulsion systems and their elements (propellers as an example), combustion chambers, and many others, are turbulent too.

In particular, a turbulent flow can be expected to exhibit all of the following specifics: disorganized, chaotic, seemingly random behavior; nonrepeatability in its structures; various and large range of length and timescales; enhanced mixing and dissipation, depending on the viscosity; three dimensionality, time dependence and rotationality; intermittency in both space and time [1].

Here, it is better to clarify that the so-called von Karman vortex street (behind a cylinder) is a vortex flow regular and coherent and cannot be referred to as a turbulent flow, **Figure 2**.

Figure 2. Von Karman streets behind a cylinder in a nonrotating 2D flow for $Re = 140$, fluorescein visualization, picture taken from [3].

The onset of turbulence can be predicted by the Reynolds number, a dimensionless parameter; it characterizes the ratio between inertial and viscous forces. Reynolds number is equal to

$$Re = (\rho \,.\, U \,.\, L)/\mu, \tag{1}$$

In this expression ρ and μ are, respectively, fluid density (kg/m^3) and dynamic viscosity (Pa.s), U is flow velocity (m/s), and L is length scale (m).

When a subvolume of fluid is characterized by excessive kinetic energy, which is higher than the dampening effect of the fluid's viscosity, in other words, when very high *Re* number is realized in that subvolume, turbulence will appear. That is why turbulence is easier for observation in low viscosity fluids, but more difficult for observation in highly viscous fluids. If Reynolds is less than a definite critical value, damping friction forces prevent turbulent movement and the flow is laminar.

In a turbulent flow, vortex structures of various sizes and frequencies could be found. Large vortex structures, influenced by the domain boundaries and the global flow field, break up into smaller structures, characterized of higher frequencies. Small vortex structures are characterized by less frequency. However, due to the flow aerodynamic character and domain boundaries, small vortices can form bigger vortex structures and vice versa. At the same time, vortices in a volume, being at continuous interaction, can exchange energy among them, change their energy levels and travel in the flow volume, changing their so-called "mixing length."

Vortices, which contain the highest amount of kinetic energy, are described by the Taylor scale. In the inertial range, the vortex breakup can be described by inertial effects, thus viscous effects are negligible. Small vortices contain a low amount of energy but contribute mostly to the dissipation. The smallest turbulent vortices are defined by the Kolmogorov microscale [1].

In general terms, in turbulent flow, unsteady vortices appear of many sizes which interact with each other, exchanging energy, as a result drag increases due to friction effects.

The level of turbulence has significant impact on the stability of boundary and shear layers. High stream turbulence scales can contribute to earlier laminar-turbulent transition. Small turbulence scales, observed in the boundary layer, are very important for the skin friction levels and can exert the separation due to adverse pressure gradients. Thus, separated boundary layer and vortices lead to turbulent fluctuations and increase the level of unsteadiness. Shear layers, between the separated and main flow, act as cores of further developed turbulence levels and turbulent scales.

Turbulence of flow through turbine channels is a prerequisite to problems in blades' streaming; they lead to less levels of aerodynamic efficiency; changes in flow regimes; significant pressure fluctuations, causing vibrations; and variable forces acting on blade surfaces, among many others. Aforementioned, lead to worse efficiency, possible blades destruction, and problems related to other processes and elements, which make part of the turbine aggregate and installation.

The aim of this work is to discuss the complexity of flows through turbine passages with a particular emphasis on turbulence and its mechanisms and to explain their effects on turbine aerodynamics and efficiency. The chapter discusses some current state of the art in regard to modeling and prediction of turbulence features, adequacy of turbulence models to achieve physically correct picture for flow parameters distribution in turbine stages. The aim of this chapter is to discuss the follow characteristics and to provide the reader to a better understanding of turbulence mechanisms, their impact on other phenomena in turbines, on the design of turbine components and on the working regimes. Also, this work could be very helpful for researchers and engineers working on prediction of transition onset, turbulence effects, and their impact on the overall turbine performance.

2. Turbulence and its relation to various processes and effects

2.1. Turbulence and interaction effects and losses

Turbulence in turbine stages depends on many aerodynamic features and flow conditions. In this chapter, the author is trying to shed light, based on previously conducted researched works, on the unsteady effects and loss mechanisms in turbines, how they depend on the turbulence effects and how the aerodynamic performance could be compromised.

In [4], an extensive review of loss generating mechanisms in turbomachinery is presented. The three principal sources of losses in turbine stage, are described as: viscous shear in boundary layers, shear layers and mixing processes; nonequilibrium processes such as shock waves and heat transfer processes.

Boundary layers are known as highly viscous regions, also could be referred to as regions of steep velocity gradients and shear stresses. In the boundary layer, a higher amount of energy losses is produced in the areas where the steepest gradients are found [5].

In [6], a parameter called "dissipation coefficient "Cd is described. Its variation for laminar and turbulent boundary layers, with Reynolds numbers in the range of $300 < Re\theta < 1000$, shows that the phenomenon of transition prediction is very important in the assessment of losses in turbomachinery boundary layers.

Stator and rotor blades interact, thus unsteady flow perturbations will appear both in stationary and rotating frames of reference. Various aerodynamic effects, such as wake shedding at the trailing edge, secondary flows in radial direction, blade vibrations, flow leakage in axial gaps, shock waves and effects at trailing edge in transonic stages, angles of attack at the leading edges, etc., can seriously affect the rotor blades and their efficiency performance. Blades loading and forces, acting on blades, are significantly lower under aforementioned conditions, leading to less efficiency [7].

Vortex shedding, as a phenomenon, will appear if flow detaches periodically from the back of a body, forming a Von Kármán vortex street. Such flow picture can occur at the trailing edge of the blade profile. The resulting frequency can be estimated based on the Strouhal number, Eq. (2).

$$St = (fe \,.\, L)/v \tag{2}$$

In Eq. (2) parameters are as follows: L is the characteristic length in (m), the excitation frequency is fe, and v is the flow velocity (m/s).

Turbulence intensity decays much slower than the velocity deficit in the wake [8]. The velocity deficit in an upstream blade wake can be perceived as an incidence variation by the downstream blade row. Turbulence levels in the wake can change the boundary layer from laminar to turbulent, leading to additional losses. The influence of wake will maintain through the blade rows far in a downstream direction.

Flow, characterized by low velocities, has normal velocity component toward the suction side of the downstream blade, thus an upstream wake with low velocities and higher turbulence levels can move toward the suction side of the blade. In a similar way, a fluid characterized by higher velocity can move toward the pressure side of the downstream blades. This movement of fluid particles has the following major effects on the downstream blade row, as described in [9]: change in boundary layer characteristics of a profile through its effect on the transition processes; affects the secondary flow generation through the blades in downstream direction; and influences the wake mixing losses due to the phenomenon of wake stretching or compression.

One of the first studies related to the interaction between streamwise vortices and downstream blades is described in [10]. A significant increase in random unsteadiness at the front part of rotor blades, in regions associated with stator secondary flow, was observed. This is a result of vortex breakdown, and is proposed that it occurred due the vortex filling and cutting, and the strong deformation of the vortex cross-sectional area, at the moment when the vortex enters into the rotor interblade channels [10]. The vortex energy is converted into energy of random fluctuations during the process of vortex breakdown. Later, more detailed mechanism for the vortex-rotor interaction, was developed, see Ref. [11]. According to that model, a moment after the vortex is separated from the rotor blade, thus disturbance will create and will start propagate along the vortex axis, at the local speed of sound, whilst simultaneously being swept downstream, at the local convection velocity. The place where vortex arises is located close to the pressure-side stagnation region [11]. In [12], for one and a half stage, it is demonstrated that flow passing through the first row of rotor blades is highly unsteady and is much influenced by the flow generated between two adjacent stator blades. Upstream, flow disturbances can convect through the rotor blades, without interaction with last mentioned, for a short time and vice versa [12].

In a passage of low aspect ratio blades, secondary flows generated in the form of streamwise vortices, are significant across the blade and can take more than 1/3 part of the blade span [13]. These vortices are convected downstream toward the next blade row where they interact with the main flow.

In [14], for first time, results of studies related to the origin and mechanisms of secondary flows generation, on the basis of the analytical modeling, **Figure 3**, are described. The incoming boundary layer is modeled as a vortex filament ab and is shown that when this filament convects in downstream blade rows to def, it produces three forms of vorticity: secondary, trailing filament, and trailing shed vorticity. The distributed secondary vorticity is result of

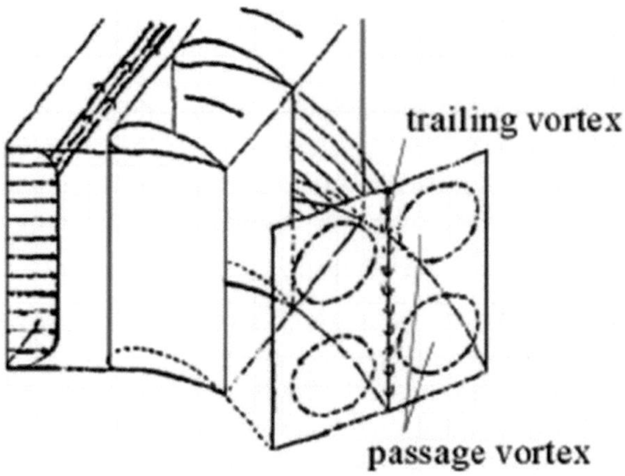

Figure 3. Model of Hawthorne for secondary flows, realized in 1955, picture taken from [14].

the turning of the inlet vortex filament; the trailing filament vorticity forms due to the differences in velocity between the concave and convex surfaces. The last one, the so-called trailing shed vorticity, is formed as a result of the spanwise variation of the blade circulation. Furthermore, it was stated that these mechanisms contribute to the overall secondary flow generation [14]. The Hawthorne's approach, applied and described in [13], is result of many assumptions; however, it can shed some light on the origin of the secondary flows. Similar models were developed and presented by other scientists [15, 16].

Every wake is initially represented as a perturbation in the uniform flow, as stated in [17, 18]. Wakes are transported with the main flow and are cut up into separate segments by the downstream blades. Inside the blade passage, the wake continues to behave as a jet, taking energy from the main flow. The velocity induced by that jet leads to generation of wake flow over the convex blade surface; the induced wake structure over the concave blade surface disappears. As a result of the blade circulation, in axial turbines, formed wake structures can stretch and shear over the blade surfaces or along the channel, they are travelling through [19].

Wake, before entering the blade passage, is subjected to "bowing" due to the higher velocities in the middle of the passage, in comparison to the near blade surfaces [20]. Moreover, wake experiences "shearing" near the suction surface and "stretching" near the pressure blade surface. It is due to the fact that the part adjacent to the convex blade surface convects more rapidly in comparison with the part adjacent to the concave surface. As a result of these processes of bowing, stretching, shearing, and distortion, the wake is moved to the convex blade surfaces and its tail is stretched in direction back to the leading edge of rotor blades. This wake transport is a prerequisite to losses generated from mixing of wakes in the downstream direction [15, 16, 19, 21, 22].

Secondary flows lead to increase in losses, formed at the end walls, and to nonuniform distribution of exit flow angles. In turbomachines, the upstream end walls are rotating relatively to a blade row, thus the inlet boundary layer is expected to be skewed. Experiments, described in [23–25], show that already mentioned skew significantly can affect the existing passage vortices and losses. The streamwise vorticity, introduced to the flow, by the skewed boundary layer at the inlet, will strengthen the streamwise vorticity, observed at the exit; all that is visible in the direction of the turbine rotation.

Existing radial gaps lead to increase in the leakage mass flow rate and causes efficiency losses. Leakage flows are formed as a result of pressure differences between concave and convex blade surfaces, dominated by formed trailing vortices, shed in the downstream direction. These vortices can reduce the local turning, performed by the blade, and can generate decrease in the extracted mechanical energy. As a consequence of the viscous effects in the tip clearance, entropy increases. The second major feature is the subsequent mixing of flow, which passes through the tip clearance gap with that coming from the main flow. Flow structure in the tip region is studied and explained by a number of researchers; see Ref. [26–29].

Low pressure, generated immediately behind the trailing edge, leads to very high losses at the trailing edges. Flow expands around the trailing edge to that low pressure values and is then recompressed after a strong shock wave, where meet flows coming from suction and pressure sides [30, 31]. The interaction between shocks and boundary layers can lead to unsteady boundary layer separations and increase in loss for transonic velocities [32, 33].

2.2. Turbulence and condensation effects in two-phase flows in turbine stages

In two-phase flows in turbine stages, the liquid phase could be presented both as water film, along the streamed blade walls, and as droplets, travelling with the main flow. The interface between primary and second phases is characterized by many aerodynamic and thermal effects.

Developed models for annular two-phase flows take into account differences and specifics for both the liquid and vapor phases and also introduce continuous and dispersed fields. Knowledge of the turbulence characteristics related to the continuous vapor phase, modified due to the presence of the droplet field, is required to introduce closure for those models. Vapor core turbulence is known to influence interfacial shear and affects transport and specifics in structure of the dispersed liquid droplet field, and ultimately interfacial shear effects and droplets dynamics and deposition are affected by vapor core turbulence.

- Condensation processes in turbomachines are random and unsteady phenomenon. Droplet growth rates show different characteristics, recognized in numerical simulations and experiments, also measured in real low-pressure steam turbines. These differences result by the large-scale temperature fluctuations, which are caused by the segmentation of blade wakes (mainly those at the trailing edges) by successive blade rows.
- In the case of two-phase flows, one has to clarify the effects of turbulence in relation to the gas and liquid phases. In other words, the gas phase turbulence can be affected in a two-way coupled manner by the dispersed phase and the droplet motion may also be influenced by turbulence. In case of liquid phase, the instantaneous flow velocity is decomposed into mean and fluctuating components:

$$u_i = u_{i-} + u_i' \quad (3)$$

$$u_i' = \Pi (2k/3)^{1/2} \quad (4)$$

where Π provides normally distributed random number, which allows the fluctuating component *ui'* to take different values around a mean velocity *ui*.

The mean component of the fluid velocity will affect the average trajectory of flow particles. Two identical particles can have different trajectories due to the specifics in flow aerodynamics, such as secondary flows, turbulence, condensation, etc. In the case of turbulent two-phase flows, the fluctuating velocity component leads to particles' trajectory deviations.

Particles with $Re < 400$ tend to suppress turbulence and its effects; particles with $Re > 400$, enhance turbulence as a result of vortex shedding [34].

In [35] turbulence measurements in flows with liquid film interface are reported; it was shown that turbulence intensities were higher than those in a pipe with wall roughness equivalent to that of the film interface.

Aerodynamic and thermal losses in turbine stages, due to the presence of second phase and turbulence effects, are as follows:

- Water film losses, originating from the movement of the water film on the blade surfaces;
- Drag loss due to the presence of coarse water droplets: small fraction of the overall losses originates from acceleration of the large coarse droplets at the trailing edges;
- Coarse water braking losses: kinetic energy of droplets decreases by impinging on the following blade row;
- Fog droplet loss due to the interphase velocity slip, entropy increases due to the viscous drag;
- Fog droplet deposition or collection loss, turbine efficiency performance is affected by decreased kinetic energy of fog droplets, merged on the blade surfaces.

The influence of direct losses in nonequilibrium flows grows with the number of stages and reaches its maximum at the end of the steam turbine. The indirect losses cause changes in aerodynamic loss by departures from equilibrium.

- Changes in flow-incidence angle, leading to increased losses, especially at the tip regions with thin leading edge blade profiles;
- Changes in profile losses due to the variations in surface shear stress, leading to transition and separation;
- Interactions between both the condensation-induced shock waves and formed boundary layer;
- Nonuniformity effects—pressure changes along streamed profiles change the working conditions for the downstream blade row.

Other specifics of the two-phase flow pass turbine surfaces are as follows:

- Formation of water film over streamed surfaces and its defragmentation due to acting viscous forces, drag forces, pressure differences, secondary flows, deviation of streamlines, water particles, trajectories of water particles hit blade surfaces;
- Depending on the axial gap, deviations of streamlines, rate of condensation and water content and specifics of pressure and velocity distribution, weaker and stronger turbulent effects are observed;
- Boundary layer separation much before the trailing edge is attained, leads to formation of droplets of various diameters, most of them are condensation nuclei in the process of blades streaming by two-phase flow [36].

From structural point of view, coarse droplets together with the turbulence impact blade surfaces, leading to erosion and efficiency decrease [36].

2.3. Turbulence and cooling in gas turbine passages

Turbine blades, as part of gas turbine aggregates, must be cooled as they are highly exerted by temperature loads, deformations and stresses. Furthermore, unsteady effects, rotationally induced forces (Corriolis and centrifugal forces), and secondary flows and change in flow parameters, among others complicate the flow aerodynamics within the interblade channels.

The flow field is characterized by higher levels of turbulence, transition effects, secondary flows, unsteadiness, particularly due to the cooling flows and effects. Turbulence results due to wakes from upstream stages; rotationally induced forces; cooling fluid (coolant) that mixes with the main stream in turbine channels; interaction effects between film cooling jets and the mainstream flow; complex flow conditions from the combustor; separation of cooling film, separation effects associated with cooling jets and cavities, etc.

Heat transfer coefficients increase by the enhancement of flow turbulence levels and boundary layer separation effects. Last mentioned are accompanied by increase in the examined pressure drop [37].

The effect of turbulence levels on heat transfer, aerodynamic performance, drag forces, skin friction and flow parameters distribution has been studied by many researchers in last decades [38–42].

High turbulence levels are provoked due to the presence of upstream wakes and rotation effects [43–45]. High levels of turbulence lead to increased temperatures over blade surfaces, particularly at leading edges where temperatures reach maximum values [46].

As an effect of the interaction between cross-flow and film cooling, higher turbulence levels present in turbine flows. These effects depend on the diameter of holes that supply the coolant jets and their exact place on turbine blade surfaces. Experiments on cooling via inclined holes, characterized by ratio ($\delta 1/d \ll 1$), show that a higher percent of turbulence levels can be achieved immediately in the downstream direction of a row of film cooling holes [47]. Turbulence effects and their increase are found even after film cooling jet dissipation.

By application of hot-wire measurement techniques is also concluded that in the case of ($\delta 1/d$<<11) jet-turbulent fluid could dominate the boundary layer in the downstream direction [48].

In [49], authors have studied how circular wall jets can increase the free-stream turbulence intensity. It is found that the increased turbulence levels in the free-stream can cause high levels of mixing and quick dissipation in the film layer.

An increase in heat transfer levels is related to moderate turbulence intensity levels and relatively small length scales recognized in the fluid.

Mainstream turbulence intensity and length scales have significant impact on the film cooling jets and their intensity in the flow, also lead to additional interactions between cooling jets and core flow and affect the effectiveness of film cooling in the downstream direction [50].

Free-stream turbulence can decrease the efficiency of film cooling, in the downstream direction and can increase it between injection holes, due to the enhanced mixing [51].

An increase in mixing of momentum, that effectively thins the boundary layer, will result in an increase in the film cooling efficiency in the downstream direction. At the same time, an increase in the rate of cooling film disintegration is a reason for less efficiency performance of the film cooling. In general, which case will be dominant depends on the "blowing ratio" [37, 52]. In [53–55], the effects of increased turbulence intensity on film cooling performance, at leading edge were examined and the effects of increased turbulence intensity are in a strong relation to the blowing ratio.

The efficiency of performed film cooling varies rapidly with the blowing ratio and is in function of the spanwise angle [56].

At higher blowing ratios and cylindrical film cooling holes, the effectiveness in downstream direction will increase if the jet angle is deflected in a close proximity to the wall.

An increase in turbulent mixing levels, due to free-stream turbulence, leads to improved spanwise mixing between holes and better film cooling performance [57].

Film layers characterized with low blowing ratios could be rapidly dispersed by increased turbulence intensity, while increasing the local heat transfer coefficient. High blowing ratio film layers were found to be relatively insensitive to increase in the turbulence intensity levels, consistent with previous studies with flat plates.

Many numerical and experimental studies have been conducted to determine the relation between turbulence and the Nusselt number. A numerical study of both heat transfer characteristics and flow field of a slot turbulent jet impinging, on a concave surface with constant heat flux, has been carried out and discussed in [58]. In [59] attention is paid on flat plate with varying curvature, results are discussed in [59]. Computational results show that maximum value of Nu is attained at the stagnation point and gains higher values as the Reynolds number increases.

http://dx.doi.org/10.5772/intechopen.68205

3. Transition modeling in turbine stages

3.1. Transition as a phenomenon: transition in turbine stages

Turbulence modeling has been a subject of intensive research for many years. Turbulent models are able to predict turbulence occurrence and its effects in turbine passages. However, in the case of modeling, with all specific features, such as stator-rotor interaction, phase changes (if any), compressible and viscosity effects, pressure gradients, among others, turbulent model lacks of applicability is not able to provide accurate predictions of turbulence effects on the mean flow characteristics. Models should be more powerful and enriched by specific terms related to existing specific aerodynamic observations due to turbulent effects.

One of the most important and most difficult turbulent phenomenon, to model and resolve, is the so-called transition.

Reynolds is the first scientist who worked on the transition phenomenon; he investigated the transition from laminar to turbulent flow by injecting a dye streak into a flow through a pipe having smooth transparent walls [1].

Prediction of the onset of boundary layer transition is one of the most important concerns in the area of fluid mechanics. There is a great interest to transition as it plays a major role in many engineering applications and raises important questions to the flow physics, also could serve as an ingesting example for determinism and chaos.

The so-called viscous instability of a laminar boundary layer was for the first time taken into account and studied by Tollmien. Under low free-stream turbulence conditions, instability is observed in the case of two-dimensional unstable Tollmien-Schlichting waves are formed and propagate in the streamwise direction. These waves lead to additional 3D aerodynamic effects to appear in the flow structure, such as peaks, stronger secondary flow effects, hairpin vortices and transition effects. Turbulent spots are formed in the regions of vorticity peaks and can develop to continuously spreading turbulence. A turbulent spot model to describe the specifics of a transitional flow is proposed in [60]. Later, turbulent spots generated over a flat plate surfaces, without imposed pressure gradients, were also visualized [61]. Recently, scientists have been working on more accurate transition length predictions, based on measurement of transition length in a field of adverse pressure gradients and of triggered turbulent spots, see Ref. [62]. It was found that spot characteristics, in the case of adverse pressure gradients, are different from those formed in the case of zero or favorable pressure gradients. Also, it became clear that in the presence of adverse pressure gradient, a spot can be formed at the center of a highly amplified transverse waves and is convected at lower velocity than under a zero pressure gradient, as discussed in [63].

Laminar to turbulent transition is proved as a phenomenon, which seriously affects the efficiency performance of various machines. The transition effects contribute to additional drag and lift forces, also heat fluxes that are crucial for overall working principles of different types of machines and installations.

A particular field, in which there is a specific interest to the transition phenomenon and its physics, is the area of turbomachinery. Transitional flows can be seen in flows past turbine blades, mainly in low pressure turbines. Transition is observed when various geometry blades are streamed at variable flow parameters and boundary conditions. Turbulence and transition effects significantly decrease the aerodynamic performance of turbine stages and must be studied and understood in detail [64].

Speaking about transition, one must stress that there are different types of transition [65]. The first one is called "natural" transition. It begins with a weak instability in the laminar boundary layer, as was described years ago by Tollmien and Schlichting [6], next proceeds through various stages of amplified instability to fully turbulent flow. The second type of transition is the "bypass"" transition, defined by Morkovin [84]. Bypass transition is caused by high levels of disturbances in the external flow (such as free-stream turbulence) and can completely bypass "natural transition." The third type of transition is the so called "separated-flow" transition. It exists in a separated laminar boundary layer and may or may not reveals some instabilities of the Tollmien-Schlichting type.

As it is related to the "bypass" transition, some researchers presumed that it could appear as an instantaneous turbulent breakdown with zero length of transitional flow. However, the bypass transition does not always exclude instability processes. Only the long region of two-dimensional wave amplification preceding the appearance of three-dimensional disturbances (spanwise periodicity) in low turbulence flow is bypassed, as clarified in detail in [63, 66]. In more detail, during the phase of the so-called "bypass transition," not all specific laminar breakdown processes would be recognized. Separated-flow transition occurs when a laminar boundary layer separates and transitions in the free shear layer, above the already formed bubble. Transition due to separated flow could develop at the leading edge and close to the place where minimum pressure on the suction surface sides is formed. This type of transition is extremely harmful for low pressure turbines and leads to early separation of the boundary layer. There is another type of transition, reverse transition, which is known as "relaminarization." Relaminarization is possible to appear at places where a previously turbulent or transitional region is affected by strong favorable pressure gradients, and as a result of that, it transfers to laminar again. Depending on the profile section geometry and the flow regime, near the leading edge, laminar flow followed by a wake-induced or shock-induced transition could be visualized. Last described phenomenon could be replaced by a relaminarization with subsequent transition to turbulence, occurring at multiple locations simultaneously [63].

The phenomenon of "separated-flow transition" could arise after the so-called boundary layer trip wires as a result of laminar separation under strong adverse pressure gradients. Thus, the flow can reattach as turbulent, forming laminar separation/turbulent-reattachment "bubble", on the surface under consideration. In gas turbine stages, the transition of separated flow could be seen in the so-called overspeed region close to the leading edge of the profile, over the convex or concave side, or both, and near the place where minimum pressure on the convex side is observed. What will be the bubble size depends on the transition process within the free shear layer and may involve all of the stages for a natural transition type. For bubbles with bigger size and characterized by low free-stream turbulence levels, flow in the

http://dx.doi.org/10.5772/intechopen.68205

bubble is dominantly laminar and instabilities could be observed [67]. Big bubbles, along the blade surfaces, produce losses and act as a prerequisite to exit flow angles deviation. Small bubble configurations are an effective way to increase the turbulence levels and can possibly control the blade aerodynamics [63].

Speaking about transition and its effects on the entire flow field, it is necessary to mention that flow passing through a turbine stage is essentially turbulent and unsteady. The nonstationary pulsations are obtained as a result from the stator-rotor interaction, mainly. Periodic phenomenon, caused by stator-rotor interaction effects, excites both the flow, passing over blade surfaces, and boundary layer characteristics [63]. This results in an increased production of the so-called turbulent spots and shifts the location of laminar-turbulent transition in the upstream direction. This laminar-turbulent transition phenomenon is known as "wake-induced transition" [65].

Figure 4, a picture of possible boundary layer development over surfaces of high pressure blade is shown, see [65]. On the suction side, it is usually expected that in the downstream direction of the initial laminar part, a boundary layer will transfer to turbulent (2) in **Figure 4**. The size of the transition zone is related to the place where transition phenomenon could be observed - in upstream direction or downstream direction of the place of minimum pressure. In the upstream direction, the zone of transition is expected to comprehend bigger area. If a laminar separation bubble occurs in the front part of the suction side (1), then the presence of high pressure gradients will force the boundary layer to develop as laminar again in downstream direction; forward transition will take place [63]. The reverse transition may appear on the suction surface [67–69]. In the case of research on film-cooled gas turbine blades, the transition is expected to appear at the places where cooling jets are injected in the main flow [63]. In downstream, a reverse transition process also could be recognized. This fact could affect the heat transfer distribution over surfaces of film-cooled blades. On the profile pressure side - if a separation bubble occurs, the reattached turbulent boundary layer may become again laminar like, (2) in **Figure 4**. In the case of lack of separation bubble, a forward transition zone, followed by a reverse one, in the rear part of the profile, could be observed, (1) in **Figure 4**.

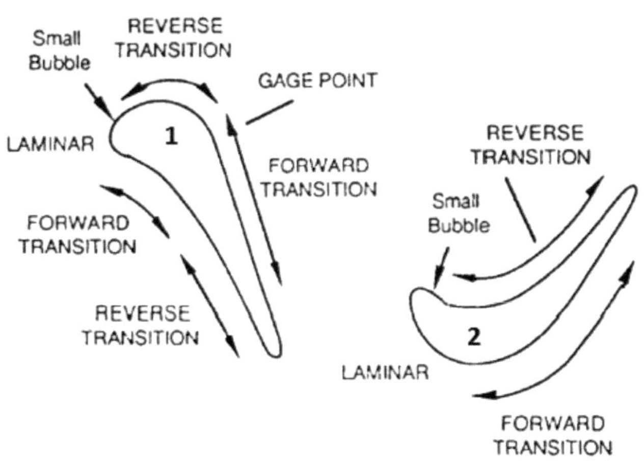

Figure 4. Boundary layer development on high pressure turbine blades [65].

In high pressure turbines, the effect of transition on losses is usually small, because the aerodynamic losses are mainly related to the turbulent flow development after the moment of transition. In low pressure turbines, the flow in interblade channels is characterized by low *Re*. Especially for gas turbines, as part of aircraft engines, the operating Reynolds numbers are low at high altitudes to begin with and a further decrease can cause separation before transition.

In regions where expansion occurs, the fluid is highly accelerated and the boundary layer has small thickness due to the favorable pressure gradients.

At high Reynolds numbers, transition occurs far in the upstream direction, flow is mainly turbulent over the profile. Near the trailing edge, in function of the blade profile geometry, the boundary layer will separate forced by turbulent levels. When Re number decreases, turbulent separation disappears and transition (the "bypass transition") moves in the downstream direction; at that moment losses are minimal. If *Re* number decreases more, laminar separation ahead of the transition region could appear. In the case of no separation, the bubble is small enough so that the flow could reattach to the blade surface. In this case, aerodynamical losses are slightly higher than the previously described case. For lower Reynolds numbers the increase of laminar shear layer and transition length, until reattachment, before the trailing edge, is no longer possible and thus a complete separation occurs, **Figure 5**.

3.2. What transition depends on?

In general, the turbulence Reynolds number ($Re_{\theta t}$) increases with increase in acceleration or with decrease in the free-stream turbulence levels. The effect of acceleration is significant for low turbulence levels. However, for turbulence levels found in gas turbine stages, it has a negligible value. In the case of high turbulence levels, the transition onset is controlled by the free-stream turbulence [65].

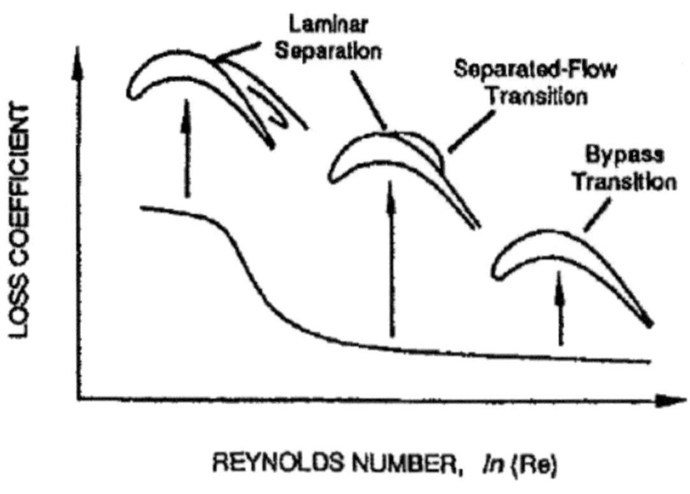

Figure 5. Transition on a low-pressure turbine airfoil at various Reynolds number.

For accelerating flows, length of transition is variable for thermal and momentum boundary layers [70, 71].

The local point where boundary layer can separate is not in a close relation to the levels of free-stream turbulence. In the case of observed increase in levels of free-stream turbulence, the separation bubble size decreases and turbulent transition point moves in upstream in regard to the flow direction. The separation bubble decreases with increase of the Reynolds number [72]. At high free-stream turbulence intensity, the streak structures can be observed in the upstream direction and are related to the place of boundary layer separation, showing that bypass transition of an attached boundary layer can be realized at the high Reynolds number. In the case of low free-stream turbulence intensity, velocity fluctuations are seen within the shear layer of the separation bubble.

The roughness over streamed surfaces is nonuniform and characterized by significant variations in streamwise and spanwise directions [73, 74]. Modifications in the behavior of the boundary layer due to the presence of definite values of surface roughness can decrease the aerodynamic efficiency [75–77] and can also increase the heat transfer [78–80]). Levels of heat transfer may also be affected by changes in material properties or in the case of eroded or broken protective coatings.

The effect of geometry curvature and/or streamlines curvature on the transition was studied and described in [81, 82]. In [81], it was found that a laminar boundary layer formed over a concave surface becomes unstable as a result of acting centrifugal forces, three-dimensional disturbances and streamwise vortices presented in the boundary layer. Liepmann in [82] showed that transition on a convex surface is only slightly delayed, but can occur earlier over the concave surface.

The increase in transition Reynolds number is caused by the Görtler vortices, which can increase velocity gradients near the wall and thus can delay the transition. For highly curved surfaces, this effect dominates that one caused by turbulence. A concave curvature can decrease or can increase the transition The Reynolds number depends on the turbulence intensity and the curvature.

Heating or cooling can seriously affect boundary layer transition at low free-stream turbulence. Heat transfer through a laminar boundary layer formed over the concave blade surface is influenced both by Taylor-Görtler vortices and the main flow turbulence levels [83]. Transition occurs when these factors surpass the tendency of boundary layer to remain laminar in the presence of higher pressure gradients. If spot production is not affected by the heat transfer, at high free-stream turbulence intensities, transition would not be observed.

Film cooling affects the boundary layer formed on the streamed surfaces of gas turbine blades. At places where cooling fluid is injected, holes are usually much larger than the boundary layer thickness, thus the injection of coolant through these holes disrupts the flow close to the surfaces and provides higher turbulence levels within the downstream developing boundary layer [63]. Therefore, it may be said that film cooling effect is to "trip" a laminar boundary layer and initiates transition to turbulence.

In the case of acceleration, sufficient to cause reverse transition in the downstream direction of the injection, the heat transfer intensity approaches that for laminar flow [69]. This implies that even though injection can initiate transition, a subsequent strong acceleration can cause the flow to become laminar again. Such a situation is common for film-cooled blades of first gas turbine stages. Heat transfer measurements on a stator vane, presented in [85, 86], indicated that the behavior of the boundary layer transition along the suction side of the vane showed dependency to the film-cooling injection place.

4. Numerical modeling of turbulence in turbine stages

Many numerical and experimental research works have been performed, and various codes and approaches have been developed and applied, for the purposes of modeling and research of origin and mechanisms of laminar-turbulent transition and how it exerts the fluid dynamics in turbine stages. However, the phenomenon in turbomachinery flows is not well understood. Transition modeling still limits the performance of nowadays CFD codes, and problems in estimation of the transition onset and extension of the transition affect the efficiency by several percent and the component life by more than an order of magnitude [87, 88]. Transition as a phenomenon and its understanding is of huge importance to arrive to appropriate design of blade geometry and increased aerodynamic performance.

A brief history of development of turbulence models dates back more than 140 years ago and is shown in [89].

Many of existing turbulence models are applied for modeling and research of turbulence effects in turbine stages. The performances of standard $k-\varepsilon$ model, RNG $k-\varepsilon$ model, realizable $k-\varepsilon$ model, SST $k-\omega$ model, and LRR Reynolds stress transport models for the purposes of research on the convective heat transfer, during slot jet impingement over flat and concave cylindrical surfaces, were evaluated against available experimental data [90, 91]). Near-wall models such as equilibrium wall function and two-layer enhanced wall treatment were applied to the boundary layer to obtain physically correct results. When the impingement surface is within the potential core of the jet, applied models overpredict the Nusselt number in the impingement region, and at the same time, Nu is not very correct in the wall jet region. The RNG $k-\varepsilon$ model, applied together with the enhanced wall treatment, also the SST $k-\omega$ model, gives better Nusselt number distribution in comparison with other models for flat plate and concave surface impingement cases. However, mean velocity profiles are not well predicted by the SST $k-\omega$ model for the concave surface impingement case. Results for velocity profiles, obtained with RNG $k-\varepsilon$ model, agree very well with the experiment. The Reynolds stress model could not give better prediction, compared to other eddy viscosity models [91].

In [58, 90], performance of various turbulence models to predict the convective heat transfer for slot jets impinged on flat and concave surfaces was under consideration. The outcomes as a result of application of more specific model are found very accurate in the case of an impingement surface placed outside of the jet core. In the case of surface placed inside the jet core, results obtained for *Nu* demonstrated larger discrepancies and variation in the impingement region, as their models overpredicted the *Nu* in that region. However, prediction of values for *Nu* is fairly accurate in the wall jet region.

Other interesting numerical studies on rib-roughened channels are related to measurements and simulations with standard $k-\varepsilon$ and nonlinear $k-\varepsilon$ turbulence models [92].

The predicted Nusselt number depends on the selected turbulence model. An improvement in the predicted Nusselt number was found when comparing the LES with a standard $k-\varepsilon$ turbulence model [93].

The standard $k-\varepsilon$ and RNG $k-\varepsilon$ models can accurately predict streaming over an impingement surface [94].

As a result of the numerical modeling and analysis on turbulent flow field and heat transfer in 3D ribbed ducts, it is found that heat transfer predictions obtained using the v^2-f cannot cover well the experimental data for the 3D ribbed duct. On the wall of the duct where ribs exist, predicted heat transfer agrees well with the experimental data for all configurations, heat transfer predictions on the smooth-side wall do not cover well the experimental data. It is mainly due to the presence of strong secondary flow structures which might not be properly simulated with turbulence models based on eddy viscosity.

The standard $k-\varepsilon$ turbulence model with wall functions cannot predict very well heat transfer, due to the presence of large separation regions in the flow field [95]. In [96], it was found that the $k-\varepsilon$ model can give only reasonable qualitative agreement with the experimental data. The application of standard $k-\varepsilon$ model to complex 3D problems is computationally expensive and leads to wrong results [97, 98]. Application of two-layer $k-\varepsilon$ with the effective viscosity model gives bad predictions for heat transfer in rotating ribbed passages [99]. Computations with low-*Re* models could give good heat transfer predictions by introducing a differential version of the Yap length scale correction term [100].

Since low-*Re* models cannot correctly capture the separation and reattachment that takes place between ribs, v^2-f and Spalart-Allmaras (S-A) were taken under consideration in many research works. The v^2-f turbulence model was successfully applied to separated flow in [100], to 3D boundary layers [101], impinging jets, [102, 103] and prediction of flow characteristics and heat transfer in 3D duct with ribs and in model configuration, resembling the tip of an axial turbine blade [104].

The model of Spalart and Allmaras [105] was proven to be robust and accurate in aerodynamic applications [106].

As discussed before, the phenomenon of transition is very complicated and depends on many parameters, such as free-stream turbulence, roughness, curvature, heat transfer and film cooling, among others, and needs specific mathematical models and turbulence closures for the purposes of its research and analysis.

Due to the strong accelerations and decelerations of flow in turbine cascades, the local value of free-stream turbulence, at the location of boundary layer transition onset, may significantly vary from first to the last turbine stage. Currently applied transition onset correlations involve data from many scientists, who have adopted different approaches to define the free-stream turbulence values.

Laminar separation bubbles can result from laminar separation followed by sufficiently early transition in the separated shear layer and subsequent turbulent reattachment. Errors in predicting the length of these bubbles will lead to failures in the blade design and wrong solutions. Early attempts at describing bubble development and separation, see [107, 108], were based on semi-empirical models. In those models, constant pressure for the region of the separated laminar shear layer, instantaneous transition, and linear variations in free-stream velocities during the phase of turbulent reattachment was assumed. An integral boundary layer computation procedure was applied, the location of the transition onset was found with correlations for separated laminar layer, in function of the momentum thickness and *Re*.

Flows with transition and separation phenomena could be modeled with application of contemporary models for accurate description of all aerodynamic effects, which are expected to be observed, together with innovative and very correct transition model. Modeling of bubble dynamics is important for the purposes of research and prediction of separated flows with transition [63].

In [109–113] various predictive techniques were described in detail in the area of turbulence in turbines and stressed on the need for application of improved and correct transition modeling.

In the literature, many papers discuss experiments and their outcomes related to turbine blades. Mostly, they present research in more global aspect, a small number of experiments provide detailed results useful for turbulence modeling. This is related to the fact that it is very difficult to obtain sufficiently thick boundary layers to perform detailed measurements on the suction and pressure sides. Many results are obtained after modeling and measurements related to research on a flat plate with a pressure gradient, imposed by the external wall [70, 114, 115] for negative pressure gradients; also, after application of Görtler vortex on the concave plate [116]. Results for the streaming effects of blade convex side, are shown in [12]. Studies in [117] discuss results of measurements on the suction surface of blade under conditions of very low Reynolds number.

There are mainly two approaches used to model bypass transition in industry [65]. The first is to apply low-Reynolds number turbulence models in which wall-damping functions implemented into the turbulent transport equations were applied to obtain the moment when boundary layer transition will occur [129, 130]. Research activities have proved that this approach cannot predict very well the influence of various factors, such as pressure gradients, free-stream turbulence, and wall roughness to predict the transition onset [131]. Damping functions, optimized to damp the turbulence in the viscous sublayer, cannot give reliable prediction of the transition when subjected different and complicated processes [132].

The second approach is application of experimental correlations related to the free-stream turbulence intensity and to the transition Reynolds number, with included boundary layer momentum thickness [65, 118]. The last approach is proved as accurate, but very challenging, - actual momentum-thickness, Reynolds numbers must be compared with their critical values, obtained from correlations, included into the mathematical model, applied for the

http://dx.doi.org/10.5772/intechopen.68205

purposes to arrive to physically correct numerical solution. There are additional difficulties related to application of unstructured mesh, not well-defined boundary layer, and various approaches to attain numerical solution [64].

5. An example of modeling of turbine stage with twisted rotor blade with three different turbulence models

The main target of this research in which geometry modeling, numerical set-up, and convergence problem solution are described in detail in [119, 120] is to define the flow parameters distribution in a 3D turbine stage with twisted rotor blade. For the purposes of the turbulence modeling, the standard $k - \varepsilon$ turbulence model, RNG $k - \varepsilon$, standard $k - \varepsilon$, for the case of research on radial gap, and RSM (Reynolds stress model) models are applied, in regard to the flow conditions.

The Reynolds stress model (RSM) is applicable for modeling effects of additional vortices, found in flow and shear stress effects over fluid particles [122].

The standard $k - \varepsilon$ model gives quite good values, especially for the turbulent kinetic energy, in the core flow see [119, 121, 122]. In [123], results show that the advantage of using the RSM in regions of flow separation; however, the main flow features were still good enough, captured by the $k - \varepsilon$ model.

The RNG model gives the highest prediction of lift and maximal lift angle [124, 127].

The $k - \varepsilon$ turbulence models are appropriate for flows characterized by high adverse pressure and intensive separation. This model allows for a more accurate near wall treatment with an automatic switch from wall function to low-Reynolds number formulation, based on grid spacing see [125–128].

In the current study, it is found that depending on the specific flow feature, under consideration, different turbulence model have to be applied.

Qualitative results are shown in **Figure 6–Figure 9**.

Figure 6 shows velocity field distribution in the case of rotating rotor blade and activated the standard $k - \varepsilon$ turbulence model.

Figure 7 presents vorticity magnitude values, in radial direction, for the turbine stage under consideration.

Numerical results for pressure distribution, in the case of applied standard $k - \varepsilon$ model, are shown in **Figure 8**.

Figure 9 visualized vortices, in radial direction, due to difference between the pressure field values for hub and shroud sections in the turbine stage, in the case of applied Reynolds stress model (RSM). The area occupied by this vortex is bigger than the one formed in the case of standard $k - \varepsilon$ turbulence model, **Figure 7**.

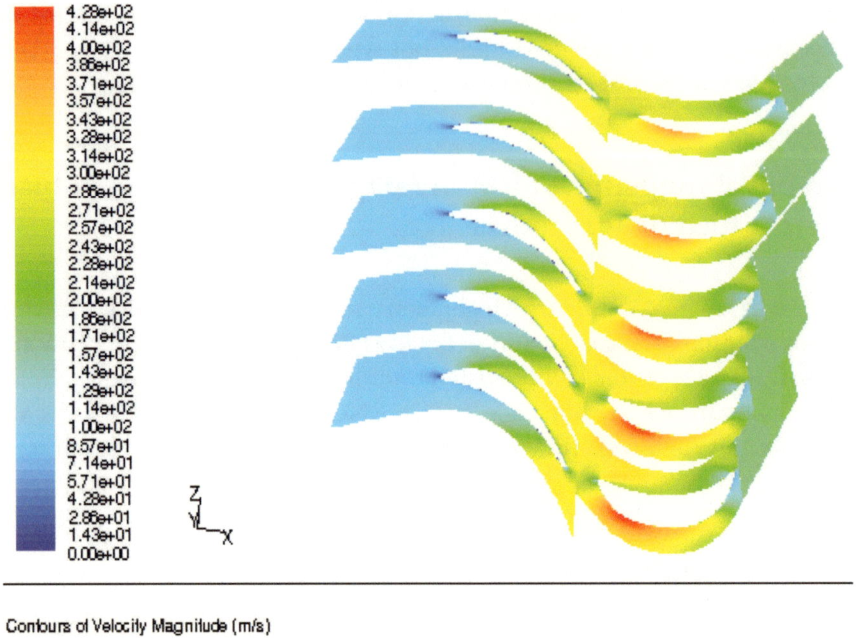

Figure 6. Flow velocity field distribution in control sections, in radial direction.

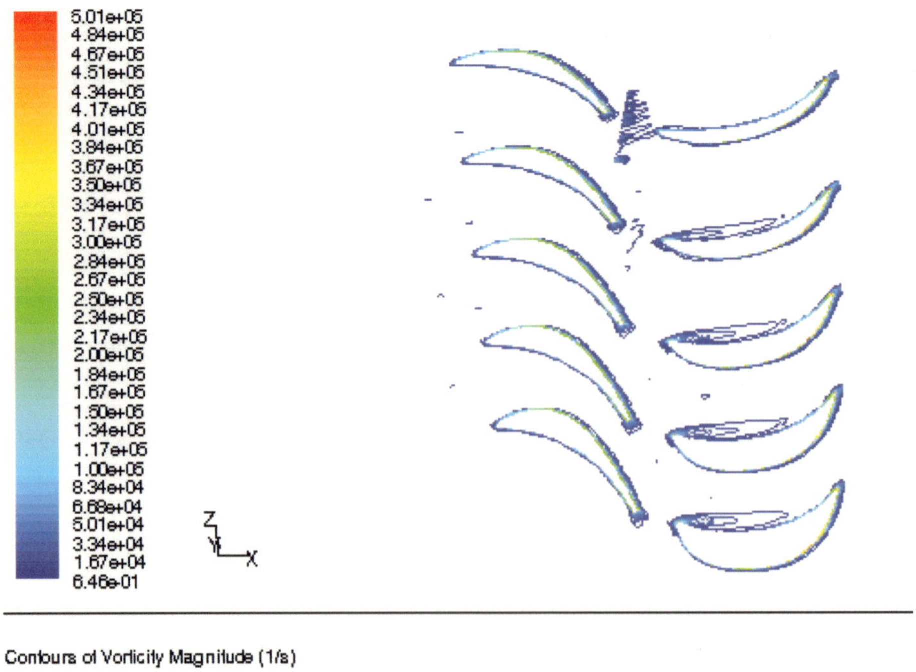

Figure 7. Vorticity magnitude values in control sections, in radial direction.

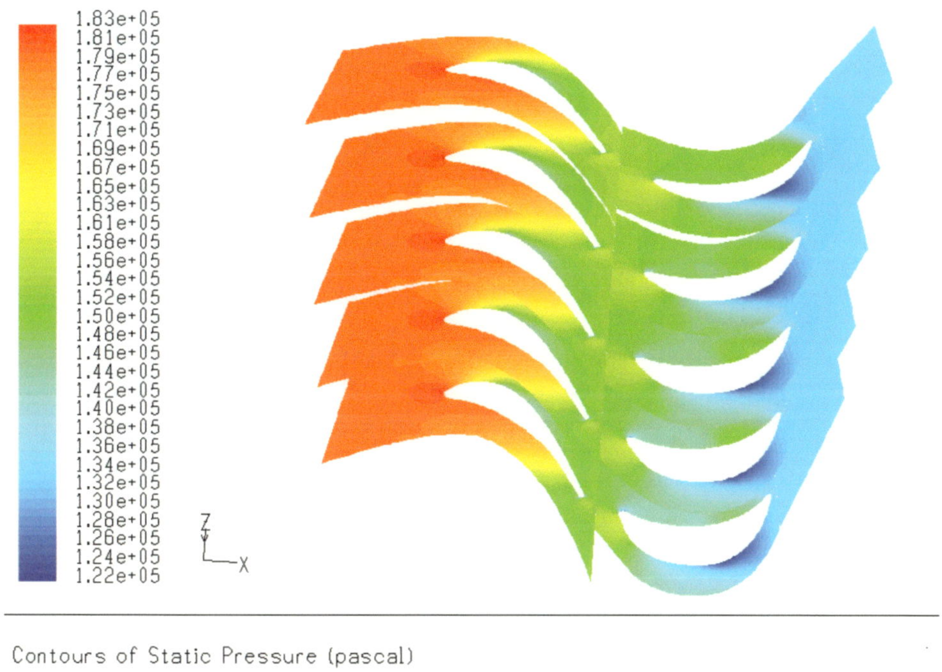

(a)

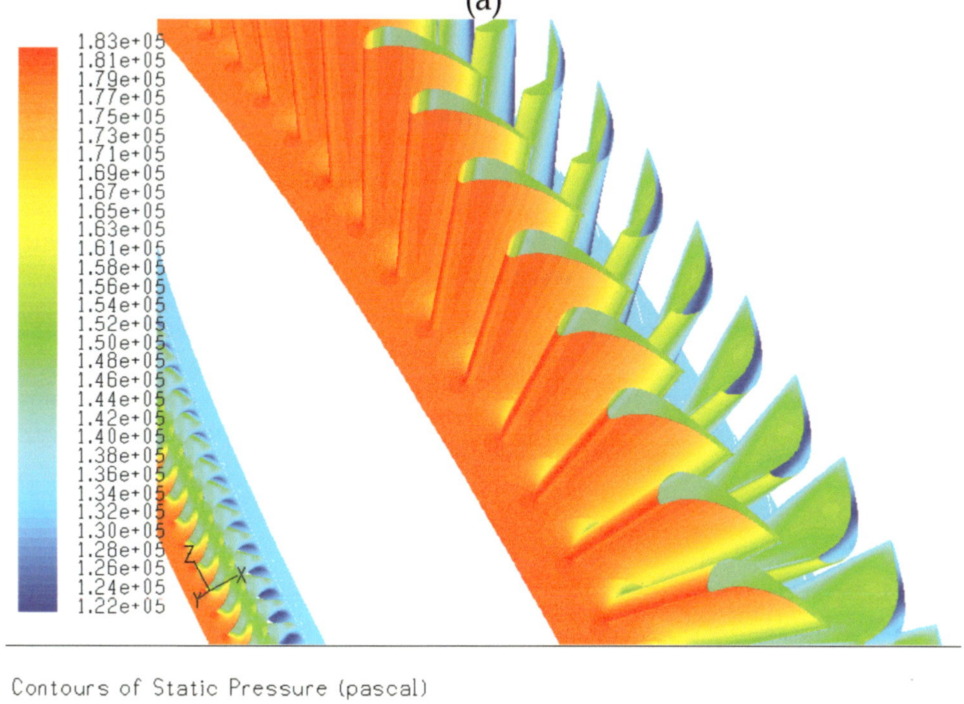

(b)

Figure 8. Static pressure distribution - by control sections (a), for the turbine blade (b) in the case of rotating rotor blade.

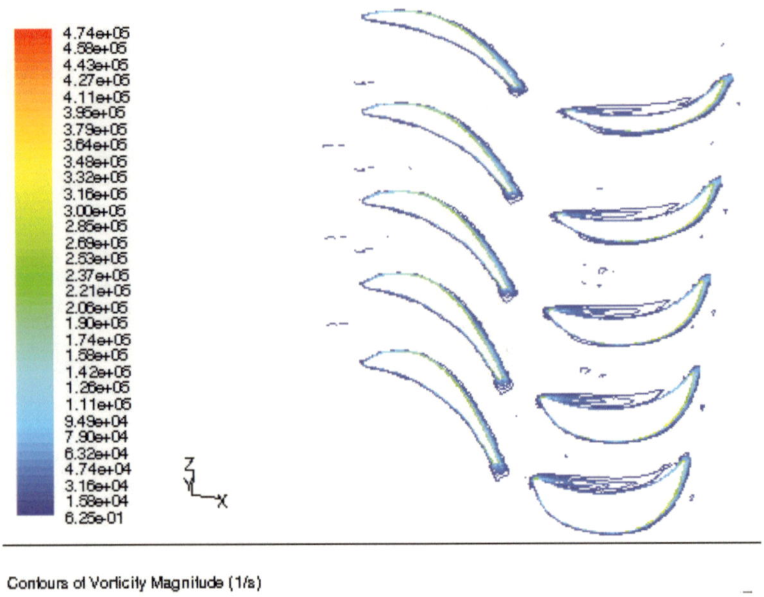

Figure 9. Vorticity magnitude values by control sections, in radial direction.

The outcomes of the performed research are as follows:

- The RNG model is acceptable to study both the shear stress and streamlines curvature effects. It presents vortices formed at the trailing edge and also provides results for aerodynamic features at the leading edge.
- In the case of applied RSM model, a relative decrease of 1.308% for turbine stage efficiency is observed. This is a result of taking into account of all pulsations and vortex structures near the wall regions, boundary layer separation, viscosity, and compressibility effects.
- The RNG $k - \varepsilon$ turbulence model leads to increased values for turbulent intensity and less turbulent viscosity. This is a prerequisite for decrease of the left-hand side term values in the momentum equations, furthermore causes relative increase of stage efficiency with 0.147%, in a comparison with the case of implemented RSM turbulence model.

6. Conclusion

On the basis of previously performed numerical and experimental works by many researchers and by the author, transition onset, and turbulence origin, their effects and impact on efficiency, flow parameters distribution, and possible blades design. Various modeling techniques, turbulence models and their application are discussed to obtain physically correct prediction of turbulence effects in flow past turbine blades. Various turbulence features, and fluid dynamics specifics, streaming of blades and their efficiency performance are discussed.

The chapter presents contemporary approaches to turbulence modeling and the adequacy of turbulence models to obtain flow characteristics, also.

This work could be very helpful for engineers working on prediction of transition onset, turbulence effects and their impact on the overall turbine performance. Moreover, it could be a basis for future research works related to innovative models and advanced numerical techniques to turbulence modeling and analysis.

Acknowledgements

This research is supported by the National Science Fund of Bulgaria and the Technical University - Varna.

Author details

Galina Ilieva Ilieva

Address all correspondence to: galinaili@yahoo.com

Technical University, Varna, Bulgaria

References

[1] McDonough JM. Introductory Lecturers on Turbulence, Physics, Mathematics and Modeling. Lexington, KY: Departments of Mechanical Engineering and Mathematics University of Kentucky. 2004/2007.

[2] Bingham R. Plasma physics: On the crest of a wake. Nature. 2007;**445**:721-722. http://www.nature.com/nature/journal/v445/n7129/fig_tab/445721a_F1.html

[3] http://wavelets.ens.fr/RESULTATS/GAELE/Website/resume_anglais.html

[4] Denton JD. Loss mechanisms in turbomachines. IGTI Gas Turbine Scholar Lecture. 1993; ASME 93-GT-435.

[5] Dawes WN. A comparison of Zero and One Equation Turbulence Models for Turbomachinery calculations. ASME Paper No. 90-GT-303. 1990.

[6] Schlichting H. Boundary Layer Theory. 6th ed. New York: McGraw-Hill; 1966.

[7] Megerle, B. Unsteady aerodynamics of low pressure steam turbines operating under low volume flow conditions. These N 6096 (2014). École Polythechnique Federale de Lausanne, presented on 16 May 2014.

[8] Raj R, Lakshminarayana B. Three-dimensional characteristics of turbulent wakes behind rotors of axial flow turbomachinery. ASME Journal of Engineering for Power. April 1976; 218-228.

[9] Chaluvadi VSP. Blade Vortex Interactions in High Pressure Steam Turbines. Girton College Department of Engineering University of Cambridge; November 2000.

[10] Binder A. Turbulent production due to secondary vortex cutting in a turbine rotor. Journal of Engineering for Gas Turbines and Power. 1985;**107**:1039-1046.

[11] Binder A, Forster W, Mach K, Rogge H. Unsteady flow interaction caused by stator secondary vortices in a turbine rotor. ASME Paper No. 86-GT-302; 1986.

[12] Sharma OP, Wells RA, Schlinker RH, Bailey DA. Boundary layer development on turbine airfoil suction surfaces. ASME Journal of Engineering for Power 1982;**104**:698-706.

[13] Harrison S. The Influence of Blade Stacking on Turbine Losses [Ph.D. thesis]. Cambridge University; 1989.

[14] Hawthorne WR. Rotational flow through cascades. Journal of Mechanics and Applied Mathematics. 1955; **8**.

[15] Adamczyk JJ. Wake Mixing in Axial Flow Compressors. ASME Paper No. 96-GT-029; 1996.

[16] Van de Wall A, Kadambi JR, Adamczyk JJ. A Transport Model for the Deterministic Stresses Associated with Turbomachinery Blade Row Interactions. ASME Paper No. 2000-GT-430; 2000.

[17] Meyer RX. The Effect of Wakes on the Transient Pressure and Velocity Distributions in Turbomachines. ASME Journal of Basic Engineering. 1958;1544-1552.

[18] Kemp NH, Sears WR. The Unsteady Forces due to Viscous Wakes in Turbomachinery. Journal of Aeronautical Sciences. 1955;**22**(7):478-483.

[19] Smith LH. Wake Dispersion in Turbomachines. ASME Journal of Basic Engineering. 1966; 688-690.

[20] Hodson HP, Dawes WN. On the Interaction of Measured Profile Losses in Unsteady Wake-Turbine Blade Interaction Studies. ASME 96-GT-494, 1996.

[21] Deregel P, Tan CS. Impact of Rotor Wakes on Steady State Axial Compressor Performance. ASME Paper No. 96-GT-253; 1996.

[22] Valkov T. Control of the Unsteady Flow in a Stator Blade Row Interacting with Upstream Moving Wakes. ASME Journal of Turbomachinery. 1995;**117**:97-105.

[23] Bindon JP. Exit Plane and Suction Surface Flows in an Annular Cascade with a Skewed Inlet Boundary Layer. International Journal of Heat and Fluid Flow. 1980;**2**(2P):57.

[24] Boletis E, Sieverding CH, van Hove W. Effect of Skewed Inlet Endwall Boundary Layer on the 3-D Flowfield in an Annular Cascade, Viscous Effects in Turbomachinery. AGARD-CP-351; 1983.

[25] Walsh JA, Gregory-Smith, DG. The Effect of Inlet Skew on the Secondary Flows and Losses in a Turbine Cascade, Efficiency Prediction Improvement. I Mech. E, Paper No. C275/87; 1987.

[26] Wadia AR, Booth, TC. Rotor-tip leakage: Part II – design optimisation through viscous analysis and experiment. ASME Journal of Engineering for Power. 1982;**104**:162.

[27] Sjolander SJ, Amrud KK. Effects of tip clearance on blade loading in a planar cascade of turbine blades, ASME Journal of Turbomachinery. 1987;**109**:237.

[28] Heyes FJG, Hodson HP, Dailey GM. The effect of blade tip geometry on the tip leakage flow in axial turbine cascades. ASME Journal of Turbomachinery. 1992;**114**:643.

[29] Moore J, Tilton JS. Tip leakage flow in a linear turbine cascade. ASME Journal of Turbomachinery. 1988; **110**:18.

[30] Bindon JP. The measurement and formation of tip clearance loss. ASME Journal of Turbomachinery. 1989;**111**:257.

[31] Yaras MI, Sjolander SA. Prediction of Tip Leakage Losses in Axial Turbines. ASME Paper No. 90-GT-154; 1990.

[32] Atkin CJ, Squire LC. A study of the interaction of a normal shock wave with a turbulent boundary layer at mach numbers between 1.3 and 1.55'. European Journal of B/Fluids. 1992;**11**(1).

[33] Scherbakov AP, Tischenko AA, Dohler SW. The loss behaviour of transonic flow in nozzle cascades under the influence of unsteady effects. In: 2nd European Conference on Turbomachinery, Fluid Dynamics and Thermodynamics, Antwerpen; 5-7March 1997.

[34] Hetsroni G. Particles-turbulence interaction. International Journal of Multiphase Flow. 1989;**15**(5):735-746.

[35] Azzopardi BJ, Teixeira JCF. Detailed measurements of vertical annular two-phase flow Part It: gas core turbulence. ASME Journal of Fluids Engineering. 1994b;**116**: 796-800.

[36] Ilieva GI. Erosion failure mechanisms in turbine stage with twisted rotor blade. Engineering Failure Analysis. 2016;**70**: 90-104.

[37] Siddique W. Design of internal cooling passages: investigation of thermal performance of serpentine passages [Doctoral Thesis]. Stockholm, Sweden: Division of Heat and Power Technology, Department of Energy Technology, Royal Institute of Technology ; 2011.

[38] Simonich J, Bradshaw, P. Effect of free-stream turbulence on heat transfer through a turbulent boundary layer. Journal of Heat Transfer. 1978;**100**: 671-677.

[39] Blair M, Werle M. The Influence of Free-Stream Turbulence on the Zero Pressure Gradient Fully Turbulent Boundary Layer. (R80-914388-12), Contract No. F49620-78-C-0064. UTCRC, East Hartford, CT; 1980.

[40] Blair M. Influence of free-stream turbulence on turbulent boundary layer heat transfer and mean profile development, parts I and II. Journal of Heat Transfer. 1983;**105** (1): 33-47.

[41] Hancock P. Effect of high free-stream turbulence in turbulent boundary layer [Ph.D. thesis]. UK: Imperial College, London University.

[42] Maciejewski P, Moffat R. Heat transfer with very high free-stream turbulence. Part I. Experimental Data. Journal of Heat Transfer. 1992;**114** (4): 827-833.

[43] Du H, Ekkad S, Han J. Effect of unsteady wake with trailing edge coolant ejection on film cooling performance for a gas turbine blade. Journal of Turbomachinery. 1999;**121** (3): 448-455.

[44] Du H, Han J, Ekkad S. Effect of unsteady wake on detailed heat transfer coefficient and film effectiveness distributions for a gas turbine blade. Journal of Turbomachinery. 1997;**120** (4): 808-813.

[45] Du H, Han J, Ekkad SV. Effect of unsteady wake on detailed heat transfer coefficient and film effectiveness distributions for a gas turbine blade. Journal of Turbomachinery. 1998;**120** (4): 808-817.

[46] Włodzimierz W. Numerical evaluation of the blade cooling for the supercritical steam turbine. Applied Thermal Engineering. 2013; **51**(1): 953-962.

[47] Pietrzyk J, Bogard D, Crawford M. Hydrodynamic measurements of jets in crossflow for gas turbine film cooling applications. Journal of Turbomachinery. 1989;**111** (2):139-145.

[48] Andreopoulos J. On the structure of jets in a crossflow. Journal of Fluid Mechanics. 1985;**157**:163-197.

[49] Jumper G, Elrod W, Rivir R. Film cooling effectiveness in high turbulence flow. Journal of Turbomachinery. 1991;**113** (3): 479-483.

[50] Kadotani K, Goldstein R. On the nature of jets entering a turbulent flow: Part B - film cooling performance. Journal of Engineering for Power. 1979c;**101** (3): 466-470.

[51] Bons J, MacArthur C, Rivir R. The effect of high free-stream turbulence on film cooling effectiveness. Journal of Turbomachinery. 1996;**118**(4): 814-825.

[52] Kodzwa Jr. PM, Eaton JK. Measurements of film cooling performance in a transonic single passage model. Report No. TF 93 June 2005, Flow Physics and Computation Division Department of Mechanical Engineering Stanford University Stanford, CA 94305-3035; 2005.

[53] Mehendale A., Han J. Influence of High Mainstream Turbulence on Leading Edge Film Cooling Heat Transfer, Journal of Turbomachinery. 1992;**114** (4), 707-715.

[54] Ou S, Mehendale A, Han J. Influence of high mainstream turbulence on leading edge film cooling heat transfer: effect of film hole location. Journal of Turbomachinery. 1992;**114** (4):716-723.

[55] Ou S, Han J. Influence of high mainstream turbulence on leading edge film cooling heat transfer through two rows of inclined film slots. Journal of Turbomachinery. 1992;**114** (4):724-733.

[56] Zhao L, et al. Numerical simulation on impingement and film composite cooling of blade leading edge model for gas turbine. Applied Thermal Engineering. 2014.

[57] Saumweber C, Schulz A, Wittig S. Free-stream turbulence effects on film cooling with shaped holes. Journal of Turbomachinery. 2003;**125** (1): 65-73.

[58] Yue-Tzu Y, Wei T-C, Wang Y-H. Numerical study of turbulent slot jet impingement cooling on a semicircular concave surface. International Journal of Heat and Mass Transfer. 2011;**54**(1): 482-489.

[59] Öztekin E. et al. Heat transfer in a turbulent slot jet flow impinging on concave surfaces. International Communications in Heat and Mass Transfer. 2013;**44**:77-82.

[60] Emmons HW. The Laminar-turbulent transition in a boundary layer: Part I. Journal of Aeronautical Science. 1951;**18**(7): 490-498.

[61] Schubauer GB, Klebanoff PS. Contributions on the mechanics of boundary-layer transition. NACA Tech. Note TN 3489; 1955.

[62] Walker GJ, Gostelow JP. Laminar-Turbulent Transition in Boundary Layers on Axial Turbomachine Blades. ERCOFTAC Bulletin. 2009;**80**:11-15.

[63] Jahanmiri M. Boundary Layer Transitional Flow in Gas Turbines, Research Report 2011:01, Göteborg, Sweden: Division of Fluid Dynamics Department of Applied Mechanics, Chalmers University of Technology; 2011.

[64] Reza Taghavi Z, Salary M, Kolaei A. Prediction of boundary layer transition based on modeling of laminar fluctuations using RANS approach. Chinese Journal of Aeronautics. 2009;**22**:113-120.

[65] Mayle RE. The role of laminar-turbulent transition in gas turbine engines. ASME Journal of Turbomachinery. 1991;**113**: 509-537.

[66] Sieverding CH. Recent Progress in the Understanding of Basic Aspects of Secondary Flows in Turbine Blade Passages, ASME Paper No. 84-GT-78; 1984.

[67] Gaster M. The structure and behavior of laminar separation bubbles. ARC R&M Report No. 3595; 1969.

[68] Hodson HP. Boundary layer and loss measurements on the rotor of an axial-flow turbine. ASME Journal of Engineering for Gas Turbines and Power. 1984;**106**: 391-399.

[69] Warren JM, Metzger DE. Heat transfer with film cooling in the presence of laminarizing mainstream. ASME Paper 72-HT-11; 1972.

[70] Blair MF. Influence of free-stream turbulence on boundary layer transition in favorable pressure gradients. ASME Engineering for Gas Turbines and Power. 1982;**104**:743-750.

[71] Sharma OP. Momentum and thermal boundary layer development on turbine airfoil suction surfaces. AIAA Paper No. 87-1918;1987.

[72] Taniguchi H, Sakai H, Funazaki K. Effects of free-stream turbulence on bypass transition of separated boundary layer on low-pressure turbine airfoils. In: Proceedings of 3rd Asian Joint Workshop on Thermophysics and Fluid Science, Matsue, Japan; 2010.

[73] Bons JP, McClain ST, Taylor RP, Rivir RB. The many faces of turbine surface roughness. ASME Journal of Turbomachinery. 2001;**123**:739-748.

[74] Taylor RP. Surface roughness measurements on gas turbine blades. ASME Journal of Turbomachinery. 1990;**112**:175-180.

[75] Kind RJ, Serjak PJ, Abbott MWP. Measurements and prediction of the effects of surface roughness on profile losses and deviation in a turbine cascade. ASME Journal of Turbomachinery. 1998;**120**:20-27.

[76] Boynton JL, Tabibzadeh R, Hudson ST. Investigation of rotor blade roughness effects on turbine performance. ASME Journal of Turbomachinery. 1992;**115**: 614-620.

[77] Suder KL, Chima RV, Strazisar AJ, Roberts WB. The effect of adding roughness and thickness to a transonic axial compressor rotor. ASME Journal of Turbomachinery. 1995;**117**: 491-505.

[78] Bons JP, McClain ST. The effect of real turbine roughness with pressure gradient on heat transfer. ASME Journal of Turbomachinery. 2003;**126**:333-441.

[79] Blair MF. An experimental study of heat transfer in a large scale turbine rotor passage. ASME Journal of Turbomachinery. 1994;**116**:1-13.

[80] Wang T, Rice MC. Effect of elevated free-stream turbulence on transitional heat transfer over dual-scaled rough surfaces. ASME GT2003-38835; 2003.

[81] Götler H. Über eine dreidimensionale Instabiltät laminar Grenzschichten an konkaven Wänden. Nachr. Wiss. Ges. Göttingen, Math Physics Klasse, Neue Folge 1940;**2**(1); see also ZAMM, 1941;**21**; 250-252.

[82] Liepmann HW. Investigations on laminar boundary layer stability and transition on curved boundaries. NACA ACR 3H30 (NACA-WR-W-107); 1943.

[83] Martin BW, Brown A. Factors influencing heat transfer to the pressure surfaces of gas turbine blades. International Journal of Heat and Fluid Flow. 1979;**1**:107-119.

[84] Morkovin MV. On the Many Faces of Transition. In: Wells CS, editor. Viscous Drag Reduction. New York: Plenum Press. 1969. pp. 1-31.

[85] Colban W, Gratton A, Thole KA, Haendler M. Heat transfer and film-cooling measurements on a stator vane with fan-shaped cooling holes. ASME Journal of Turbomachinery. 2006;**128**: 53-61.

[86] Colban W, Thole KA, Haendler M. Experimental and computational comparisons of fan-shaped film cooling on a turbine vane surface. ASME Journal of Turbomachinery. 2007;**129**:23-31.

[87] Elsner W. Transition modelling in turbomachinery. Journal of Theoretical and Applied Mechanics. 2007;**45**(3):539-556.

[88] Elsner W, Vilmin S, Drobniak S, Piotrowski W. Experimental analysis and prediction of wake-induced transition in turbomachinery. ASME Paper no. GT2004-53757, ASME TURBO EXPO 2004, Vienna.

[89] http://documentslide.com/documents/1-intro-turbulent-flows.html

[90] Sharif MAR, Mothe KK. Evaluation of turbulence models in the prediction of heat transfer due to slot jet impingement numerical heat transfer, part B: Fundamentals. International Journal of Computation and Methodology, 2009;**55**:(4):273-294.

[91] Ali ARA, Janajreh I. Numerical simulation of turbine blade cooling via jet impingement. Energy Procedia. 2015;**75**: 3220-3229.

[92] Bredberg, J. Turbulence Modelling for Internal Cooling of Gas-Turbine Blades. Goteborg, Sweden: Department of Thermo and Fluid Dynamics, Chalmers University of Technology; 2002.

[93] Acharya S, Dutta S, Myrum TA, Baker RS. Periodically developed flow and heat transfer in a ribbed duct. International Journal of Heat and Mass Transfer. 1993;**36**:2069-2082.

[94] Isman MK, Pulat E, Etemoglu AB, Can M. Numerical investigation of turbulent impinging jet cooling of a constant heat flux surface. Numerical Heat Transfer, Part A. 2008;**53**:1109-1132.

[95] Simoneau RJ, Simon FF. Progress towards understanding and predicting heat transfer in the turbine gas path. International Journal of Heat Fluid Flow. 1992;**14** (2):106-128.

[96] Stephens MA. Computation of flow and heat transfer in a rectangular channel with ribs. AIAA Paper 95-0180; 1995.

[97] Gatski TB, Speziale CG. On explicit algebraic stress models for complex turbulent flows. Journal of Fluid Mechanics. 1993;**254**:59-78.

[98] Speziale C. Comparison of explicit and traditional algebraic stress models of turbulence. AIAA Journal. 1997;**35** (9):1506-1509.

[99] Iacovides H, Raisee M. Recent progress in the computation of flow and heat transfer in internal cooling passages of turbine blades. International Journal of Heat Fluid Flow. 1999;**20**:320-328.

[100] Durbin PA. Separated flow computations with the k–f– v^2 model. AIAA Journal. 1995; **33**:659-664.

[101] Parneix S, Durbin PA, Behnia M. Computation of 3-D turbulent boundary layers using the v^2-f model. Flow, Turbulence Combustion. 1998;**60**:19-46.

[102] Behnia M, Parneix S, Durbin P. Prediction of heat transfer in an axisymmetric turbulent jet impinging on a flat plate. International Journal of Heat Mass Transfer. 1998;**41**:1845-1855.

[103] Behnia M, Parneix S, Shabany Y, Durbin PA. Numerical study of turbulent heat transfer in confined and unconfined impinging jets. International Journal of Heat Fluid Flow. 1999;**20**:1-9.

[104] Metzger DE, Bunker RS, Chyu RK. Cavity heat transfer on a transverse grooved wall in a narrow flow channel. Journal of Heat Transfer. 1989;**111**:73-79.

[105] Spalart PR, Allmaras SR. A one-equation turbulence model for aerodynamic flows. AIAA;92-0439; 1992.

[106] Bardina JE, Huang PG, Coakley TJ. Turbulence modeling validation, testing, and development. NASA TM110446; 1997.

[107] Horton HP. A Semi-Empirical Theory for the Growth and Bursting of Laminar Separation Bubbles. ARC CP 1073; 1969.

[108] Roberts WB. Calculation of laminar separation bubbles and their effects on airfoil performances. AIAA Journal. 1980;**18**: 25-31.

[109] Lakshminarayana B. An assessment of computational fluid dynamic techniques in the analysis and design of turbomachinery: - The 1990 freeman scholar lecture. ASME Journal of Fluids Engineering. 1991;**113**: 315-352.

[110] Simoneau RJ, Simon FF. Progress towards understanding and predicting heat transfer in the turbine gas path. International Journal of Heat Fluid Flow. 1993;**14**:106-128.

[111] Simon FF, Ashpis DE. Progress in Modeling of Laminar to Turbulent Transition on Turbine Vanes and Blades. NASA Technical Memorandum No. 107180; 1996.

[112] Dunn MG. Convective Heat Transfer and Aerodynamics in Axial Flow Turbines. ASME Paper No. 2001-GT-0506; 2001.

[113] Yaras MI. Measurements of the Effects of Freestream Turbulence on Separation Bubble Transition. ASME Paper No. GT-2002-30232; 2002.

[114] Keller FJ, Wang T. Flow and Heat Transfer Behavior in Transitional Boundary Layers with Streamwise Acceleration. ASME 94-GT-24; 1994.

[115] Kestoras MD, Simon TW. Effect of Free-Stream Turbulence Intensity on a Boundary Layer Recovering from Concave Curvature Effects. ASME 93-GT-25; 1993.

[116] Peerhossaini, H, Wesfreid JE. Experimental study of the Görtler-Taylor instability. In: Wesfreid JE, Brand H, Mannville P, Albinet G, Boccara N, editors. Propagation in Systems for From Equilibrium. Berlin: Springer; 1988. pp. 399-412.

[117] van Treuren KW, Simon T, von Koller M, Byerley AR, Baughn JW, Rivir R. Measurements in a turbine cascade flow under ultra low Reynolds number conditions. Journal of Turbomachinery, Transaction of ASME. 2002;**124**:100-106.

[118] Abu-Ghannam BJ, Shaw R. Natural transition of boundary layers-the effects of turbulence, pressure gradient, and flow history. Journal of Mechanical Engineering Science. 1980;**22**(5):213-228.

[119] Ilieva GI. Modelling, Research and Analysis of 3D Real Flow in Turbine Stages with Complex Geometry [Dissertation Thesis]. Varna: Technical University, 2009.

[120] Ilieva G. Numerical Modeling and Research of 3D Turbine Stage. Engineering Applications of Computational Fluid Dynamics. (Series of Advanced Structured Materials) 2015;**44**:103-126.

[121] Buntić Ogor I, Dietze S, Ruprecht A. Numerical Simulation of the Flow in Turbine-99 Draft Tube. In: Proceedings of the third IAHR/ERCOFTAC Workshop on Draft Tube Flow; 8-9 December 2005; Porjus, Sweden.

[122] Menter F. Turbulence Modeling for Turbomachinery Applications, Lucerne: QNET-CFD Meeting; 2002.

[123] Villalpando F, Reggio M, Ilinca A. Assessment of turbulence models for flow simulation around a wind turbine airfoil. Modelling and Simulation in Engineering. 2011;**2011**, Article ID 714146, 8 pages http://dx.doi.org/10.1155/2011/714146

[124] Ilieva GI, Iosifov RD. Geometry modeling features for 3D turbine cascade with twisted rotor blade in GAMBIT. Acta Universitatis Pontica Euxinus. 2005;**5**(2):7-12 2005; ISSN 1312-1669.

[125] Menter FR. Two-equation eddy-viscosity turbulence models for engineering applications. AIAA Journal. 1994;**32** (8):1598-1605.

[126] Menter FR. Zonal Two Equation k-w Turbulence Models for Aerodynamic Flows. AIAA Paper #93-2906, 24th Fluid Dynamics Conference; July 1993.

[127] Jošt D, et al. Numerical flow simulation and efficiency prediction for axialturbines by advanced turbulence models. In 2012 IOP Conference Series: Earth and Environmental Science. 2012. 15 062016 doi:10.1088/1755-1315/15/6/062016

[128] Cleak JGE, Gregory-Smith DG. Turbulence modeling for secondary flow prediction in a turbine cascade. Journal of Turbomachinery. 1992;**114**(3):590-598. doi:10.1115/1.2929183

[129] Savill AM. Some Recent Progress in the Turbulence Modelling of Bypass Transition. In: So RMC, Speziale CG, Launder BE, editors. Near-Wall Turbulent Flows. Elsevier Science Publishers B.V.; 1993. pp. 829-848.

[130] Savill AM. One-Point Closures Applied to Transition: Turbulence and Transition Modelling. Dordrecht: Kluwer Academic Publisher; 1996. pp. 233-268.

[131] Suzen YB, Huang PG, Hultgren, LS, et al. Predictions of separated and transitional boundary layers under low-pressure turbine airfoil conditions using an intermittency transport equation. Journal of Turbomachinery. 2003:**125**(3):455-464.

[132] Reynolds O. An experimental investigation of the circumstances which determine whether the motion of water shall be direct or sinuous, and of the law of resistance in parallel channels. Philosophical Transactions of the Royal Society of London A. 1883;**174**:935-982.

Chapter 7

RANS Modelling of Turbulence in Combustors

Lei-Yong Jiang

Additional information is available at the end of the chapter

http://dx.doi.org/10.5772/intechopen.68361

Abstract

Turbulence modelling is a major issue, affecting the precision of current numerical simulations, particularly for reacting flows. The RANS (Reynolds-averaged Navier-Stokes) modelling of turbulence is necessary in the development of advanced combustion systems in the foreseeable future. Therefore, it is important to understand advantages and limitations of these models. In this chapter, six widely used RANS turbulence models are discussed and validated against a comprehensive experimental database from a model combustor. The results indicate that all six models can catch the flow features; however, various degrees of agreement with the experimental data are found. The Reynolds stress model (RSM) gives the best performance, and the Rk-ε model can provide similar predictions as those from the RSM. The Reynolds analogy used in almost all turbulent reacting flow simulations is also assessed in this chapter and validated against the experimental data. It is found that the turbulent Prandtl/Schmidt number has a significant effect on the temperature field in the combustor. In contrast, its effect on the velocity field is insignificant in the range considered (0.2–0.85). For the present configuration and operating conditions, the optimal turbulent Prandtl/Schmidt number is 0.5, lower than the traditionally used value of 0.6–0.85.

Keywords: turbulence modelling, RANS, momentum and scalar modelling, combustor

1. Introduction

Turbulence is one of the principal unsolved problems in physics today [1] and its modelling is one of the major issues that affect the precision of current numerical simulations in engineering applications, particularly for reacting flows. Turbulence is characterized by irregularity or randomness, diffusion, vortices and viscous dissipation, and involves a wide range of time and length scales. Despite the rapid development of computing power, large eddy simulations are limited to benchmark cases with relatively simple geometries, while direct numerical simulations of turbulent flows remain practical only at low Reynolds numbers [2–4]. It is particularly true for turbulent reacting flows. Even without turbulence, combustion is a complicated process and can consist of hundreds of species and thousands of element reactions, where numerical difficulties occur [5]. Consequently, it is necessary to utilize turbulence models for the development of advanced combustion systems in the foreseeable future.

Much effort has been made to the development of turbulence modelling in the last six decades. Advances focused on constant density flows have been reviewed or described by a number of researchers [6–8] and brought up to date for the second momentum closure in reacting flows [9]. Various algebraic, one- and two-equation turbulence models were systematically evaluated [2, 3] against a number of well-documented non-reacting flows, including free-shear, boundary-layer and separated flows. Some guidelines regarding applications of these models were provided. Recently, six eddy-viscosity and two variants of Reynolds stress turbulence models were used to study the flow field around a ship hull [10]. It was found that the two Reynolds stress models were able to reproduce all the salient features and the predicted Reynolds stresses and turbulence kinetic energy were in good agreement with the experimental results. Despite the considerable progresse in turbulence modelling, no universal turbulence model is available for all flows at the current time. Therefore, it is important to understand advantages and shortcomings of these models and select the best one for defined engineering problems.

In contrast to turbulence momentum transfer modelling, studies on turbulence scalar transfer modelling are limited, but are of great engineering interest. From the 1970s (CFD pioneer work) up to present, in almost all turbulent reacting flow simulations, the Reynolds analogy concept has been used to model turbulence scalar transfers (mixture fraction, species and energy or temperature). In this approach, the turbulent Prandtl (Prt) and Schmidt (Sct) numbers are used to link the turbulence scalar transfers to the momentum transfer that is calculated by a selected turbulence model. The main advantage of this approach is that the turbulence scalar transfers can be effectively computed from the modelled momentum transfer without solving a full-second moment closure for both momentum and scalar transportations.

The Reynolds analogy concept was first postulated over a century ago on the similarity between wall shear and heat flux in boundary layers [11]. This original hypothesis has been considerably amended and applied to general 3D (three-dimensional) turbulent heat and species transfers [12, 13]. Recently, its applications to high-Mach-number boundary layers [14], turbine flows [15] and film cooling [16] have been studied.

In most turbulent reacting or mixing flow simulations, it has become a common practice to set Prt = Sct [17]. Traditionally, a constant value of Prt = Sct = 0.6–0.85 has been used in jet and gas turbine flows [18–21], and these values are consistent with numerous measurements performed in the 1930s–1980s [12, 13]. However, low Prt and Sct numbers from 0.20 to 0.5 have been used by a large number of authors for simulating gas turbine combustors. Effort was made to validate a two-dimensional finite difference code against a number of isothermal and reacting flow measurements [22], and a value of Prt = Sct = 0.5 was recommended for recirculation zone simulations. The numerical results of five RQL (rich burn, quick quench and lean burn) low-emission combustor designs were calibrated against CARS (coherent anti-Stokes

http://dx.doi.org/10.5772/intechopen.68361

Raman spectroscopy) temperature measurements, and good agreement was found by using Prt = Sct = 0.2 [23]. An entire combustor from the compressor diffuser exit to the turbine inlet was successfully studied, and a low value of 0.25 was used for Prt and Sct since it consistently demonstrated better agreement with the fuel/air mixing results [24]. Moreover, the turbulence scalar mixing of a gaseous jet issued into a cross airflow was investigated, and in comparison with the available experimental data, Prt = Sct = 0.2 was recommended [25]. The above examples suggest that for reacting flow modelling, the scalar transfer modelling or Reynolds analogy has to be investigated.

Although there are a large number of publications on numerical simulations of practical combustion systems [for example, 20–24, 26–29], systematic assessment and validation of turbulence models in combustor flow fields against well-defined comprehensive experimental results are rare.

This chapter focuses on most widely used turbulence models in practical engineering, that is RANS (Reynolds-averaged Navier-Stokes) models, including the Reynolds stress model (RSM), a second moment closure and five popular two-equation eddy-viscosity models, the standard k-ε, renormalization group (RNG) k-ε, realizable k-ε (Rk-ε), standard k-ω and shear-stress transport (SST) k-ω model. The contents are based on the author's experience and publications accumulated over many years on turbulent reacting flow studies, related to gas turbine combustion systems [30–36].

A benchmark case, a model combustor, is used as technology demonstration. Although the model combustor geometry is simple, the complex phenomena, such as jet flows, wall boundary layers, shear layers, flow separations and reattachments, as well as recirculation zones, are involved, which are fundamental features in practical combustion systems. In addition, because the model combustor geometry is much simpler than practical combustors, its boundary conditions can be well defined. More importantly, a comprehensive experimental database is available, and then the assessment of the above issues is appropriate.

In the following sections, firstly, the governing equations, turbulence models and Reynolds analogy are discussed and then the other physical models and experimental measurements are briefly described, followed by the benchmark results. Finally, a few conclusions are highlighted.

2. Governing equations, turbulence models and Reynolds analogy

2.1. Governing equations

The first-moment Favre-averaged conservation equations for mass, momentum, species, mixture fraction and total enthalpy may be expressed in a coordinate-free form as [37–39]

$$\nabla \cdot (\overline{\rho}\, \tilde{\mathbf{V}}) = 0 \tag{1}$$

$$\nabla \cdot (\overline{\rho}\, \tilde{\mathbf{V}}\tilde{\mathbf{V}}) = -\nabla\, \overline{p} + \nabla \cdot \mathbf{T} - \nabla \cdot (\overline{\rho\, \mathbf{v}''\mathbf{v}''}) \tag{2}$$

$$\nabla \cdot (\overline{\rho}\, \tilde{\mathbf{V}} \tilde{Y}) = \nabla \cdot (\rho D_i \nabla \tilde{Y}_i) - \nabla \cdot (\overline{\rho\, \mathbf{v}'' Y''_i}) + \omega_i \tag{3}$$

$$\nabla \cdot (\overline{\rho}\, \tilde{\mathbf{V}} \tilde{f}) = \nabla \cdot (\rho D \nabla \tilde{f}) - \nabla \cdot (\overline{\rho\, \mathbf{v}'' f''}) \tag{4}$$

$$\nabla \cdot (\overline{\rho}\, \tilde{\mathbf{V}} \tilde{h}) = \nabla \cdot \left(\frac{\mu}{Pr} \nabla \tilde{h}\right) - \nabla \cdot (\overline{\rho\, \mathbf{v}'' h''}) + S_h \tag{5}$$

where $\tilde{\mathbf{V}}$ stands for the mean velocity vector, $\overline{\rho}$ represents the mean density, $\mathbf{v}''$ is the fluctuation velocity vector, $\mathbf{T} = \mu[\nabla \tilde{\mathbf{V}} + (\nabla \tilde{\mathbf{V}})^T] - \frac{2}{3}\mu \nabla \cdot \tilde{\mathbf{V}}\, \mathbf{I}$ represents the viscous stress tensor with $\mathbf{I}$ a unit tensor and μ molecular viscosity, Y_i stands for the mass fraction of the ith species, f denotes the mixture fraction, $\overline{p}$ is the pressure, h stands for the total enthalpy, $\overline{\rho\, \mathbf{v}'' \mathbf{v}''}$ is the Reynolds stresses and D and Pr are molecular diffusivity and Prandtl number, respectively. Note that in all equations, the symbols with straight overbars are time-averaged variables, and the symbols with curly overbars stand for Favre-averaged variables.

The source terms in Eqs. (2)–(5) should be modelled or determined in order to close these equations. A combustion model is chosen to compute the species source term, ω_i in Eq. (3). In Eq. (5), the energy source term, S_H contains radiation heat transfer and viscous heating. Turbulence momentum transfer or Reynolds stresses, $\overline{\rho\, \mathbf{v}'' \mathbf{v}''}$, in Eq. (2) are calculated by a selected turbulence model, whilst turbulence scalar transfers in Eqs. (3)–(5), $\overline{\rho\, \mathbf{v}'' Y''}$, $\overline{\rho\, \mathbf{v}'' f''}$ and $\overline{\rho\, \mathbf{v}'' h''}$ are computed by means of Reynolds analogy. That is

$$-\nabla \cdot \overline{\rho\, \mathbf{v}'' \phi''} \cong \nabla \cdot \left(\frac{\mu_t}{\Gamma t} \nabla \tilde{\phi}\right) \tag{6}$$

where ϕ stands for species, mixture fraction or enthalpy, μ_t is the turbulence viscosity that is computed from the chosen turbulence model and Γt represents the turbulence Prandtl (Prt) or Schmidt (Sct) number. Note that in Eq. (6), the turbulence scalar transfer coefficients, $\mu_t/\Gamma t$, are simply the products of the turbulence momentum transfer coefficient (μ_t) and $1/\Gamma t$.

2.2. Turbulence models

The main features of these turbulence models are outlined here, and detailed description and formation of each model can be found in the references mentioned below. The Boussinesq hypothesis is utilized to model Reynolds stresses for the five two-equation eddy-viscosity turbulence models,

$$-\overline{\rho}\; \overline{u_i'' u_j''} = \mu_t \left(\frac{\partial U_i}{\partial x_j} + \frac{\partial U_j}{\partial x_i}\right) - \frac{2}{3}\delta_{ij}\left(\mu_t \frac{\partial U_k}{\partial x_k} + \rho\, \overline{k}\right) \tag{7}$$

Using this approach, the turbulence viscosity, μ_t, for high Reynolds number flows is given by the expression for the standard k-ε, RNG, Rk-ε models,

$$\mu_t = \frac{\overline{\rho}\, C_\mu k^2}{\varepsilon} \tag{8}$$

where C_μ is a constant and the values of turbulence kinetic energy, k, and dissipation rate, ε, are calculated from a pair of transport differential equations. The detailed description of the standard k-ε turbulence model is given in Ref. [40], which represents a pioneer work in turbulence modelling.

The RNG turbulence model is originated from a re-normalization group theory [41]. The major difference between the standard k-ε and RNG models is that the coefficient of the destruction term in the turbulence dissipation rate equation is not a constant, but a function of flow mean strain rate and turbulence parameters of k and ε. Moreover, an analytical formula to account for variations of turbulent Prandtl and Schmidt numbers for the energy and species equations is provided. These modifications have made this model more responsive to the effects of strain rate and streamline curvature than the standard k-ε model. Using this model, good agreement between the numerical and experimental results is observed for the isothermal flow over a backward facing step [41].

The main improvement of the Rk-ε turbulence model is that Reynolds stresses comply with physics. That is, turbulence normal stresses always remain positive and shear stresses obey Schwarz inequality [42]. In addition, instead of a constant, C_μ in Eq. (8) is a function of the turbulence parameters and mean strain and rotation rates. The turbulence dissipation rate equation is obtained from the dynamic equation of the mean-square vorticity fluctuation at high Reynolds numbers. Some advantages have been observed with this model for flows with separations and recirculation zones, as well as jet spread rates over the standard k-ε model [42, 43].

The details of the standard k-ω and SST models are given in Refs. [2, 3, 44]. The corresponding turbulence viscosity for high Reynolds number flows is obtained by the following two expressions, respectively,

$$\mu_t = \frac{\overline{\rho}\, k}{\omega} \tag{9}$$

$$\mu_t = \frac{\overline{\rho} k}{\omega} \frac{1}{\max\left[1, \dfrac{SF}{a\omega}\right]} \tag{10}$$

In Eqs. (9) and (10), a pair of transportation differential equations is used to obtain turbulence kinetic energy, k, and specific dissipation rate, ω; F equals zero in the free-shear layer and one in the near-wall region; 'a' is a constant and S stands for the strain rate magnitude.

For the above two-equation models, the linear relationship of Reynolds stresses with mean strain rate and isotropic eddy viscosity is presumed, as implied in Eqs. (7)–(10). For turbulent flow simulations with the RSM, a transportation differential equation is solved for each Reynolds stress component in the flow field. Therefore, it is expected that this second-moment closure model is more 'applicable' than the two-equation, eddy-viscosity models. To convert Reynolds stress equations into a closed set of equations, unknown terms must be modelled by mean flow variables and/or Reynolds stresses [45].

2.3. Reynolds analogy

By reducing the three-dimensional conservation equations (2)–(5) to two-dimensional steady boundary flows and neglecting the streamwise pressure gradient, molecular viscous terms and source terms, Eqs. (11) and (12) are obtained and the rationale and limitation of Reynolds analogy can be revealed.

$$\bar{\rho}\tilde{u}\frac{\partial \tilde{u}}{\partial x} + \bar{\rho}\tilde{v}\frac{\partial \tilde{u}}{\partial y} \approx \frac{\partial}{\partial y}\left(\mu_t \frac{\partial \tilde{u}}{\partial y}\right) \tag{11}$$

$$\bar{\rho}\tilde{u}\frac{\partial \tilde{\phi}}{\partial x} + \bar{\rho}\tilde{v}\frac{\partial \tilde{\phi}}{\partial y} \approx \frac{\partial}{\partial y}\left(\frac{\mu_t}{\Gamma t} \frac{\partial \tilde{\phi}}{\partial y}\right) \tag{12}$$

In Eqs. (11) and (12), the turbulence viscosity concept is applied to both streamwise momentum and scalar transfers. With $\Gamma t = 1$, the two equations become identical. That is, under appropriate boundary conditions, the solution of all these flow parameters can be obtained from a single-partial differential equation or the momentum and scalar fields are similar. For wall boundary flows, the original form of Reynolds analogy can be deduced [15],

$$\frac{2St}{c_f} = \frac{(h/\rho\, C_p\, U_\infty)}{(\tau_w/\rho\, U_\infty^2)} \approx 1 \tag{13}$$

where $c_f = \tau_w/(0.5\rho U_\infty^2)$ represents the wall friction coefficient, and $St = h/(\rho\, C_p U_\infty)$ denotes the Stanton number. From Eq. (13), the turbulence heat transfer coefficient can be calculated from the measured pressure loss owing to friction in the flow.

The above analysis suggests that the Reynolds analogy method can be used to adequately calculate turbulence scalar transfers in a boundary type of flows, such as free jets, wall boundary layers and shear layers, where the effects of the streamwise pressure gradient, viscous and source terms are minor. However, it should be cautious to apply it to general complex three-dimensional flows. Its failure to disturbed turbulent boundary flows has been reported by a number of authors [46–48].

2.4. Other physical models and numerical methods

For combustion modelling, the laminar flamelet, probability density function (PDF) and eddy-dissipation (EDS) models were considered. The previous benchmark study on combustion models indicated that for mixing-control dominated diffusion flames the temperature and velocity fields could be fairly well captured by these three combustion models [33].

A major advantage of the flamelet model over the probability density function and eddy-dissipation models is that detailed more realistic chemical kinetics can be incorporated into turbulent reacting flows [49]. For the present case, the propane-air chemical mechanism from Ref. [50] was used. This mechanism was consisted of 228 element reactions, and 31 chemical species, i.e. O_2, N_2, CO_2, H_2O, C_3H_8, CH_4, H_2, CHO, CH_2O, CH_2CO, CH_3CO, CH_3CHO, C_2H, C_2H_2, C_2H_3, C_2H_4, C_2H_5, C_2H_6, C_3H_6, O, OH, H, H_2O_2, HO_2, CO, CH, CH_2, CH_3, C_2HO, $N^*C_3H_7$ and $I^*C_3H_7$.

The PDF combustion model is based on the mixture fraction approach with an assumption of fast chemistry [51]. It offers some advantages over the EDS or EDS-finite-rate models and allows intermediate (radical) species prediction, more thorough turbulence-chemistry coupling and dissociation effect. Eighteen species were considered for the PDF model, including C_3H_8, CO_2, H_2O, O_2, N_2, CO, HO, H, O, H_2, C_3H_6, C_2H_6, C_2H_4, CH_4, CH_3, CH_2, CH and C(s). The selection of these species was based on the basic chemical kinetics and requirements for pollutant predictions [52]. As full chemical equilibrium gave considerable errors in temperature on the rich side of hydrocarbon flames [53, 54], to avoid this, a partial equilibrium approach was applied in the rich flame region. When the instantaneous equivalence ratio exceeded 1.75, the combustion reaction was considered extinguished and unburned fuel coexisted with reacted products.

The EDS model is widely accepted in diffusion flame modelling [53]. For this model, the reaction rate is governed by turbulent mixing, or the large-eddy mixing timescale, k/ε [55]. In the present case, the heat radiation from the hot gas mixture to the surroundings was computed by the discrete ordinates model [56], where the local species mass fractions were used to calculate the absorption coefficient of gaseous mixture in the flow. At all wall boundaries, an enhanced wall boundary approach was utilized, where the traditional two-layer zonal model was improved by smoothly combining the viscous sub-layer with the fully turbulent region.

The specific heat of species was calculated by polynomials as a function of temperature. For the case of the flamelet and PDF models, the polynomials were determined from the JANAF tables [57]; while in the case of the EDS model, the polynomials from Rose and Cooper [58] were used, where the chemical dissociation was considered. For other thermal properties such as molecular viscosity, thermal conductivity and diffusivity, the values of air at 900 K were used.

A segregated solver with a second-order-accuracy scheme from a commercial software, Fluent, was used to resolve the flow fields. The results were well-converged, and the normalized residuals of the flow variables were about or less than 10^{-5} for all test cases. The axial velocities monitored in shear layers of the flow fields were unchanged for the first four digits. A LINUX cluster with eight nodes and 64-GB RAM/node was employed to perform all numerical simulations.

3. Benchmark experimental measurements

A series of experimental measurements on a diffusion flame model combustor were carried out at the National Research Council of Canada (NRCC). The results provided a comprehensive database for the evaluation and development of various physical models, including mean and fluctuation velocity components, mean temperature, wall temperature, radiation heat flux through walls, as well as species concentrations [59].

The test apparatus and model combustor are shown in **Figure 1** where all dimensions are in mm. The model combustor consists of the air and fuel inlet section, combustion chamber and contracted exhaust section. Fuel entered the combustion chamber through the centre of the bluff body, while air flowed into the chamber around a disc flame-holder. The combustor was mounted on a three-axis traversing unit with a resolution of ±100 μm. Fuel was commercial grade propane, and air was from a dry air supply. Both air and fuel flow rates were regulated by Sierra Side-Trak mass-flow controllers with 2% accuracy of the full scale (100 l/min for fuel and 2550 l/min for air).

A 25.4-mm thick fibre blanket of Al_2O_3 was wrapped around the combustion chamber in order to reduce the heat losses through the combustor walls. The optical and physical access to the combustion chamber was through four windows. The viewing area of the windows measured 17 mm in width, 342 mm in length and 44–388 mm from the disk flame-holder in the axial direction. Interchangeable sets of stainless steel and fused silica windows were used, the former for physical probing with gas sampling probes, radiometers and thermocouples, and the latter for optical probing with a laser Doppler anemometer (LDA).

Both two- and three-component LDA systems operating in a back-scattering mode were used to measure flow velocities. The restricted optical access in the lower section of the combustion chamber forced the use of a single fibre head to measure axial and tangential velocities, and a

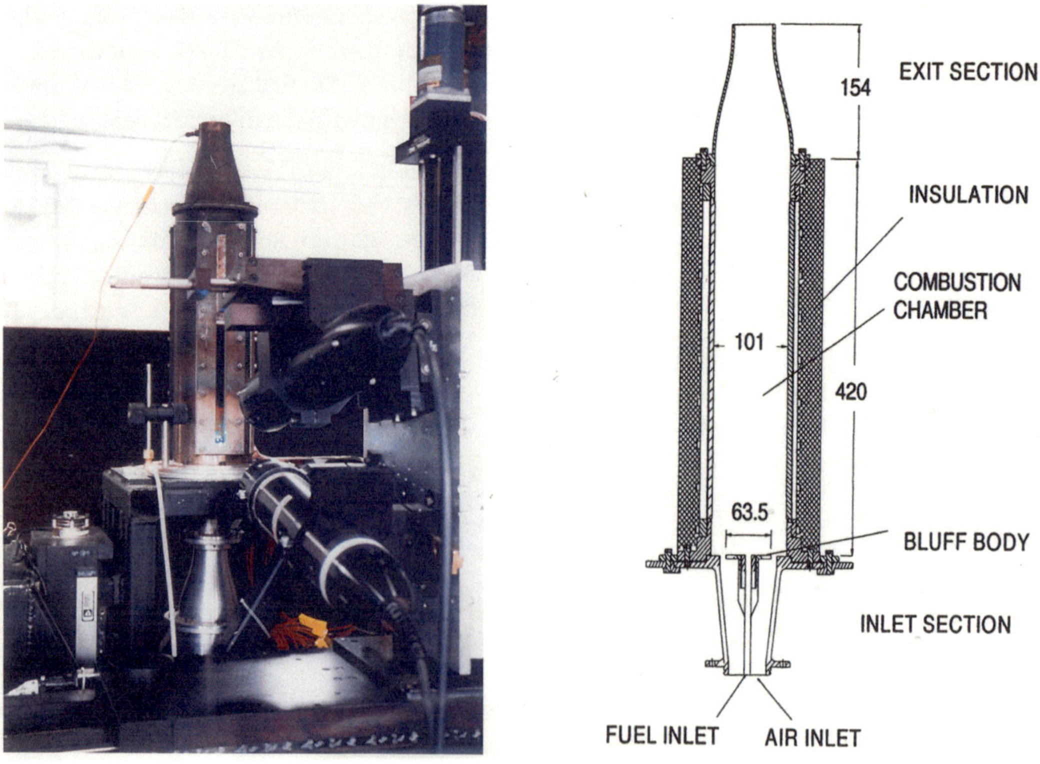

Figure 1. The test apparatus and model combustor.

complete three-component LDA system was applied in the upper section of the chamber. An uncoated 250-μm diameter, type 'S' thermocouple held by a twin-bore ceramic tube was used to measure gas temperatures in the flow field. The wall temperatures were measured by the thermocouples embedded in and flush with the wall. Gas sample was obtained by a sampling probe and the species were measured using a Varian model 3400 gas chromatograph. The measured major species were CO, CO_2, H_2O and C_3H_8. In addition, minor species fractions, such as CH_4 and C_2H_2, were also obtained. NO_x and NO were collected through the same probe but analysed using a Scintrix NO_x analyser.

4. Application of RANS turbulence models

4.1. Computational domain and boundary conditions

The computational domain covers the entire combustor flow field from the fuel and air inlets to the exhaust exit, as shown in **Figure 2**. The two-dimensional quadrilateral meshes were generated because of the axisymmetric geometry. To resolve the recirculation region, fine grids were created behind the flame-holder in the combustion chamber. Fine grids were also laid in the shear layers between the fuel and air jets and recirculation region, and the gap between the flame-holder edge and air inlet section wall as well. In the solid stainless steel wall and ceramic blanket regions, coarse grids were generated. A number of meshes were created and tested to check mesh independence issue. Finally, a mesh with 74,100 cells was used for most of the simulations. The skewness in the flow-field domain was less than 0.2 and the aspect ratio was less than 12 for 99.5% cells. Effort was made to keep the wall parameter, y^+ ($\sqrt{\tau_w/\rho}\, y \,/\, \upsilon$), in the desired range of 30–60.

The air and fuel flow rates were 550 and 16.2 g/s, respectively, and the corresponding overall equivalence ratio was 0.46. The inlet temperature for both flows was 293 K. The Reynolds number based on the flame-holder diameter and air entry velocity was 1.9×10^5. To estimate Reynolds stress components and turbulence dissipation or specific-dissipation rates at the fuel and air inlets, the turbulence intensity of 10% and hydraulic diameters were specified. The effect of the inlet turbulence intensity assignment on the flow field was examined by comparing the simulation results from three inlet turbulence intensity settings, 2, 5 and 10%. The effect is only observable for turbulence variables, for example, the change in turbulent kinetic energy was seen in the fuel inlet path and a small portion at $x \approx 80$ mm along the combustor axis, with a maximum difference of 2.3%. For the mean flow variables along the combustor axis, such as temperature and axial velocity, the deviation from the experimental data was negligibly small.

At the upstream edges of the ceramic insulation and combustion chamber and at the inlet section walls, the room temperature of 293 K was defined. Along the outer boundary of the ceramic insulation, a linear temperature profile from 294 to 405 K was assigned. The temperature of the external boundary of the contract section was set to 960 K. The same temperature was given to the downstream edge of the combustion chamber since the metal heat resistance was much smaller than the ceramic insulation. For the downstream edge of the insulation, a

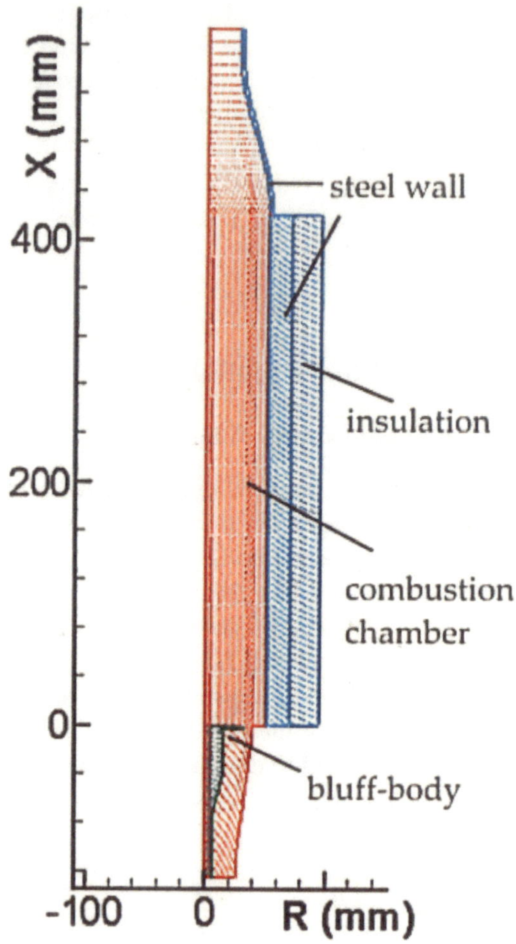

Figure 2. Computational domain.

linear temperature profile from 960 to 405 K was defined. Finally, the atmospheric pressure was set at the combustor exit.

4.2. Velocity distributions

The predicted distributions of velocity, temperature and species inside the combustor chamber are presented in the following sections. These results are obtained with the flamelet combustion model and an optimized turbulent Prandtl/Schmidt number of 0.5 (please see Section 5). The advantages and limitations of the six turbulence models can be found by comparing the numerical results with the experimental data.

The numerical results of axial velocity contours and flow path lines for six turbulence models are shown in the upper halves of six plots in **Figure 3**, respectively, while the experimental data with the zero axial velocity lines specified are displayed in the lower halves. Owing to the limited number of measured data points, no flow path lines are drawn for the experimental

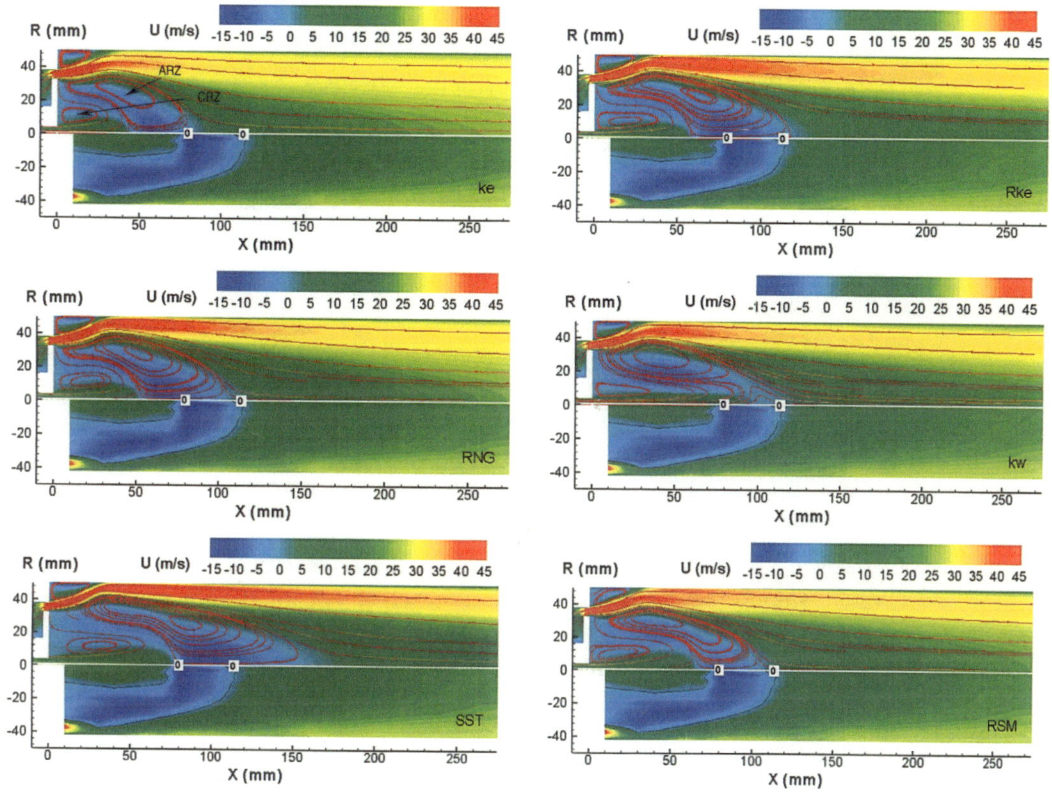

Figure 3. Axial velocity contours and flow path lines.

plots. It is noted that all models can catch the main flow features or patterns in the combustion chamber. There are two recirculation zones behind the flame-holder, i.e. a central recirculation zone (CRZ) created by the central fuel jet flow and an annular recirculation zone (ARZ) induced due to the annular air jet flow. The CRZ is completely buried inside the ARZ region, which indicates that the laminar and turbulent diffusions across the ARZ are only mechanisms for fuel transportation into the main flow field. Each recirculation zone is divided to two regions by the zero axial velocity line, and the gas mixture moves downstream in one region and moves upstream in the other. In addition, at the upper left corner of the combustion chamber, another small recirculation zone is formed due to the same reason, flow passage increases suddenly.

Various degrees of agreement with the experimental data are illustrated among the six models for predicting the reattachment points or lengths of the ARZ and CRZ. As shown in the first plot for the standard k-ε model, both ARZ and CRZ lengths are considerably under-predicted. This type of shortcoming is also pointed out by other researchers for non-reacting flow studies [60, 61]. The Rk-ε model illustrates superior performance over the standard k-ε model. It can properly predict the ARZ length and give a moderate result for the CRZ length. The RNG model underestimates the ARZ length slightly, but the CRZ length considerably. The results from the two k-ω models are poorer than those from the Rk-ε and RNG models. The k-ω model underestimates the ARZ length, while overestimates the CRZ length. In terms of the SST model, it considerably over-predicts the ARZ length though it gives a good result for the CRZ length. The RSM model, as shown in the last plot of **Figure 3**, illustrates the best performance, where both ARZ and CRZ lengths are satisfactorily provided.

A few parameters that may be valuable to combustion emission and stability studies can be obtained from the above RSM results. It is found that the gas mixture flow rate re-circulated in the ARZ is equal to 5.5% of the total inlet airflow, and the ARZ length is 1.7 times longer than the diameter of the flame-holder.

The axial velocity profiles along the combustor axis are illustrated in **Figure 4** for the six turbulence models and compared with the experimental data. As shown in the figure, the peak value of measured negative axial velocity is −10 m/s, located at $x \approx 80$ mm. The error bars representing 4% measurement accuracy are included in the figure. Considerable differences are found in the upper stream region from 10 to 80 mm for the k-ε and k-ω models, and in the downstream region from 80 to 360 mm for the k-ε and SST models. The RSM, Rk-ε and RNG models reasonably well predict the axial velocities along the central axis. Generally, the RSM gives the best results among the six turbulence models, which is consistent with the fact that only the RSM can adequately calculate both recirculation zones. For the five two-equation models, the Rk-ε illustrates better performance than the other four models.

The axial velocity profiles from $x = 20$ to 240 mm are displayed in **Figure 5** at selected seven cross-sections. Among these sections, three pass through the recirculation zones, one nearly cuts through the annular stagnation point and the remaining three are located further downstream. The features and magnitudes predicted from these models are generally close to the experimental measurements, but considerable differences are found at $x = 40 - 200$ mm sections for the k-ε model, and $x \geq 160$ mm downstream sections for the SST model. The best performance at the three upstream sections is given by the SST model because it adequately

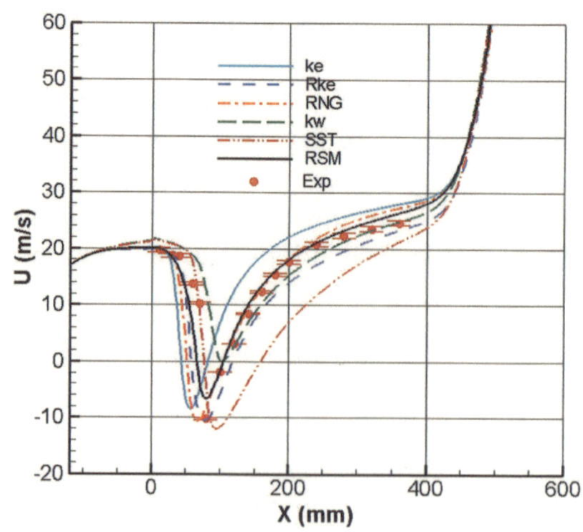

Figure 4. Axial velocities along the combustor central axis.

http://dx.doi.org/10.5772/intechopen.68361

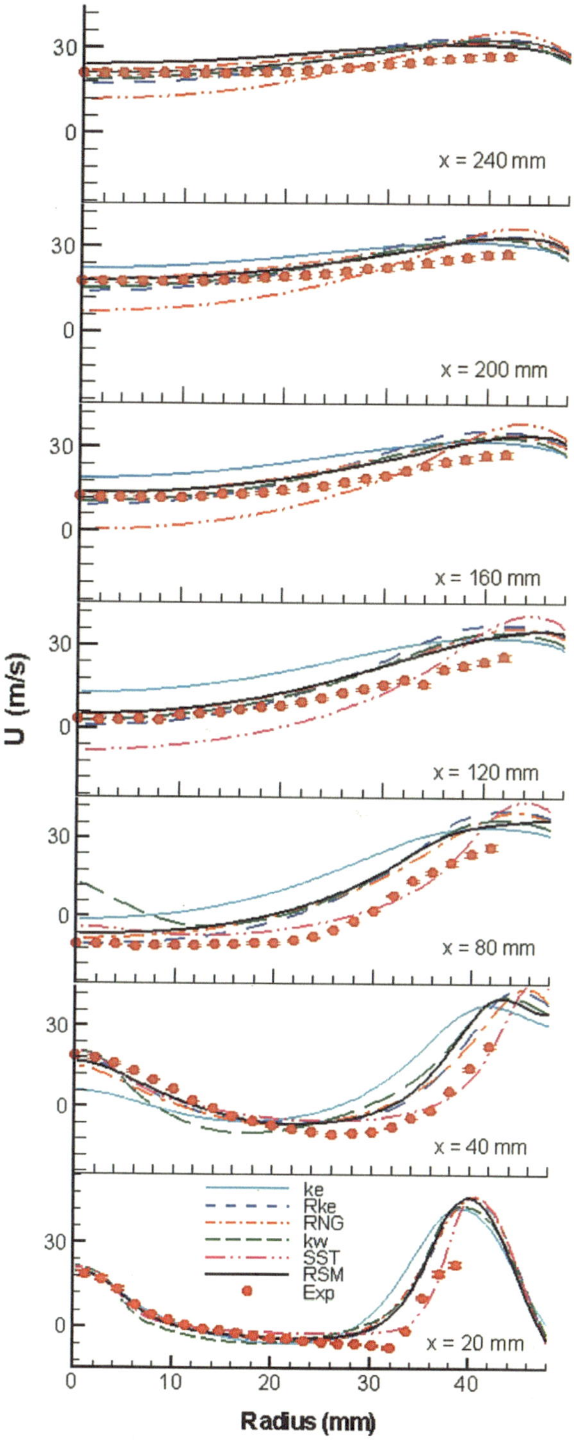

Figure 5. Axial velocity profiles at cross-sections, $x = 20$–240 mm.

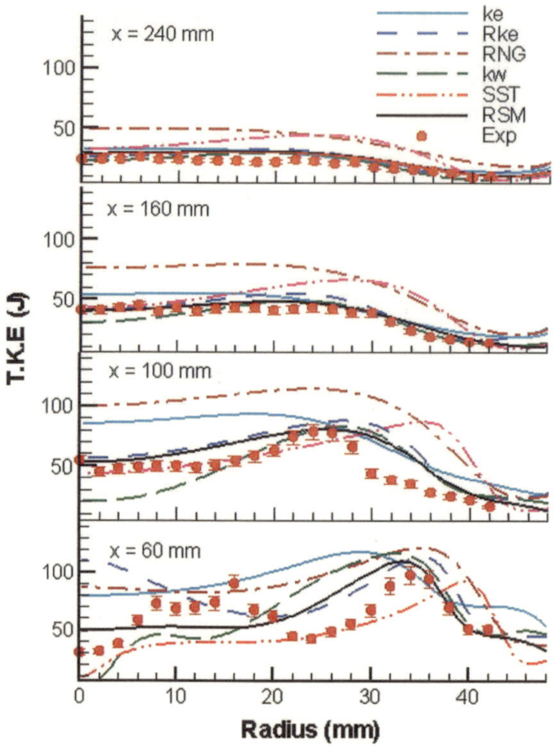

Figure 6. Turbulence kinetic energy profiles at cross-sections, $x = 60–240$ mm.

estimates the central recirculation zone, as shown in **Figure 3**. The results of the Rk-ε and RNG models are close to those obtained from the RSM. The improvement of the RSM over the Rk-ε model is observed in the regions of $x = 120–200$ mm and $R = 30–40$ mm.

The comparisons between the numerical and experimental results for the turbulence kinetic energy at four cross-sections from $x = 60$ to 240 mm are presented in **Figure 6**, and the measurement accuracy of 8% is shown by error bars. Again, only the RSM provides encouraging predictions at all sections. In addition, the promising results of three normal turbulence stresses and one shear stress from the RSM model are also obtained. The numerical results agree reasonably well with the experimental data, particularly for the shear-stress profiles. Please see Ref. [31] for detail.

In **Figure 6**, the turbulence kinetic energy is considerably overestimated by the RNG model at all sections and the k-ε model at the upstream sections. For the Rk-ε and k-ω models, except for the k-ω model at section $x = 100$ mm, reasonable agreement with the experimental data is observed. The SST model fairly well predicts the turbulence kinetic energy in the central area, and however overestimates its value away from the combustor axis. These comply again with the above observation shown in **Figure 3** that the SST model can properly calculate the CRZ, but fails to assess the ARZ. Note that since the $x = 60$ mm section crosses both recirculation zones where the two peak regions of turbulence kinetic energy are located, it represents a challenging task for numerical prediction. Unfortunately, none of these models can properly capture the central peak value.

In short, in terms of velocity flow-field prediction, the RSM is superior over the five two-equation models and in general, the Rk-ε model illustrates better performance than the other four two-equation models.

4.3. Temperature distributions

The upper halves of **Figure 7** present the temperature contour results of the six turbulence models. The stoichiometric line of the mean mixture fraction is superimposed in each plot, and it starts from the edge of the disk flame-holder, passes through the high-temperature region and ends at the combustor axis. As illustrated in the figure, the flame is ignited at the downstream end of the flame-holder edge and propagates downstream over the annular recirculation zone envelope. Along the envelope, the gaseous mixture of fuel and hot gas re-circulated from downstream mixes with the fresh air from the inlet section and the combustion takes place. During the experimental testing, a carbon deposit was observed at the disk edge of the flame-holder, which is consistent with the numerical prediction.

The comparisons of the numerical results with the experimental data in the lower halves of **Figure 7** have shown that the size and location of the high-temperature region are in good agreement with the experimental data for the RSM and Rk-ε models, while the RSM performance is a little better than the Rk-ε model. As shown in the figure, the standard k-ε and RNG models underestimate high-temperature region and the high-temperature regions are shifted upstream. On the contrary, the k-ω and SST models considerably overestimate high-temperature region, and the high-temperature regions are shifted downstream.

Figure 8 quantitatively compares the calculated temperature profiles along the combustor central axis with the experimental data, where the measurement error is about 5%. The calculated trends are close to the experimental values along the combustor central axis from 50 to 350 mm. However, in the central portion, the experimental profile is almost flat, while the predicted profiles display peak values. Overall, the better performance is given by the RSM and Rk-ε models among the six models. The k-ε and RNG models predict higher temperature than the experimental data in the upstream area and lower temperature in the downstream area. On the contrary, the k-ω and SST models underestimate the temperature in the upstream and significantly overestimate it in the downstream.

As indicated in **Figures 7** and **8**, the calculated temperatures are higher than the measured values in the centre region from $x \approx 140$ to 250 mm, and the maximum deviation is about 150 K. Three reasons are anticipated. Firstly, as mentioned earlier, the temperature was measured by thermocouples. Due to the radiation and conduction heat losses from the thermocouple, the measurement error could be higher than 100 K at the locations where the flow velocity was low and the gas temperature was high [62]. Secondly, the temperature probe could modify local flow structure, thus increases local turbulent mixing and causes an increase in local temperature [62]. Thirdly, the turbulence kinetic energy (**Figure 6**) may not be properly calculated in

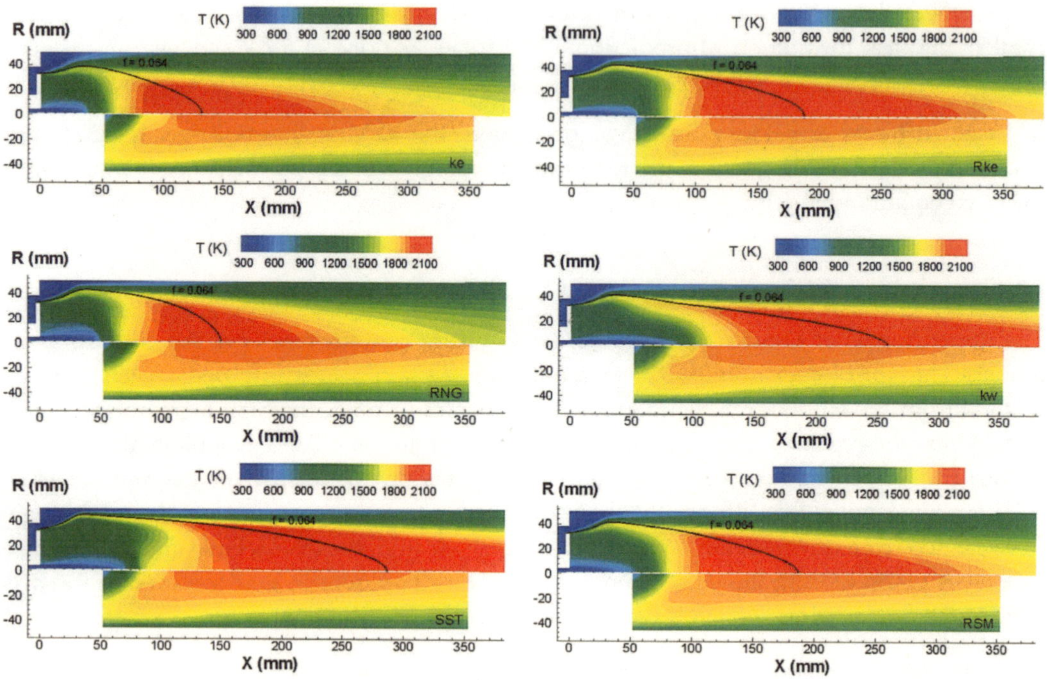

Figure 7. Temperature contours and flow path lines.

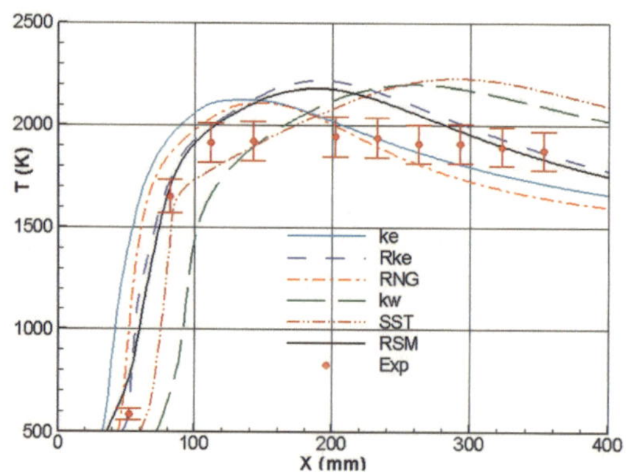

Figure 8. Temperature profiles along the combustor central axis.

the local region, and consequently the combustion and temperature calculation could be affected.

The temperature profiles at seven cross-sections, $x = 52$–353 mm are presented in **Figure 9** for the six models. For the RSM and Rk-ε models, the predicted results agree reasonably well with the experimental data, besides the most upstream section and the small local area near the combustor wall. The RNG model illustrates the similar performance at sections from $x = 82$ to 293 mm.

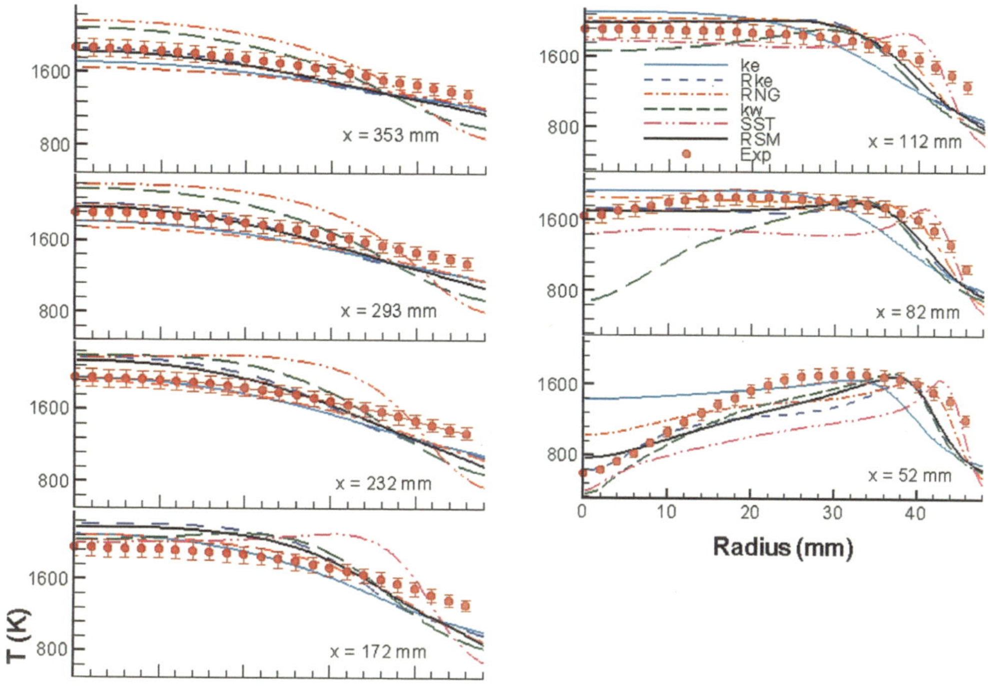

Figure 9. Temperature profiles at cross-sections, $x = 52$–353 mm.

For the other three models, poor performance is found at upstream sections from $x = 52$ to 112 mm for the k-ε model, $x = 82$, 232–353 mm sections for the k-ω model, and most sections for the SST model.

Similar to the velocity predictions observed above, in general, the predicted temperature results from the RSM and Rk-ε models fairly agree with the experimental data.

4.4. Species distributions

Figure 10 presents the CO_2 mole fraction profiles at five cross-sections, from $x = 21$ to 171 mm for six turbulence models, and compares them with the experimental data. These results are obtained with the flamelet combustion model (31 species) and the estimated error for species measurements is about 5%, as shown by error bars in the figure. The five cross-sections are chosen to show the main trends of chemical reactions, that is two passing through the central recirculation zone, two across the annular recirculation zone and the last one after the recirculation region (see **Figure 3**).

Carbon dioxide is one final major species of propane-air combustion. The Rk-ε and RSM predictions agree fairly well with the experimental data, except for the central region at $x = 111$ mm and the middle region at $x = 81$ mm, as illustrated in **Figure 10**. Poor prediction

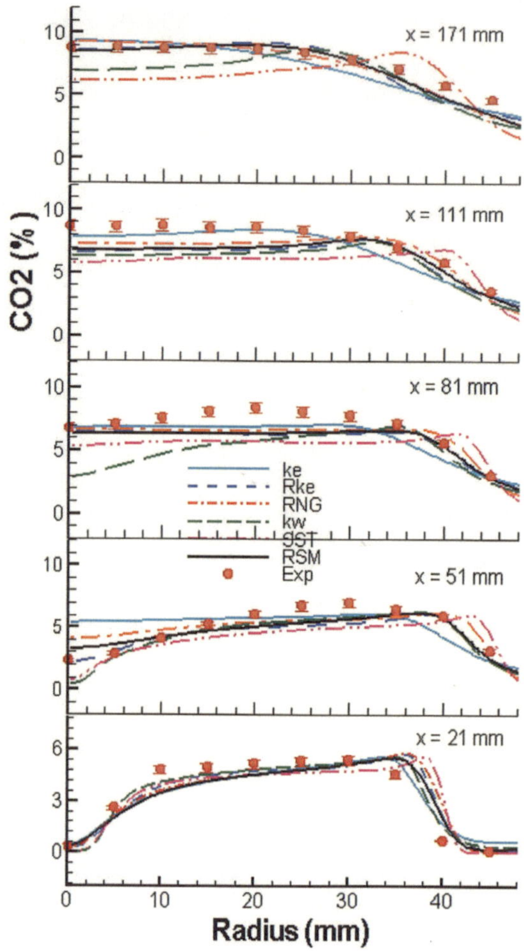

Figure 10. CO_2 profiles at cross-sections, $x = 21–291$ mm.

is found at $x = 51$ mm section for the k-ε model, $x = 81$ and 171 mm sections for the k-ω model and $x = 171$ mm section for the SST model.

Carbon monoxide, CO, is one major immediate species in hydrocarbon fuel combustion. The CO radial profiles are represented in **Figure 11**, where the numerical results are compared with the experimental data. Similar to the CO_2 case, all models properly estimate the CO profile at the most upstream section, except for the SST model showing a small bump at $R \approx 38$ mm. The RSM and Rk-ε predictions agree fairly with the experimental data at $x = 51$ and 171 mm. It is found that the two models over-predict the CO mole fraction in the central region, while under-predict the CO_2 at the two other sections, $x = 81$ and 111 mm. However, the sum of CO_2 and CO of the predicted results is close to the sum of the experimental data for CO_2 and CO. This indicates that the calculated oxidization of CO at these two sections is lag behind. Poor prediction is found again for the k-ε model at section $x = 51$ mm, k-ω model at section $x = 171$ mm and SST model at sections $x = 111$ and 171 mm.

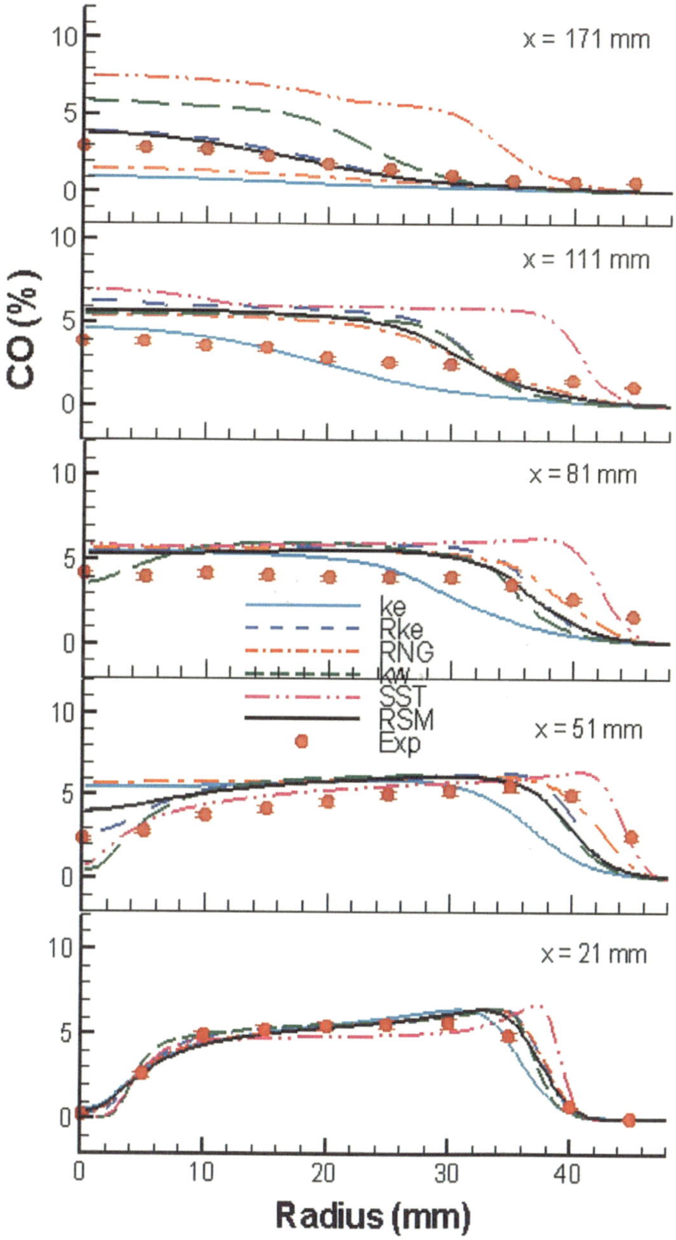

Figure 11. CO profiles at cross-sections, $x = 21–291$ mm.

In brief, the Rk-ε and RSM results are consistent with the experimental data except for some local regions, and better than the other models. The species predictions are encouraging in general.

The above qualitative and quantitative comparisons of velocity, temperature and species distributions inside the combustor between the numerical results and experimental database clearly indicate that the RSM model, a second-moment closure, is better than the eddy-viscosity models. This agrees with the findings from other authors, such as Ref. [10] for a non-reacting flow and Ref. [27] for a reacting flow. Furthermore, the Rk-ε model illustrates better performance than other four two-equation models. Instead of the RSM, utilization of the Rk-ε model for practical gas turbine combustor simulations can avoid some numerical problems, such as stability and time-consuming issues.

For the SST model, it can provide good solutions in many non-reacting flows, such as the NACA 4412 air foil, backward-facing step and adverse pressure gradient flows [44]. However, the model considerably overestimates the high-temperature region and annular recirculation zone in the combustion chamber and this type of result is also found in the simulations of a real-world gas turbine combustor [36]. Two reasons are anticipated. Firstly, the testing cases used for model validation are isothermal or almost isothermal flows, and the considerable thermal expansion and chemical reaction may not be adequately accounted for in the model [3, 44]. This may justify that the features in the central recirculation zone can be appropriately estimated by the SST model, as seen from **Figures 3**, **5** and **6**, because the temperature is low in the central recirculation zone, as seen from **Figure 7**. Secondly, multiple large vortices or recirculation regions play an important role in fuel-air mixing and combustion management in the combustion chamber, and this type of flow is more complex than single vortex flows, such as the backward-facing step flow. For example, in the present case, the whole central recirculation zone is buried inside the annular recirculation zone.

5. Application of Reynolds analogy

Most of these results are obtained with the RSM turbulence model and PDF combustion model. By comparing the numerical results with the experimental database, the Reynolds analogy approach can be assessed and the optimized turbulence Prandtl/Schmidt number for the combustor flow-fields can be identified.

5.1. Velocity distributions

The predicted axial velocity contours are illustrated in **Figure 12** for Lt = Prt = Sct = 0.85, 0.50 and 0.25, respectively, and compared with the experimental data. The flow patterns in the combustion chamber are well captured in all three plots and two reattachment points or lengths of the two recirculation zones are properly predicted.

As mentioned earlier, turbulence scalar transfers are calculated based on the modelled turbulent momentum transfer, and however the former may also affect the latter since they are coupled. As shown in **Figure 12**, the effect of Γt on velocity field for $\Gamma t = 0.85$–0.25 is minor. For $\Gamma t = 0.25$, the length and volume of the annular recirculation zone are slightly reduced in comparison with those from $\Gamma t = 0.50$ and 0.85.

The similar trends for the axial velocity profiles along the combustor central axis, and the axial velocity, turbulence kinetic energy and shear-stress profiles at a number of cross-sections are also observed [32]. These results indicate that the effect of Γt variation on the predicted velocity

http://dx.doi.org/10.5772/intechopen.68361

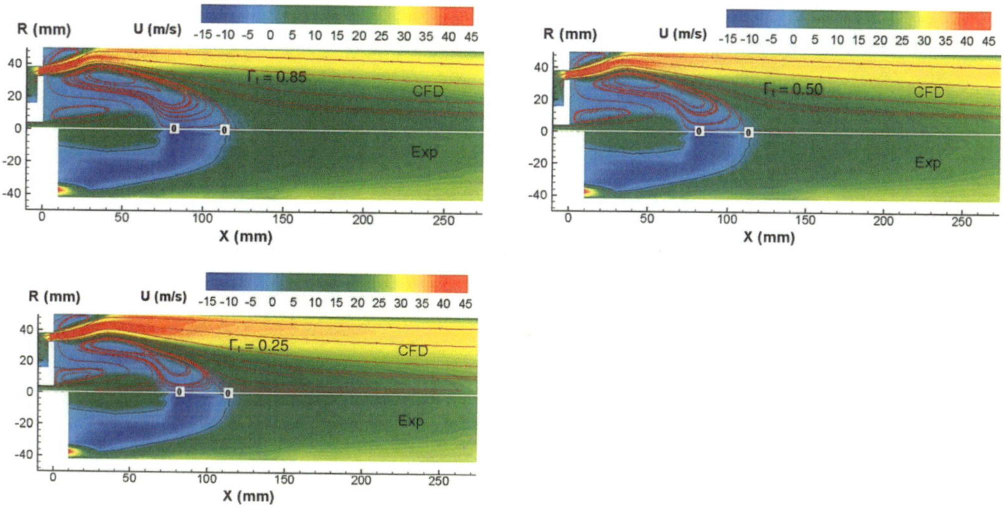

Figure 12. Axial velocity contours and flow path lines, $\Gamma t = 0.85$, 0.50 and 0.25.

field is minor, and the predicted velocity fields agree fairly well with the experimental data for $\Gamma t > 0.35$.

5.2. Temperature distributions

The temperature contours for $\Gamma t = 0.85$, 0.50 and 0.25 are presented and compared with the experimental database in **Figure 13**. Strong turbulent mixing makes the temperature in the recirculation region relatively uniform. Along the envelope of the annular recirculation zone, combustion occurs intensively. In comparison with the experimental data, for the large Γt number of 0.85, the high-temperature region is enlarged and shifted downstream, while for the small Γt number of 0.25, the high-temperature region is considerably reduced and shifted upstream. The best results are observed for $\Gamma t = 0.50$. Generally, as Γt decreases, the turbulence transfers of fuel into the airflow and then the chemical reaction are enhanced. As a result, the high temperature region is smaller and shifts upstream. **Figure 13** shows that the predicted maximum temperature in the high temperature region is higher than the measured values, which has been explained earlier.

Figure 14 illustrates the effect of Γt on the predicted flame length. In the upper plot, the stoichiometric line of the mean mixture fraction ($\tilde{f} = 0.0603$) is used to signify the flame length, and in the lower plot, the OH mole fraction contour lines are employed. The effect of Γt on the flame volume or length is clearly observed. Both the flame volume and length are substantially decreased as Γt decreases from 0.85 to 0.25, and the flame length reduces more than three times from 365 to 110 mm.

The predicted temperature profiles along the combustor central axis are compared with the experimental data in **Figure 15**. The limited effect of Γt is shown in the upstream region ($x < \sim 80$ mm); the predicted values for $\Gamma t = 0.50$ and 0.85 agree well with the measured data.

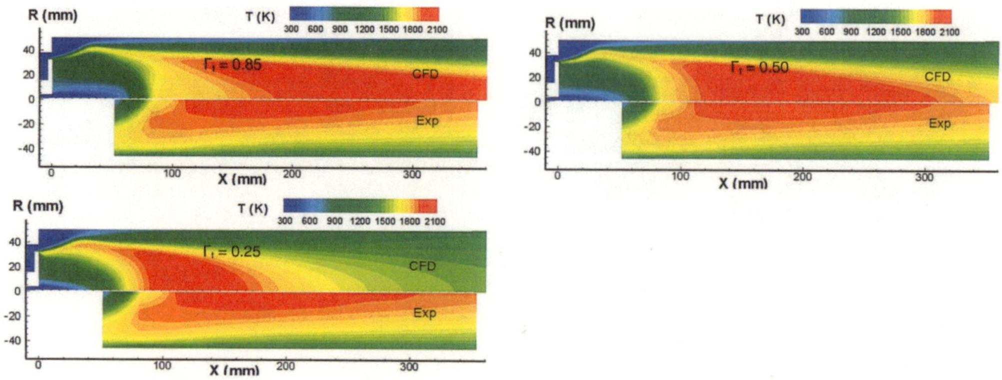

Figure 13. Temperature contours, $\Gamma t = 0.85, 0.50, 0.25$.

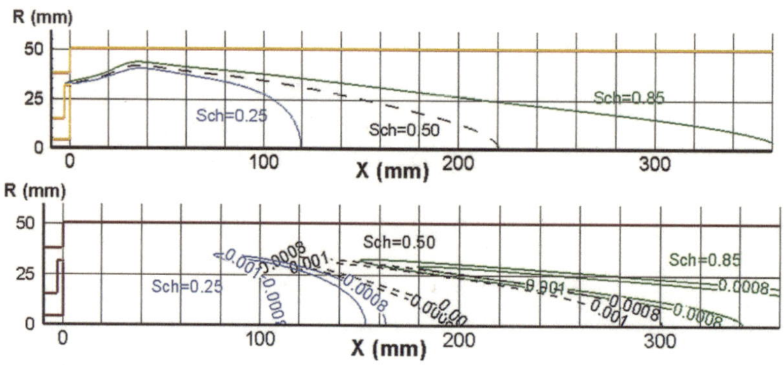

Figure 14. Variation of predicted flame length with Γt.

Conversely, the strong effect of Γt is observed in the downstream region. This is because the fuel distribution or chemical reaction in the upstream region is mainly controlled by the size and location of the annular recirculation zone, which is created by complex flow interactions among the central fuel jet, annular airflow and two recirculation zones. In other words, the flow field is dominated by convection. However, the turbulent diffusion or transfer in the downstream region plays a major role in the fuel spreading away from the axis of symmetry, as shown in **Figure 12** where the flow path lines are almost parallel to each other. As shown in **Figure 15**, $\Gamma t = 0.50$ gives the best results although the predicted profile shows a peak in the middle portion, while the measurements tend to be flat.

Figure 16 presents the temperature profiles for $\Gamma t = 0.85$, 0.50 and 0.25 at five cross-sections from $x = 52$ to 233 mm. The temperature profiles become flatter or the fuel spreading becomes faster as Γt decreases at all sections. Generally, the predicted results for $\Gamma t = 0.50$ are close to the measured data, except for the local regions around the combustor wall. In these near-wall regions, the temperature is underestimated and it may indicate the under-prediction of fuel spreading in these regions. Poor prediction is found for $\Gamma t = 0.25$ at sections $x = 82$ and 233 mm, and $\Gamma t = 0.85$ at section $x = 52$ mm. The effect of Γt is strong at all five sections.

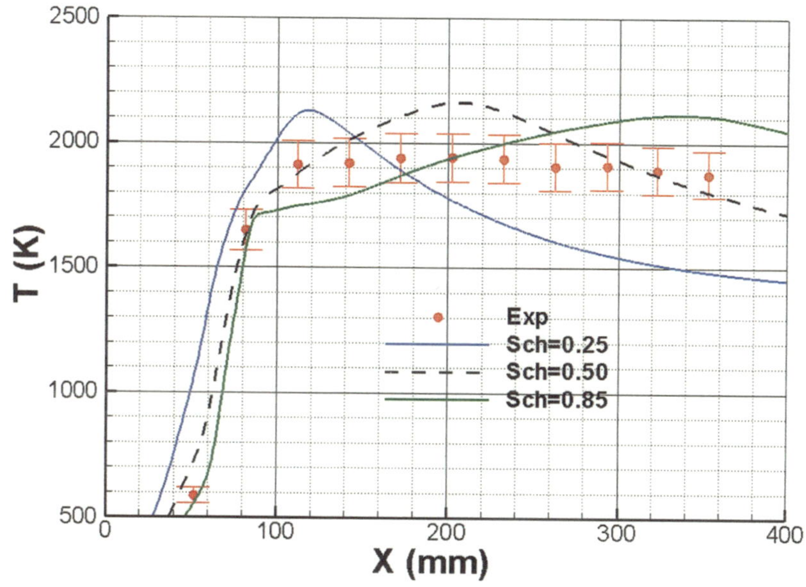

Figure 15. Temperature profiles along the combustor central axis.

5.3. Wall temperature distribution

Variation of combustor wall temperature with Γt is shown in **Figure 17**, and the numerical results are evaluated against the experimental data. Unsurprisingly, the predicted wall temperature increases as Γt decreases. The best agreement with the experimental data is obtained with $\Gamma t = 0.35$ although the wall temperature is a little overestimated in the upstream region and underestimated in the downstream region. Note that this Γt number is different from the preferred value of 0.50 for the temperature prediction inside the combustor as discussed earlier. It may indicate that instead of a constant number, varying Γt should be considered in turbulent reacting flow simulations.

In order to thoroughly assess the Reynolds analogy issue, numerical simulations were also carried out with the eddy-dissipation (EDS) and laminar flamelet combustion models. A large amount of numerical results and figures were generated, with a Γt increment of 0.05 or even 0.01. The trends and magnitudes of velocity, turbulence kinetic energy, Reynolds stresses and temperature distributions are similar to those obtained from the PDF combustion model. Although the results are not presented in this chapter, the optimized Γt numbers are given in **Table 1** for the purpose of comparison.

As shown in the above results, the optimal Γt number for the temperature prediction inside the combustor is 0.50 for all three combustion models. This number is different from 0.20 [23] and 0.25 [24] for gas turbine combustor studies, and 0.20 for a cross-jet flow simulation [25]. However, it is the same as recommended by Syed and Sturgess [22] for recirculation zone simulations.

Two important facts are revealed from the above all examples. Firstly, the Γt number optimized is smaller than the conventionally accepted value of 0.6–0.85. Secondly, most likely the

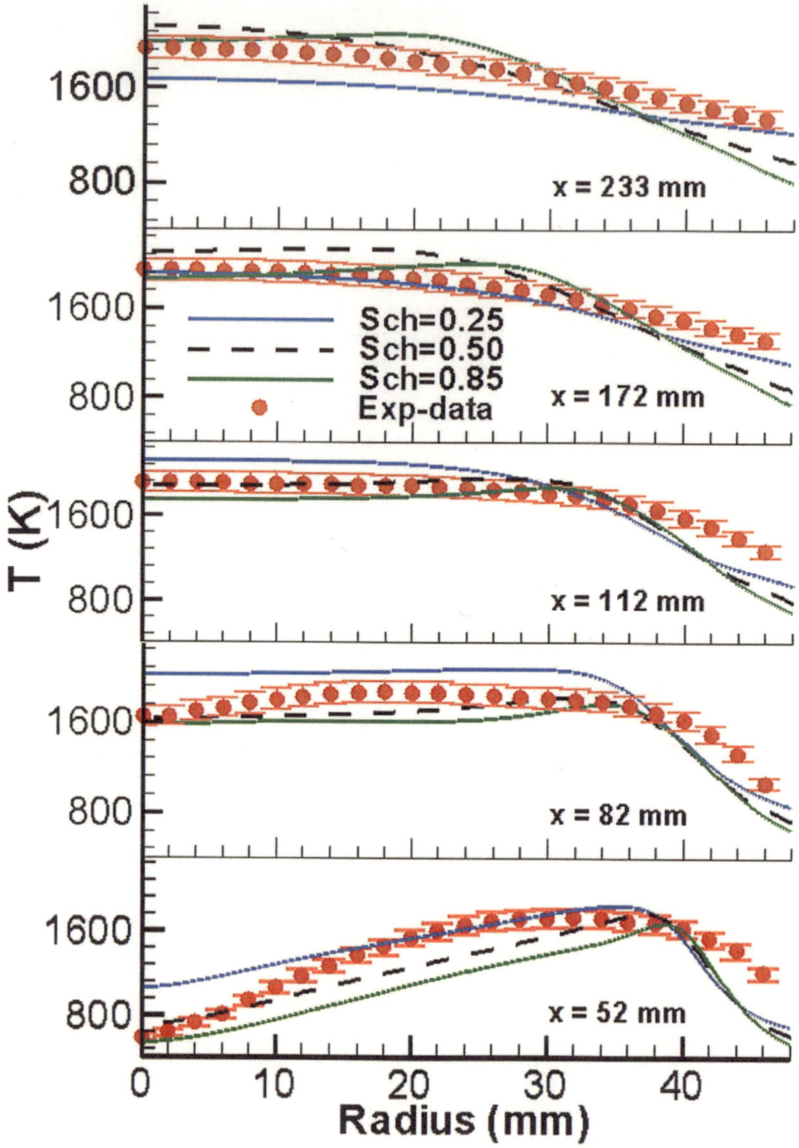

Figure 16. Temperature profiles at sections $x = 52–233$ mm.

combustor configuration and possibly the operating conditions affect the optimal value of *Γt*. Therefore, a priori optimization of *Γt* is preferred in order to confidently predict temperature and species distributions inside combustors.

These observations may be attributed to the following reasons. Firstly, theoretically, Eqs. (11)–(13) are only valid for boundary layer flows, where the streamwise pressure gradient, viscous and source terms can be neglected. Certainly, its application to complex turbulent reacting flows is questionable. Secondly, the experimental numbers of *Γt* (∼0.7) are measured from fully developed boundary or pipe flows [12, 13, 63], which are quite different from practical three-dimensional

http://dx.doi.org/10.5772/intechopen.68361

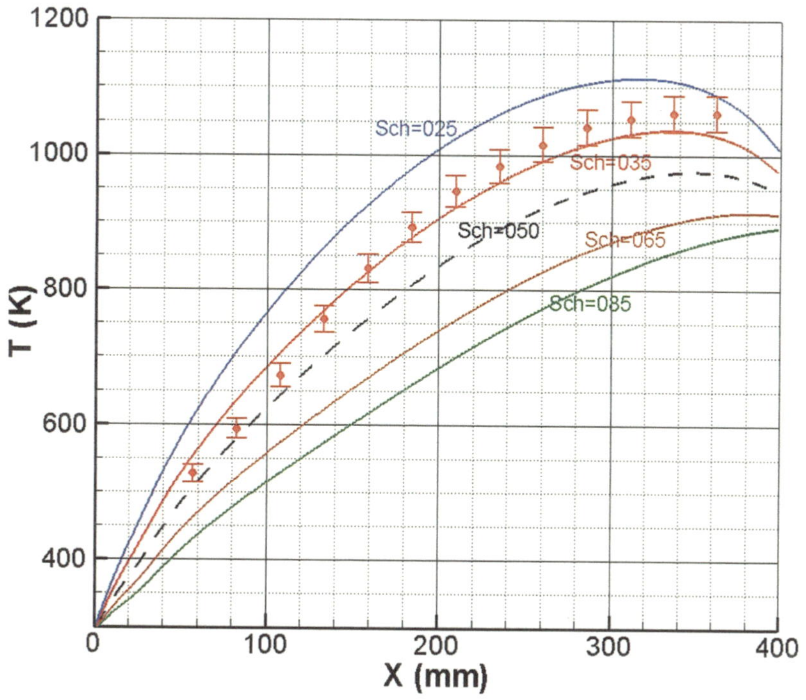

Figure 17. Temperature profiles along the combustor wall.

	Flamelet	PDF	EDS
T in chamber	0.50	0.50	0.50
T at wall	0.40	0.35	0.5

Table 1. Optimal Prandtl/Schmidt number.

turbulent reacting flows. In terms of the relative strength between the turbulence momentum and scalar transfers for the numerical methods and models used today, a low value of Γt is true, and it may change with flow configurations.

Thirdly, the gradient-type diffusion assumption used in Eqs. (11)–(13) has been questioned by a large number of researchers, in particular for turbulent energy and heat transfer. As pointed out in Ref. [12], to adequately model turbulence scalar transfers, not only the gradient-based diffusion from small-scale turbulence, but also the convection effect from large-scale turbulent motion should be taken into consideration. This may imply that the gradient-based diffusion method is appropriate for turbulent boundary flows, but it may not be proper for complex flows, such as practical combustion systems.

As a summary, the Reynolds analogy approach has been applied to general three-dimensional flow-field simulations from the 1970s. In order to accurately predict the turbulence scalar transfers without a prior optimization, the improvement of the current approach is necessary, and new ideas should be considered.

6. Conclusions

Turbulence modelling is one of major issues, which affects the precision of current numerical simulations in engineering applications, particularly for reacting flows. To systematically study and validate various physical models, a series of experimental measurements have been carried out at the National Research Council of Canada on a model combustor, and a comprehensive database has been obtained. The combustor simulations with the interior and exterior conjugate heat transfers have been carried out with six turbulence models, i.e. the standard k-ε, re-normalization group k-ε, realizable k-ε, standard k-ω, shear-stress transport (SST) and Reynolds stress models. The laminar flamelet, PDF and EDS combustion models and the discrete ordinate radiation model as well are also used.

All six turbulence models can capture the flow features or patterns; however, for the quantitative predictions of velocity, temperature and species fields, different levels of performance are revealed. The RSM model gives the best performance, and it is the only one that can accurately predict the lengths of both recirculation zones and offer reasonable prediction on the turbulence kinetic energy distribution in the combustor. In addition, the performance of the Rk-ε model is better than other four two-equation models, and it can give similar results as those from the RSM under the present configuration and operating conditions.

The effect of the turbulent Prandtl/Schmidt number on the flow field of the model combustor has also been numerically studied. In this chapter, some of the results obtained with turbulent Prandtl/Schmidt number varying from 0.85 to 0.25 have been presented and discussed. It has a strong effect on the temperature fields, particularly downstream in the combustor. This is also true for the temperature profile along the combustor wall. On the contrary, its effect on the velocity field is limited.

For all three combustion models, the optimal Γt is 0.5 for temperature prediction in the combustor, while for predicting temperature at the combustor wall the optimal value alters from 0.35 to 0.50. With $\Gamma t = 0.50$, except for some local regions, the velocity, temperature and major species fields in the combustor are fairly well simulated.

As a final point, considering the foundation and shortcoming of the Reynolds analogy, to accurately predict temperature and species distributions in turbulent reacting flow fields without an optimization of turbulent Prt and Sct numbers, the Reynolds analogy approach should be enhanced and new ideas should be considered.

Author details

Lei-Yong Jiang

Address all correspondence to: lei-yong.jiang@nrc-cnrc.gc.ca

Gas Turbine Laboratory, Aerospace, National Research Council of Canada, Ottawa, Ontario, Canada

References

[1] Tennekes H, Lumley JL. A First Course in Turbulence. 2nd ed. Massachusetts: The MIT Press; 1983

[2] Wilcox DC. Turbulence Modelling for CFD. 2nd ed. La Canada California: DCW Industries Inc.; 2002

[3] Wilcox DC. Turbulence Modelling: An Overview. AIAA paper, 2001-0724, 2001

[4] Jiang LY, Campbell I. An attempt at large eddy simulation for combustor modelling. The Open Aerospace Engineering Journal. 2012;**5**:1–10. DOI:10.1115/GT2010-22257

[5] Dagaut P, Reuillon M, Boettner JC, Cathonne M. Kerosene combustion at pressures up to 40 atm: Experimental study and detailed chemical kinetic modelling. In: 25th Symposium (Int.) on Combustion; 1994. pp. 919–926

[6] Reynolds WC, Cebeci T. Calculation of turbulent flows. Topics in Applied Physics. 1976;**12**:193–229

[7] Marvin JG. Turbulence modelling for computational aerodynamics. AIAA Journal. 1983; **21**:941–955. DOI: 10.2514/3.8182

[8] Launder BE. Second-moment closure: Present and future? International Journal of Heat and Fluid Flow. 1989;**10**:282–299. DOI: 10.1016/0142-727X(89)90017-9

[9] Jones WP. Turbulence modelling and numerical solution methods for variable density and combustion flows. In: Libby PA, Williams FA editors. Turbulent Reacting Flows. New York: Academic Press; 1994:211–304

[10] Kim SE, Rhee SH. Assessment of eight turbulence models for a three-dimensional boundary layer involving crossflow and streamwise vortices. AIAA paper 2002-0852, 2002

[11] Reynolds O. On the Extent and action of the heating surface for steam boilers. Manchester Literary and Philosophical Society. 1874;**14**:7–12

[12] Hinze JO. Turbulence. New York: The McGraw-Hill Book Company Inc.; 1987. pp. 372–753

[13] White FM. Heat and Mass Transfer. New York: Addison-Wesley Publishing Company; 1988. pp. 320–641

[14] Suraweera MV, Mee DJ, Stalker RJ. Reynolds analogy in high-enthalpy and high-Mach-number turbulent flows. AIAA Journal. 2006;**44**:917–919

[15] Bons J. A critical assessment of Reynolds analogy for turbine flows. Transactions of the ASME. Journal of Heat Transfer. 2005; **127**:472–485

[16] Jones TV, Walton TW. Reynolds analogy in film cooling. In: 10th International Symposium on Air Breathing Engines; Nottingham, England; 1991. pp. 1263–1268

[17] Libby PA, Williams FA. Fundamental aspects and a review. In: Libby PA, Williams FA, editors. Turbulent Reacting Flows. New York: Academic Press; 1994. pp. 1–61

[18] Lubbers CL, Brethouwer G, Boersma BJ. Simulation of the mixing of a passive scalar in a round turbulent jet. Fluid Dynamics Research. 2001;**28**:189–208

[19] Spalding DB. Concentration fluctuations in a round turbulent free jet. Chemical Engineering Science. 1971; **26**:95–107

[20] Bai XS, Fuchs L. Sensitivity study of turbulent reacting flow modelling in gas turbine combustors. AIAA Journal. 1995;**33**:1857–1864

[21] Lai MK. CFD analysis of liquid spray combustion in a gas turbine combustor. ASME IGTI paper, 1997-GT-309; 1997

[22] Syed SA, Sturgess GJ. Validation studies of turbulence and combustion models for aircraft gas turbine combustors. In: Lauder BE, Humphrey JAC, editors. Proceeding of Momentum and Heat Transfer Processes in Recirculating Flows. Vol. 13. ASME HTD, ASME Winter Annular Meeting, Chicago, Illinois, USA; 1980. pp. 71–89

[23] Kaaling H, Ryden R, Bouchie Y, Ansart D, Magre P, Guin C. RQL combustor development including design, CFD calculations, cars measurements and combustion tests. In: The 13th International Symposium Air Breathing Engines; Chattanooga, TN, USA; 1997. ISABE paper, 97-7069

[24] Crocker DS, Nickolaus D, Smith CE. CFD modelling of a gas turbine combustor from compressor exit to turbine inlet. ASME IGTI paper; 1998-GT-184; 1998

[25] He G, Guo Y, Hsu AT. The effect of Schmidt number on turbulent scalar mixing in a jet-in-crossing flow. International Journal of Heat and Mass Transfer. 1999;**42**:3727–3738

[26] Cadorin M, Pinelli M, Vaccari A, Calabria R, Chiariello F, Massoli P, Bianchi E. Analysis of a micro gas turbine fed by natural gas and synthesis gas: Test bench and combustor CFD analysis. ASME IGTI paper, 2011-GT-46090; 2011

[27] Turrell MD, Stopfod PJ, Syed KJ, Buchanan E. CFD simulation of the flow within and downstream of high-swirl lean premixed gas turbine combustor. ASME IGTI paper 2004-53112; 2004

[28] Mongia HC. Aero-thermal design and analysis of gas turbine combustion systems: Current status and future direction. AIAA paper 98-3982; 1998

[29] Alkabie H. Design methods of the ABB Alstom power gas turbine dry low emission combustion system. Journal of Power and Energy. 2000; **214:** 293–315

[30] Jiang LY. A critical evaluation of turbulence modelling in a model combustor. ASME Journal of Thermal Science and Engineering Applications. 2013;**5:**031002

[31] Jiang LY, Campbell I. Turbulence modelling in a model combustor. Canadian Aeronautics and Space Journal. 2007;**53:**47–57

[32] Jiang LY, Campbell I. Reynolds analogy in combustor modelling. International Journal of Heat and Mass Transfer. 2008;**51**:1251–1263

[33] Jiang LY, Campbell I. Combustion modelling in a model combustor. Journal of Aerospace Power. 2007;**22**:694–703

[34] Jiang LY, Campbell I. Radiation benchmarking in a model combustor. Journal of Engineering for Gas Turbine and Power. 2009;**131**:011501

[35] Jiang LY, Campbell I. A critical evaluation of NOx modelling in a model combustor. Journal of Engineering for Gas Turbines and Power. 2005;**127**:483–491

[36] Jiang LY, Corber A. Assessment of combustor working environments. International Journal of Aerospace Engineering. 2012;**2012**:1–6. DOI:10.1155/2012/217463, 2012.

[37] Ferziger JH, Peric M. Computational Methods For Fluid Dynamics. New York: Springer-Verlag; 2002. pp. 1–10

[38] Moore JG, Moore J. Functional Reynolds Stress Modelling. USA: Pocahontas Press Inc.; 2006

[39] ANSYS Inc. Fluent 13.0 Documentation. 10 Cavendish Court, Lebanon, NH, USA; 2010

[40] Launder BE, Spalding DB. Lectures in Mathematical Models of Turbulence. UK: Academic Press; 1972

[41] Yakhot V, Orszag SA. Renormalization group analysis of turbulence: I. Basic theory. Journal of Scientific Computing. 1986; **1**:1-51–

[42] Shih TH, Liou WW, Shabbir A, Yang Z, Zhu J. A new k-ε eddy-viscosity model for high Reynolds number turbulent flows—Model development and validation. Computers Fluids. 1995;**24**:227–238

[43] Kim SE, Choudhury D, Patel B, M.D. Salas et al. (eds.) Modeling Complex Turbulent Flows, Computations of complex turbulent flows using the commercial code fluent. Proceedings of the ICASE/LaRC Interdisciplinary Symposium on Modeling Complex Turbulent Flows. Hampton, Virginia. 1997:259–276

[44] Menter FR. Two-equation eddy-viscosity turbulence models for engineering applications. AIAA Journal. 1994;**32:**1598–1605

[45] Launder BE, Reece GJ, Rodi W. Progress in the development of a Reynolds-stress turbulence closure. Journal of Fluid Mechanics. 1975;**68**:537–566. DOI: 10.1017/S0022112075001814

[46] Choi KS, Orchard DM. Turbulence management using riblets for heat and momentum transfer. Experimental Thermal and Fluid Science. 1997;**15**:109–124. DOI: 10.1016/S0894-1777(97)00047-2

[47] de Souza F, Delville J, Lewalle J, Bonnet JP. Larger scale coherent structures in a turbulent boundary layer interacting with a cylinder wake. Experimental Thermal and Fluid Science. 1999;**19**:204–213. DOI: 10.1016/S0894-1777(99)00022-9

[48] Vogel JC, Eaton LK. Combined heat transfer and fluid dynamic measurements downstream of a backward-facing step. Journal of Heat Transfer. 1985;**107**:922–929

[49] Peters N. Laminar diffusion flamelet models in non-premixed turbulent combustion. Progress in Energy and Combustion Science. 1984;**10**:319–339

[50] Stahl G, Warnatz J. Numerical investigation of time-dependent properties and extinction of strained methane- and propane-air flamelets. Combustion and Flame. 1991;**85**:285–299

[51] Jones WP, Whitelaw JH. Calculation methods for reacting turbulent flows: A review. Combustion and Flame. 1982;**48**:1–26.

[52] Glassman I. Combustion. New York: Academic Press Inc.; 1987

[53] Bilger RW. Turbulent jet diffusion flames. Progress in Energy and Combustion Science. 1976;**1**:87–109

[54] Sivathanu YR, Faeth GM. Generalized state relationships for scalar properties in non-premixed hydrocarbon/air flames. Combustion and Flame. 1990;**82**:211–230. DOI: 10.1016/0010-2180(90)90099-D

[55] Magnussen BF, Hjertager BH. Mathematical models of turbulent combustion with special emphasis on soot formation and combustion. In: 16th International Symposium on Combustion; 15–20 August, 1976. pp. 719–729

[56] Raithby GD, Chui EH. A finite-volume method for predicting a radiant heat transfer in enclosures with participating media. Journal of Heat Transfer. 1990; **112:**415–423

[57] M W Chase; National Institute of Standards and Technology (U.S.) NIST, NIST-JANAF Thermochemical Tables. 4th ed. Washington, DC, American Chemical Society ; Woodbury, N.Y. : American Institute of Physics for the National Institute of Standards and Technology; 1998.

[58] Rose JW, Cooper JR. Technical Data on Fuel. New York: John Wiley & Sons; 1977

[59] Campbell I, Logan DL. An Experimental Study of a Combusting Flow Past a Confined Bluff Body. Canada: Combustion Institute, Canadian Section; 1997

[60] Widmann JF, Charagundla SR, Presser C. Aerodynamic study of a vane-cascade swirl generator. Chemical Engineering Science. 2000;**55**:5311–5320

[61] Xu D, Khoo BC. Numerical simulation of turbulent flow in an axisymmetric diffuser with a curved surface center-body. International Journal of Numerical Methods for Heat & Fluid Flow. 1988;**82**:45–255

[62] Sislian JP, Jiang LY, Cusworth RA. Laser Doppler velocimetry investigation of the turbulence structure of axisymmetric diffusion flames. Progress in Energy and Combustion Science. 1988;**14**:99–146. DOI: 10.1016/0360-1285(88)90001-9

[63] Panchapakesan NR, Lumley JL. Turbulence measurements in axi-symmetric jets of air and helium. Part 2, helium jet. Journal of Fluid Mechanics. 1993;**246**:225–247. DOI: 10.1017/S0022112093000096

Chapter 8

Testing Physical and Mathematical Criteria in a New Meandering Autocorrelation Function

Charles R.P. Szinvelski, Lidiane Buligon,
Michel Baptistella Stefanello, Silvana Maldaner,
Debora R. Roberti and Gervásio Annes Degrazia

Additional information is available at the end of the chapter

http://dx.doi.org/10.5772/67920

Abstract

An alternative formulation for the low wind speed-meandering autocorrelation function is presented. Employing distinct theoretical criteria, this mathematical formulation, from a physical point of view, is validated. This expression for the meandering autocorrelation function reproduces well-observed wind-meandering data measured in a micrometeorological site located in a pampa ecosystem area (South Brazil). The comparison shows that the alternative relation for the meandering autocorrelation function is suitable to provide meandering characteristic parameters. Employing MacLaurin's series expansion of a lateral dispersion parameter that represents cases in which turbulence and oscillatory movements associated to the meandering events coexist, a new formulation for the turbulence/meandering dissipation rate has been presented.

Keywords: wind meandering, meandering autocorrelation function, physical and mathematical criteria, turbulent energy spectrum, meandering period

1. Introduction

The autocorrelation function is a physical quantity important both for diffusion investigation and to provide basic relations concerning the turbulence processes in the planetary boundary layer (PBL). The employment of autocorrelation functions derived from observed data for distinct movement patterns in the PBL allows to evaluate relevant quantities used to understand complex phenomena in geophysical flow. Therefore, the utilization of the autocorrelation function in the Taylor statistical diffusion theory provides directly the dispersion parameters and also formulations for the turbulence dissipation rate, which are utilized, respectively, in Eulerian (Gaussian diffusion plume) and Lagrangian dispersion models.

Simulating turbulent diffusion and species scalar and vector transport in low wind speed situations is a challenging problem. For conditions, in which the mean wind speed tends to be negligible [1], it is very difficult to identify a defined mean wind direction. In this case, large-scale motions, such as submeso motions, control, in a dominate manner, the atmospheric contaminant dispersion [2]. These large horizontal wind oscillations are called meandering and are responsible for the fact that measured autocorrelation functions of the horizontal wind velocity components show a looping behavior, characterized by the presence of negative lobes [3]. This looping pattern with the presence of negative lobes in autocorrelation functions, which characterize the meandering observed phenomenon, can be well reproduced by particular mathematical formulations. With this good fit, it is possible to estimate the principal physical variables that specify the meandering flow [4, 5].

Recently, Moor et al. [6] proposed to utilize a new mathematical formulation to fit the observed low wind speed-meandering autocorrelation functions. This functional form for the horizontal wind velocity components u and v may be written as follows:

$$R(\tau) = \frac{\cos\left(\frac{m\tau}{(m^2+1)T_L}\right)}{\left(1 + \frac{\tau}{(m^2+1)T_L}\right)^2} \tag{1}$$

where τ is the time lag, T_L is the decorrelation time scale, and m is the loop parameter. The mathematical formulation, as provided by Eq. (1), is composed of the product of the binomial function (describing the autocorrelation function for the turbulent degrees of freedom [7]) by the cosine (describing the influence of the submeso motions in the autocorrelation function [8]). Therefore, this formulation presents a heuristical flexibility that allows describing the observed characteristics of the atmospheric multiple-scale interaction. Eq. (1) can also be written in a distinct way, namely

$$R(\tau) = \frac{\cos(q\tau)}{(1 + p\tau)^2} \tag{2}$$

with

$$p = \frac{1}{(m^2 + 1)T_L} \tag{3}$$

and

$$q = \frac{m}{(m^2 + 1)T_L} \tag{4}$$

q and p are hybrid relations described in terms of the turbulent (T_L) and meandering parameters (m) and are originated from the Frenkiel autocorrelation function [9]. Furthermore, a meandering characteristic time scale (meandering period) can be represented as

http://dx.doi.org/10.5772/67920

$$T^* = \frac{2\pi}{q} \tag{5}$$

and will be calculated by

$$T^* = \frac{2\pi\,(m^2+1)T_L}{m} \tag{6}$$

The purpose of this study is to verify that the expression given by Eq. (1) fills the mathematical and physical requirements that must be fulfilled by an autocorrelation function. An additional aim is to employ the best-fitting curves, obtained from Eq. (1), to calculate the loop parameter and the meandering characteristic time scale. The present investigation also provides a new relation for the turbulence/meandering dissipation rate.

2. Discussion of the physical and mathematical criteria to the autocorrelation function

In this part of the chapter, we use four established criteria [10] to validate the autocorrelation function, as given by the mathematical formulation described in Eq. (1). Therefore, these criteria are applied to stationary and homogeneous turbulence, which are mathematical descriptions of respective physical requirements for the validation of the autocorrelation function applied to the studies of turbulence. Eq. (1) is described in terms of a parameter m, which indicates the intensity of the meandering phenomena. This parameter is responsible to introduce variations in the usual results that are those normally expected in the classical forms that represent the autocorrelation functions. In sequence, the four criteria are discussed in detail as follows:

(I) $R(\tau)$ is an even function, and it has the following properties:

a. $|R(\tau)| \leq= R(0)$, it is limited in the origin and its maximum value 1 occurs in $\tau = 0$. This fact agrees with the theoretical consideration presented in [11, 12]. The authors suggest that R must osculate the theoretical parabolic form of R in the origin ($\tau \approx 0$).

b. The next property takes into account the qualitative behavior to large values of autocorrelation time. In this case, $R(\tau) \to 0$ when $|\tau| \to \infty$. In Eq. (1), this consideration results of confront theorem applied in $|R(|\tau|)| \leq (1+p|\tau|)^{-2}$;

c. To define the integral time scale, it is necessary to calculate the integral $\int_0^\infty R(\tau)d\tau$, namely the integral must be convergent. Thus,

$$\int_0^\infty R(\tau)d\tau \leq \left| \int_0^\infty R(\tau)d\tau \right| \leq \int_0^\infty |R(\tau)|d\tau = \frac{1}{p} = (m^2+1)T_L \tag{7}$$

is a finite value.

(II) $R(\tau) \in C^0(\mathfrak{R})$, wherever the smoothness is not valid in τ because the lateral derivatives in origin are not defined. This fact yields the angular functional form of the autocorrelation function to $\tau = 0$, and its physical implication is that this autocorrelation function is not suitable to investigate the viscous microstructure of turbulence [13]. Therefore, to practical applications, where the hypothesis of fully developed turbulence is admitted ($Re \gg 0$), the validity of this criteria is not relevant by the fact that the energy-containing and the characteristics scales of eddies in this region are not relevant in the scalar diffusion process.

(III) As already mentioned, the integral time scale T_L is well defined and its value is obtained as $\int_0^{\infty} R(\tau)d\tau$. Here, it is necessary to make some considerations:

- the integral resolution is[1]

$$\int_0^{\infty} \frac{\cos(q\tau)}{(1+p\tau)^2} d\tau = (1+m^2)T_L\left[1 - m\left(\sin(m)Ci(m) - \cos(m)\left(Si(m) - \frac{\pi}{2}\right)\right)\right] \\ = (1-\beta_m)(1+m^2)T_L \tag{8}$$

- to $m = 0$, the result obtained is T_L, in agreement with [7] and attends the given interpretation of [10] to these criteria, and, following [14], this result is expected because the argument of the autocorrelation function has dimensions of length inverse and after integration recovers dimensions of integral scale;

- but to $m \neq 0$. There is the factor $(1 - \beta_m)$, which goes to zero when m goes to infinity. This fact tells us that it does not have significant correlation to large values of m;

Therefore, using the above considerations, this result suggests the following interpretation: the case $m = 0$ will establish the autocorrelation time T_L to a fully developed turbulence hypothesis. On the other hand, when $m \neq 0$, the time scale will be proportional to T_L by the following correction factor $(1 - \beta_m)$ yielding

$$T_{Lm} = (1-\beta_m)(1+m^2)T_L \tag{9}$$

(IV) To validate these criteria, basically, we must verify the following relation [15]:

$$E(\omega) = \frac{2\langle u^2\rangle}{\pi}\int_0^{\infty} R(\tau)\cos(\omega\tau)d\tau \approx \omega^{-2} \tag{10}$$

[1] $Ci(m)$ is the cosine integral function and $Si(m)$ is the sine integral function.

http://dx.doi.org/10.5772/67920

In this case, applying the Fourier transform in Eq. (1) results

$$\frac{E(\omega)}{\langle u^2\rangle T_L} = \frac{(m^2+1)}{\pi}\Big\{2 - A\Big[\cos(A)\Big(\frac{\pi}{2} - Si(A)\Big) + \sin(A)Ci(A)\Big] - B\Big[\cos(B)\Big(\frac{\pi}{2} - Si(B)\Big) + \sin(B)Ci(B)\Big]\Big\} \tag{11}$$

with $A = (m^2+1)(\omega T_L) + m$ and $B = (m^2+1)(\omega T_L) - m$.

Eq. (11) does not express the form ω^{-2}, explicitly, but this behavior is founded in its graphic representation (see **Figure 1**).

The present analysis, based on mathematical and physical criteria extensively discussed in the literature, showed that the meandering autocorrelation function, as given by Eq. (1), can be employed to investigate the physical patterns associated to the meandering multiple-scale flow.

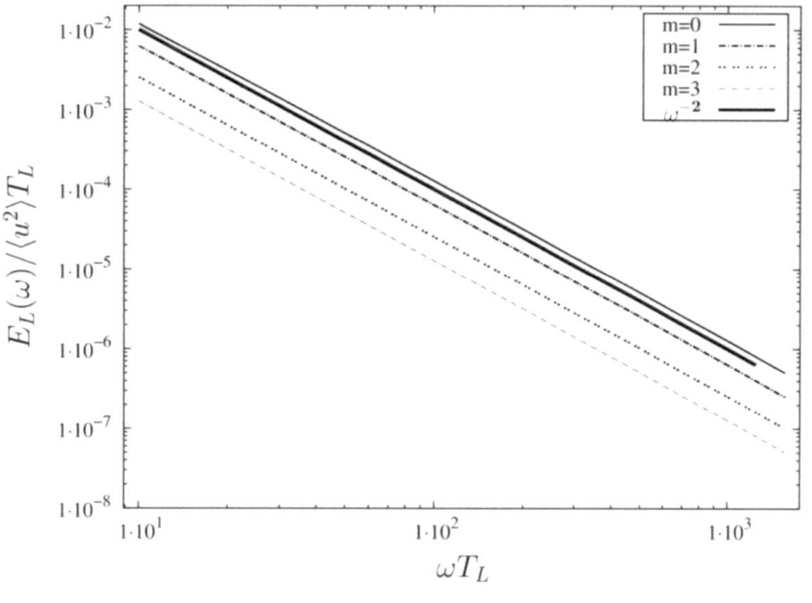

Figure 1. Turbulent energy spectrum from the autocorrelation functions (Eq. (1)) for distinct values of the m parameter.

3. Reproducing observed meandering autocorrelation functions

In this part of the study, we use meandering data observed in a low wind speed PBL to calculate the experimental autocorrelation functions. These experimental meandering autocorrelation functions were obtained from the following relation:

$$R(\tau) = \frac{\langle u(t)u(t+\tau)\rangle}{\langle u^2\rangle} \quad (12)$$

where $\langle u^2\rangle = \sigma_u^2$ is the variance of the velocity. These observed functions are fitted by Eq. (1). Therefore, Eq. (1) is evaluated as a mathematical representation to reproduce the observed negative lobes in the meandering autocorrelation functions and utilized to calculate the loop parameter and the meandering period.

The low wind speed data were collected at the Federal University of Santa Maria (Rio Grande do Sul, Brazil) meteorological site. The wind velocity components were sampled at a frequency of 10 Hz by a sonic anemometer installed at a height of 3 m located in a pampa ecosystem area.

Figures 2 and **3** establish the comparison among the autocorrelation function obtained from the low wind speed data set (continuous line) and the correspondent best fit (dashed line)

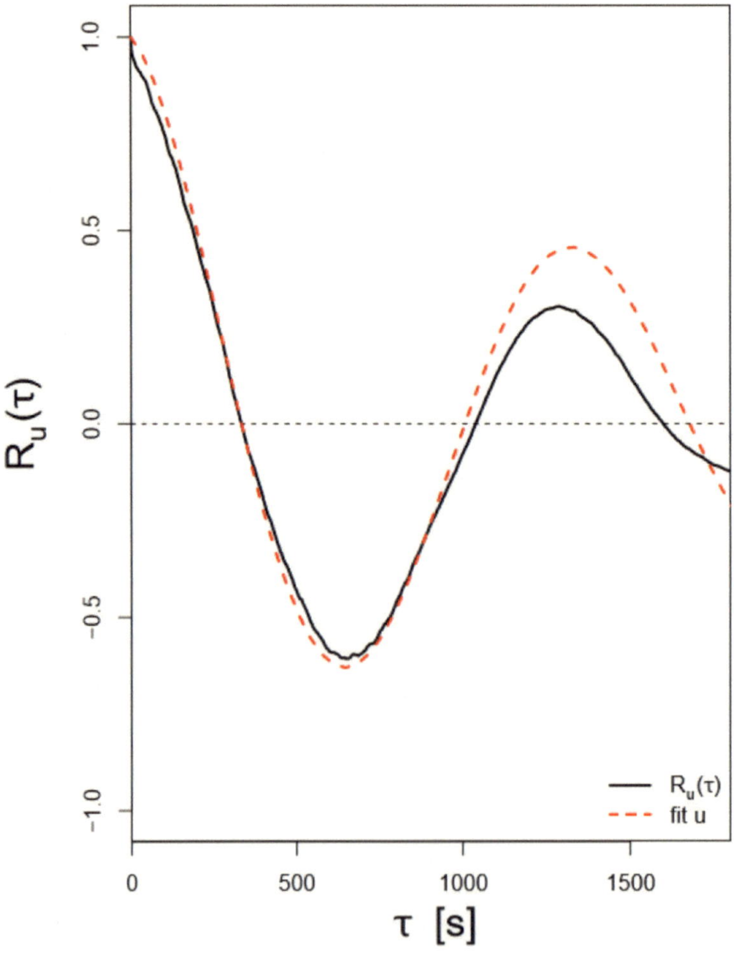

Figure 2. Autocorrelation function for the horizontal wind-velocity component u.

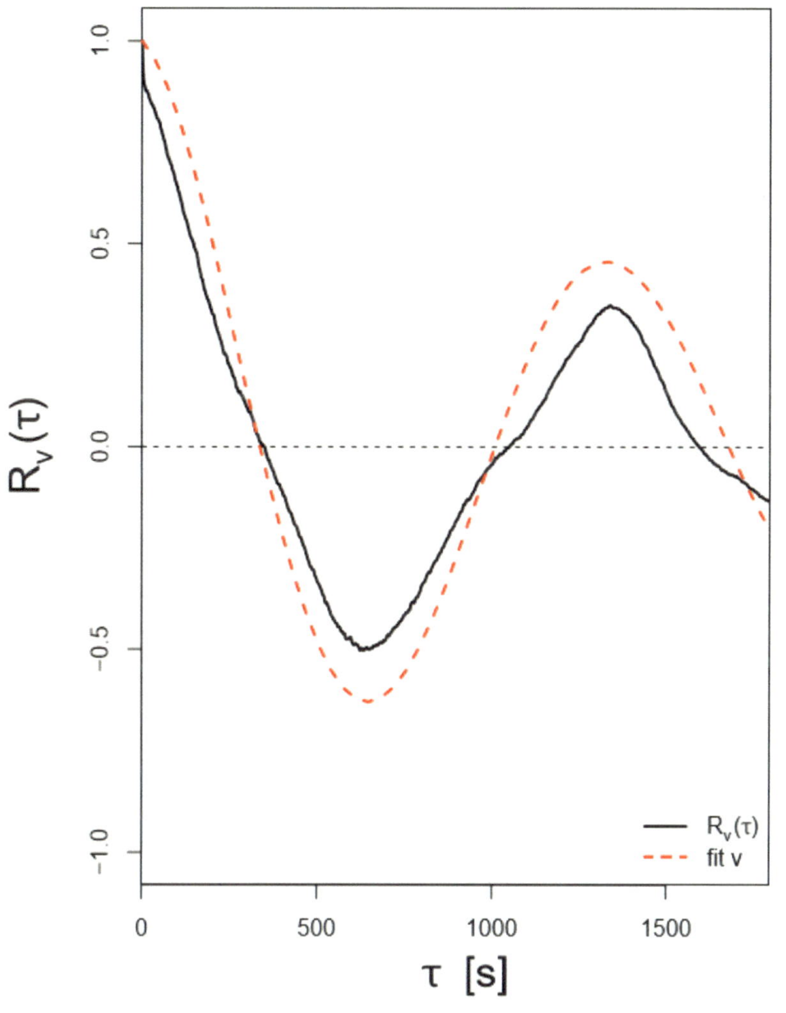

Figure 3. Autocorrelation function for the horizontal wind-velocity component v.

provided from Eq. (1). These figures were calculated from 1-h time series presenting the mean horizontal speed ($\overline{u} = 0.96$ ms^{-1}). They show that the autocorrelation function, provided by the mathematical formulation (Eq. (1)), represents fairly well the observed form of the meandering autocorrelation functions including their negative lobes. This oscillatory behavior, generating negative lobes in the autocorrelation functions, calls attention for the presence of multiple scales in the horizontal wind-meandering flow. The action of the submeso [8] and turbulent scale motions provoke, for short-time intervals, large variations in the horizontal wind direction.

Employing Eqs. (3)–(5) and a best-fit curve condition, it is possible to calculate, respectively, the magnitudes for the looping parameter and meandering period (a complete discussion about this development can be seen in [1]). These values are presented in **Table 1**.

The magnitudes for m_u and m_v in **Table 1** agree very well with the results obtained by Moor et al. [6] that employed wind-meandering data which were collected in a nocturnal stable PBL in a Brazilian Amazon Large Scale Biosphere-Atmosphere Project. The meandering periods T_u^* and T_v^* in **Table 1** present similar magnitudes with the mean values found by [4]. The characteristic values of the meandering phenomenon are exhibited in **Table 1**, which can be used in Eqs. (6) and (9) to obtain the new turbulent scales, yielding $T_{L,m_u} \approx 70$ s and $T_{L,m_v} \approx 62$ s. These magnitudes of $T_{L,m}$ are in agreement with the results estimated by Moor et al. [6].

$T_u^*(s)$	$T_v^*(s)$	m_u	m_v
1337	1337	5.2	4.2

Table 1. Meandering period and loop parameter values for the u and v components.

4. Turbulence/meandering dissipation rate

The lateral dissipation parameter σ_y is a statistical quantity fundamental for the dispersion modeling and for the derivation of turbulence/meandering dissipation rate functional form [16, 17]. From the Taylor statistical diffusion theory, this relevant parameter may be described as

$$\sigma_y^2(t) = 2\sigma_v^2 \int_0^t (t-\tau) R_v(\tau) d\tau \tag{13}$$

where σ_v is the standard deviation of the turbulent lateral velocity.

Taylor [18] considered an exponential form for the autocorrelation function and first derived an expression for σ_y^2

$$\sigma_y^2(t) = 2\sigma_v^2 T_{Lv}^2 \left(\frac{t}{T_{Lv}} - 1 + e^{\frac{-t}{T_{Lv}}} \right) \tag{14}$$

where t is the travel time of the fluid particle, T_{Lv} is the lateral Lagrangian integral time scale (defined in Criteria III).

Tennekes [16] used Eq. (14) and inertial subrange similarity arguments [15, 12] to derive the following fundamental expression for the turbulence dissipation rate ε:

$$\varepsilon = \frac{2}{C_0} \frac{\sigma_v^2}{T_{Lv}} \tag{15}$$

where C_0 is the Kolmogorov constant.

Following the above development and employing the autocorrelation function as given by Eq. (1) into Eq. (13) yields

http://dx.doi.org/10.5772/67920

$$\frac{\sigma_y^2(t)}{2\sigma_v^2} = -\frac{\cos(qt)}{p^2} + \frac{1+pt}{p^2} - \frac{q(1+pt)}{p^3}\left[\sum_{n=0}^{\infty}\left(\frac{(pt)^{2n+2}}{2n+2} - \frac{(pt)^{2n+3}}{2n+3}\right)\sum_{k=0}^{n}\frac{(-1)^k m^{2k+1}}{(2k+1)!}\right]$$
$$-\frac{1}{p^2}\left[\sum_{n=0}^{\infty}\left(\frac{(pt)^{2n+1}}{2n+1} - \frac{(pt)^{2n+2}}{2n+2}\right)\sum_{k=0}^{n}\frac{(-1)^k m^{2k}}{(2k)!}\right] \quad (16)$$

For $t \ll T_{Lv}$, MacLaurin's series expansions of the meandering and turbulent dissipation parameter (Eq. (16)) can be approximated by

$$\sigma_y^2(t) = \sigma_v^2 t^2 - \frac{2p\sigma_v^2}{3} t^3 \quad (17)$$

The confrontation of Eq. (17) with the Taylor statistical diffusion theory (Eq. (13)), for dispersion periods that are very smaller than the Lagrangian integral time scale, demonstrates that the negative expression in Eq. (17) represents a term that reduces the hybrid dispersion parameter. This negative contribution results in the suppression of a number of degrees of freedom of the turbulent field associated to the high-frequency harmonics [17]. As a consequence, it is reasonable to relate the term $\frac{2p\sigma_v^2}{3} t^3$ to the Kolmogorov spectrum high-frequency eddies. This comparison was accomplished by [16] employing the Lagrangian structure function, the Lagrangian autocorrelation function, and the inertial subrange Lagrangian turbulent spectrum. Tenneke's derivation provides the following relationship for the lateral dispersion parameter in terms of inertial subrange quantities:

$$\sigma_y^2(t) = \sigma_v^2 t^2 - \frac{C_0 \varepsilon}{6} t^3 \quad (18)$$

The comparison of Eq. (17) with Eq. (18) leads to the following fundamental relation:

$$\varepsilon = \frac{4}{(m^2+1)} \frac{\sigma_v^2}{C_0 T_{Lv}} \quad (19)$$

The new expression, as given by Eq. (19), maintains the basic argument that turbulent motion is dissipated at a rate proportional to the kinetic energy and inversely proportional to the memory effect associated to the energy-containing eddies. Therefore, Eq. (19) is described in terms of the looping parameter m. Such a parameter, associated to the meandering phenomenon, becomes the new dissipation rate distinct of those applied to the purely turbulent cases. The looping parameter m determines the magnitudes of the dissipation rate. Hence, the dissipation rate increases when the turbulence is dominant. On the other hand, the dissipation rate decreases when submeso motions, associated to the meandering phenomenon, control the geophysical flow.

5. Conclusion

In this study, a mathematical relation to represent observed meandering autocorrelation functions, derived from the heuristic arguments, is tested and validated using well-known mathematical and physical criteria. Therefore, it is employed to describe experimental wind-meandering data. From this approach and utilizing best-fitting curves, it was possible to obtain characteristic values for the loop parameter and the meandering period. These quantities, which characterize the wind-meandering phenomenon, are shown in **Table 1**.

An important aspect in the present development has been the derivation of Eq. (11) which represents a theoretical formulation to model observed wind-meandering spectral data. Thus, Eq. (8) is able to describe the effect of the submeso and turbulent scales and hence allow to identify the low frequencies associated to the meandering spectral peaks.

Another relevant result is provided by Eq. (19). Therefore, this expression is able to evaluate dissipation rates generated from complex flows. Complex flows, such as meandering motions, are those in which there are interactions between distinct circulations characterized by movements presenting different spatial and time scales. As a consequence, Eq. (19) allows to describe this atmospheric phenomenon type.

Acknowledgements

The authors would like to thank the CNPq (Conselho Nacional de Desenvolvimento Científico e Tecnológico) and CAPES (Coordenação de Aperfeiçoamento de Pessoal de Nível Superior) for partial financial support of this work.

Appendix A

In this appendix, the general criteria to validate mathematical formulation for the autocorrelation function are presented. These criteria are a summary of physical and mathematical requirements applied in the statistical treatment of the turbulence founded in the classical literature [10, 13].

In particular, Manomaiphiboon and Russell [10] presented four criteria for the validation of the ACF to homogeneous and stationary turbulence, as follows:

(I) $R(\tau)$ is limited at origin (neighborhood of origin) $|R(\tau)| \leq 1 = R(0)$. Besides, $\lim_{|\tau| \to \infty} R(\tau) = 0$ and $\int_0^{\infty} |R(\tau)| d\tau < \infty$.

In fact, the generic form of the autocorrelation function is given by the Eq. (12)

following $R(0) = 1$, which indicates that the maximum correlation is $\tau = 0$ and, consequently to any other correlation time, is given by $|R(\tau)| \leq 1$.

On the other hand, the condition that $\int_0^{\infty} |R(\tau)| d\tau$ must be finite will permit to obtain the Lagrangian integral scale time, which will be defined in Criteria III.

(II) $R(\tau)$ is smooth in τ. At origin $\left.\frac{dR(\tau)}{d\tau}\right|_{\tau=0} = 0$ and $\left.\frac{d^2R(\tau)}{d\tau^2}\right|_{\tau=0} < 0$.

The mathematical properties described by Criteria II result of the Taylor series expansion of $R(\tau)$ in the neighborhood of origin ($\tau \approx 0$),

$$R(\tau) = \sum_{n=0}^{\infty} \left.\frac{d^n R(\tau)}{d\tau^n}\right|_{\tau=0} \frac{\tau^n}{n!}. \tag{20}$$

Due to the parity of $R(\tau)$ (stationary turbulence), the odd-order derivatives are null, in particular, $\frac{dR(\tau)}{d\tau}|_{\tau=0} = 0$. Therefore, from Eq. (20), it results as follows:

$$R(\tau) = 1 + \frac{1}{2}\left.\frac{d^2R(\tau)}{d\tau^2}\right|_{\tau=0} \tau^2 + O(\tau^4) \tag{21}$$

and, consequently, $\frac{1}{2}\frac{d^2R(\tau)}{d\tau^2}|_{\tau=0} < 0$. The test of the second derivative in an interval centered at the origin ($\tau \approx 0$) ensures a representation of $R(\tau)$ by quadratic polynomial and, at $\tau = 0$, agrees with the first criteria.

Following [12], to introduce a time scale τ_L, which contains temporal values close to the origin, the Taylor's autocorrelation function can be approximated by

$$R(\tau) \approx 1 - \frac{\tau^2}{\tau_L^2} \tag{22}$$

meaning an osculating parabola $R(\tau)$ at origin. This expression defines the time scale τ_L by the relation

$$\frac{1}{\tau_L^2} = -\frac{1}{2}\left.\frac{d^2R(\tau)}{d\tau^2}\right|_{\tau=0}. \tag{23}$$

(III) The Lagrangian integral time scale given by $T_L = \int_0^{\infty} R(\tau) d\tau$ is limited and well defined. It is a direct consequence of Criteria I.

(IV) From Wiener-Khintchin theorem [19], the $R(\tau)$ and $E(\omega)$ are expressed by a pair of Fourier cosine transform:

$$R(\tau) = \frac{1}{\langle u^2 \rangle} \int_0^{\infty} E(\omega) \cos(\omega\tau) d\omega \tag{24}$$

and

$$E(\omega) = \frac{2\langle u^2 \rangle}{\pi} \int_0^\infty R(\tau) \cos(\omega\tau) d\tau \tag{25}$$

where ω is the turbulent frequency.

According to the inertial subinterval theory (K41) [20], $E(\omega)$ can be expressed by $E(\omega) = \kappa\bar{\epsilon}\omega^{-2} \propto \omega^{-2}$, to $1 \ll \omega T_L \ll \frac{T_L}{\tau_\eta}$, where κ is the dimensional universal constant, $\bar{\epsilon}$ is the average rate of turbulent energy dissipation, and τ_η is the Kolmogorov time scale.

Author details

Charles R.P. Szinvelski, Lidiane Buligon, Michel Baptistella Stefanello, Silvana Maldaner, Debora R. Roberti and Gervásio Annes Degrazia*

*Address all correspondence to: gervasiodegrazia@gmail.com

Universidade Federal de Santa Maria, Santa Maria, Brazil

References

[1] Mortarini L, Anfossi D. Proposal of an empirical velocity spectrum formula in low-wind speed conditions. Quarterly Journal of the Royal Meteorological Society. 2015;**141**:85-97

[2] Anfossi D, Alessandrini S, Trini Castelli S, Ferrero E, Oettl D, Degrazia G. Tracer dispersion simulation in low wind speed conditions with a new 2-D Langevin equation system. Atmospheric Environment. 2006;**40**:7234-7245

[3] Steeneveld GJ, Holtslag AAM. Meteorological aspects of air quality. In: Air Quality in the 21st Century. New York: Nova Science Publishers; 2009. pp. 67-114

[4] Mortarini L, Stefanello M, Degrazia G, Roberti D, Castelli ST, Anfossi D. Characterization of wind meandering in low-wind-speed conditions. Boundary-Layer Meteorology. 2016;**161**(1):165-182. DOI: 10.1007/s10546-016-0165-6

[5] Cava D, Mortarini L, Giostra U, Richiardone R, Anfossi D. A wavelet analysis of low-wind-speed submeso motions in a nocturnal boundary layer. Quarterly Journal of the Royal Meteorological Society. 2017;**143**:661-669. DOI: 10.1002/qj.2954

[6] Moor L, Degrazia G, Stefanello M, Mortarini L, Acevedo O, Maldaner S, Szinvelski C, Roberti D, Buligon L, Anfossi D. Proposal of a new autocorrelation function in low wind speed conditions. Physica A. 2015;**438**:286-292

[7] Phillips P, Panofsky HA. A re-examination of lateral dispersion from continuous sources. Atmospheric Environment. 1982;**16**:1851-1860

[8] Mahrt L. Stably stratified atmospheric boundary layers. Annual Review of Fluid Mechanics. 2014;**46**:23-45

[9] Frenkiel FN. Turbulent diffusion: mean concentration distribution in a flow field of homogeneous turbulence. Advances in Applied Mechanics. 1953;**3**:61-107

[10] Manomaiphiboon K, Russell AG. Evaluation of some proposed forms of Lagrangian velocity correlation coefficient. International Journal of Heat and Fluid Flow. 2003;**24**(5):709-712

[11] Tennekes H, Lumley JK. A First Course in Turbulence. Massachusetts: The MIT Press; 1972

[12] Hinze JO. Turbulence. New York: McGraw-Hill; 1975

[13] Tennekes H. The exponential the Lagragian correlation function and turbulent diffusion in the inertial subrange. Atmospheric Environment. 1979;**13**(11):1565-1568

[14] McComb WD. The Physics of Fluid Turbulence. Oxford: Clarendon Press; 1990

[15] Kolmogorov AN. The local structure of turbulence in incompressible viscous fluid for very large Reynolds numbers. Doklady Akademii Nauk SSSR. 1941:299-303

[16] Tennekes H. Similarity relations, scaling laws and spectral dynamics. In: Atmospheric Turbulence and Air Pollution Modelling. Netherlands: Springer; 1984. pp. 37-68

[17] Degrazia G, Acevedo OC, Carvalho J, Goulart A, Moraes OLL, Campos Velho HF, Moreira DM. On the universality of the dissipation rate functional form and of the autocorrelation function exponential form. Atmospheric Environment. 2005;**39**(10):1917-1924

[18] Taylor GI. Diffusion by continuous movements. Proceedings of the London Mathematical Society. 1921;**20**:196-211

[19] Gardiner CW. Handbook of Stochastic Methods for Physics, Chemistry and the Natural Sciences. Berlin: Springer-Verlag; 1985. 442 p

[20] Kolmogorov AN. The local structure of turbulence in incompressible viscous fluid for large Reynolds number. Doklady Akademii Nauk SSSR; 1941;**30**(301):9-13

Chapter 9

Underwater Optical Wireless Communication Systems: A Concise Review

Lydia K. Gkoura, George D. Roumelas,
Hector E. Nistazakis, Harilaos G. Sandalidis,
Alexander Vavoulas, Andreas D. Tsigopoulos and
George S. Tombras

Additional information is available at the end of the chapter

http://dx.doi.org/10.5772/67915

Abstract

Underwater optical wireless communications (UOWC) have gained a considerable interest during the last years as an alternative means for broadband inexpensive submarine communications. UOWC present numerous similarities compared to free space optical (FSO) communications or laser satellite links mainly due to the fact that they employ optical wavelengths to transfer secure information between dedicated point-to-point links. By using suitable wavelengths, high data rates can be attained. Some recent works showed that broadband links can be achieved over moderate ranges. Transmissions of several Mbps have been realized in laboratory experiments by employing a simulated aquatic medium with scattering characteristics similar to oceanic waters. It was also demonstrated that UOWC networks are feasible to operate at high data rates for medium distances up to a hundred meters. However, it is not currently available as an industrial product and mainly test-bed measurements in water test tanks have been reported so far. Therefore, extensive research is expected in the near future, which is necessary in order to further reveal the "hidden" abilities of optical spectrum to transfer broadband signals at higher distances. The present work summarizes the recent advances in channel modeling and system analysis and design in the area of UOWC.

Keywords: underwater optical wireless communications (UOWC), channel modeling, propagation phenomena, oceanic turbulence, experimental studies

1. Introduction

The present work summarizes the recent advances in channel modeling and system analysis and design in the area of underwater optical wireless communications (UOWC). UOWC have gained a considerable interest during the last years as an alternative means for broadband inexpensive submarine communications.

The technology that is mostly used nowadays among divers, ships, etc. is mainly based on acoustic wave transmission. However, it is restricted to several hundreds of kbps at ranges of a few kilometers, supports slow data rates, and distresses marine mammals such as dolphins and whales. On the other hand, traditional electromagnetic (EM) waves are highly attenuated in water due to both absorption and scattering. Therefore, they can be used only for relatively short-range real-time applications in the order of a few centimeters to a few meters.

EM waves in the visible spectrum, for example, optical signals between 400 and 700nm, propagate faster in water than acoustic ones. However, the optical transmitters and receivers must be located at a short distance, since turbulence and multiscattering effects significantly deteriorate the performance. Chromatic dispersion also causes temporal broadening of the optical pulses. In contrast to the above impairments, seawater shows a decreased absorption in the blue/green region of the visible spectrum. Hence, using suitable wavelengths, high data rates can be attained. Moreover, some recent works showed that broadband links can be achieved over moderate ranges. Transmissions of several Mbps have been realized in laboratory experiments by employing a simulated aquatic medium with scattering characteristics similar to oceanic waters. It was also demonstrated that UOWC networks are feasible to operate at high data rates for medium distances up to a hundred meters.

UOWC present numerous similarities compared to free space optical (FSO) communications or laser satellite links mainly due to the fact that they employ optical wavelengths to transfer secure information between dedicated point-to-point links. It constitutes an alternate and effective transmission technique for underwater communications, instead of the traditional acoustic one. However, it is not currently available as an industrial product and mainly test-bed measurements in water test tanks have been reported so far. Therefore, extensive research is expected in the near future which is necessary in order to further reveal the "hidden" abilities of optical spectrum to transfer broadband signals at higher distances.

In the current chapter, we outline the main characteristics of UOWC channel models based on theoretical and experimental results, the phenomena limiting their performance, and the key research trials reported so far for this novel branch of optical wireless systems. A brief review of acoustic and EM communication techniques is also presented, which helps the readers better understand the different ways UOWC operates.

2. Wave transmission in the aquatic medium

2.1. Acoustic waves

Until today, underwater transmission is attained by employing acoustic waves. This technology supports, relatively, low data rates for medium distances and does not ensure the link security. Furthermore, the information signal delay is quite increased. It is a legacy technology, and even it works at long distances, it can only establish low speed transmissions. Therefore, it is inappropriate for broadband transmission driven by the current and ongoing communication demands.

Although sound travels decently through air, it travels much better through water. The speed of sound in clear air is 340 m/s, which is almost four times smaller than the speed of sound in water, which is 1500 m/s. However, it is much smaller than the speed of light and EM waves.

The speed at which sound travels through water is highly dependent on the temperature, pressure, and salinity of the water. While the sound waves travel through water, the energy is absorbed by the aquatic medium and can be transformed to other forms, such as heat [1]. Roughly speaking, the main impairments of underwater acoustic transmission are presented in the sequel.

Path Loss: The attenuation or path loss in an underwater channel is frequency dependent and expressed as [2]

$$A(l,f) = l^k \alpha(f)^l \tag{1}$$

where l is the distance between the transmitter and the receiver, f is the frequency of the signal, k is the propagation constant with values usually between 1 and 2, and $a(f)$ is the absorption coefficient, which depends on frequency. The cause for the frequency dependence is the energy absorption of the pressure waves and the spreading loss that increases by the distance. The seawater absorption coefficient is expressed by the sum of chemical relaxation processes and absorption from pure water [3]

$$\alpha = \frac{A_1 P_1 f_1 f^2}{f_1^2 + f^2} + \frac{A_2 P_2 f_2 f^2}{f_2^2 + f^2} + A_3 P_3 f^2 \tag{2}$$

where f_1 and f_2 are for the relaxation process in boric acid and magnesium sulfate parameters, P_1 and P_2 are parameters for pressure, and A_1, A_2, and A_3 are constants.

Multipath: The multipath propagation is a common problem for underwater acoustic links and causes the acoustic signals to reach the receiver following different multiple paths. In general, reflected waves from the bottom, from the surface, or from other objects are received from the transmitter. Moreover, sound speed variations with depth create wave refraction [4, 5]. As a result, the receiver detects more than one pulses with a different amplitude phase and instant of arrival [6]. An efficient expression for the channel impulse response is the following [7]

$$c(\tau, t) = \sum_P A_P(t)\delta(\tau - \tau_p(t)) \tag{3}$$

where $A_P(t)$ is the path amplitude and $\tau_p(t)$ is the path delay, both time dependent.

The transmission range, the depth, the geometry of the channel, the frequency, and the sound speed profile are the main factors that create multipath effects. As a result, horizontal channels (according to the direction of the wave with respect of the ocean bottom) exhibit longer multipath spreads, whereas vertical channels have little time dispersion [8]. Multipath propagation can severely deteriorate the acoustic signal as it generates intersymbol interference (ISI) [9].

Noise: Noise is created by machines which are in use or by shipping activities usually in high traffic areas. Another kind of noise is ambient noise, which is mostly related to hydrodynamics caused by the movement of water from one place to another like currents or tides. Weather effects on the sea like winds, rains, storms, atmospheric turbulence, thermal noise, seismic, and biological phenomena can also induce noise [2, 8]. The primary sources of noise in shallow waters seem to be created by ships and snapping shrimps [10].

Doppler spread: Doppler frequency spread is caused by the relative motion between the transmitter and the receiver. ISI is also related to Doppler spread mostly in high data rate transmissions where adjacent symbols interfere at the receiver. The Doppler effect can be estimated by the ratio $a=U_r/c$ where U_r is the relative velocity between the transmitter and the receiver, and c is the signal propagation speed. The Doppler factor is about $a=10^{-4}$ due to the relatively low frequency of the acoustic signals inside water (1500 m/s) [2, 8, 11].

Even at short ranges, the acoustic channel is limited to low data rates under Mbps. For medium ranges (1–10 km), the data rate drops to approximately 10 kbps, and eventually at very long ranges (>100 km), the data rate is less than 1 kbps [9, 12, 13]. Therefore, the acoustic wave transmission cannot satisfy the needs of new demanding technologies because of the inability to achieve high data rate communications in real-time operation.

2.2. EM waves

EM waves in the RF band can be used in order to achieve faster wireless transmission. They are also unaffected by temperature and depth. Attenuation is the main effect of water in all EM waves, due to both absorption and scattering. EM wave behavior in freshwater and seawater is quite different and that is due to the fact that sea water is a high-loss medium. Propagation velocity and the absorptive loss of EM waves can be described as functions of carrier frequency [14, 15].

EM waves inside water are highly frequency dependent and proportional to the square root of frequency. This is the main reason for using low frequencies VLF and ELF. VLF band consists of radio waves between 3 and 30 kHz, can achieve data rates of 300 bit/s, and is able to penetrate seawater at a depth of 20 m. ELF band (3–300 Hz) can penetrate seawater at depths of hundreds of meters, something quite useful for communications between submarines.

To sum up, the main limitations of EM waves are the big antenna size needed in freshwater and the high attenuation for seawater [16]. Available commercial products for underwater in RF band can achieve bit rates of 100 bps for a range of several decades of meters. 1–10 Mbps within 1 m range has been reported, as well [17].

2.3. Optical waves

Optical wireless communications are a relatively new technology providing many serious advantages, such as the very high rates of data transmission, secure links, very small and light

http://dx.doi.org/10.5772/67915

size of the transceiver components, including apertures, low installation and operational cost, and no need of licensing fees and tariffs, since the optical band is not included in the telecommunications regulations [18]. Optical wireless uses modulated optical beams in order to establish short, medium or long communications. Unfortunately, because the propagation medium is the free space, the performance and the reliability of these systems depend mainly on the weather conditions between the receiver and the transmitter [19, 20].

EM waves in the visible spectrum (400–700nm) present an alternative way to provide broadband communications in the water. They propagate faster in water (300,000,000m/s) than the acoustic ones (~340m/s in air–~1500m/s in water), which is about 200,000 times faster than sound travels through water. That is the main reason why they have gained a considerable interest during the last years to serve as a broadband (10–100Mbps), safe (non-interceptable), and reliable complement to legacy acoustic underwater communications systems [6].

In general, optical signals are highly absorbed in water, and this is one of the main disadvantages; the other one is the optical scattering by all the particles existing inside the sea. However, seawater shows a decreased absorption in the blue/green region of the visible spectrum. Thus, using suitable wavelengths, for instance in the blue/green region, high speed connections can be attained according to the type of water (400–500nm for clear to 300–700nm for turbid water conditions). Minimum attenuation is centered near 0.460μm in clear waters and shifts to higher values for dirty waters approaching 0.540μm for coastal waters [21, 22].

Roughly speaking, the power received $P(z)$, given initial power P_0, propagating through a medium of thickness z is estimated by the Beer's Law as follows

$$P(z) = P_0 e^{-c(\lambda)z} \tag{4}$$

$c(\lambda)(m^{-1})$ is the extinction coefficient expressing the total attenuation occurred by the propagation through the water. According to the Haltrin's model in Ref. [23], the total attenuation can be described as the sum of absorption and scattering. For a completely absorbing or completely scattering medium, the total attenuation coefficient in Eq. (4) can be replaced with the absorption coefficient, that is, a, or scattering coefficient, that is, b, respectively. The product cz is also referred as attenuation length, and it contributes on the reduction of the received power by a factor of $\exp(-1)$, or ~63% [24]. Based on the above, we have

$$c(\lambda) = \alpha(\lambda) + \beta(\lambda) \tag{5}$$

where $a(\lambda)$ is the absorption coefficient, $b(\lambda)$ is the scattering coefficient, and λ is the wavelength.

Beer's Law provides a limited applicability as it describes only the attenuation due to absorption and single scattering events. In reality, however, many cases of multiple scattering may occur. Also it presumes that the source and receiver are in exact alignment with each other, and it can be applied only in Line-of-Sight (LOS) communication scenarios. Moreover, Beer's Law ignores temporal dispersion [25].

More accurate expressions have to take the link geometry into account. For instance, assuming that the transmitter and receiver are positioned in a LOS configuration, the received power can be estimated by [24]

$$P_R(t, r, z) = P_T(t)D_T L_w(t, r, z)D_R \tag{6}$$

where $P_R(t, r, z)$ is the received power dependent on time t, lateral displacement from the beam axis r, and range z, $P_T(t)$ represents the transmitted power, D_T is the aperture and divergence of the optical source, and D_R is the photoreceiver aperture and field of view. The channel loss term, L_W (t, r, z), characterizes the spatial and temporal characteristics of light propagation in seawater.

3. UOWC propagation phenomena

3.1. Aquatic medium characteristics

The aquatic medium contains almost 80 different elements, dissolved or suspended in pure water, with different concentrations. Some of them are listed below [26, 27]

- Various dissolved salts such as NaCl, $MgCl_2$, etc, which absorb light at specific wavelengths and induce scattering effects.
- Detrital and mineral components, for example, sand, metal oxides, which contribute to both absorption and scattering.
- Colored dissolved organic matters such as fulvic and humic acids which affect absorption, mainly in blue and ultraviolet wavelengths.
- Organic matters such as viruses, bacteria, and organic detritus which add backscattering, especially in the blue spectral range.
- Phytoplankton with chlorophyll-A which strongly absorbs in the blue-red region and scatters green light.

Since chlorophyll absorbs the blue and red wavelengths and the particles strongly contribute to the scattering coefficient, we can use its concentration C (in mg/m^3) as the free parameter to calculate the absorption and scattering coefficients [23, 28].

The exact type of water plays a significant role in the estimation of the amount of chlorophyll concentration and consequently the amount of absorption and scattering for a specific geographic location. A classification system for the clarity of water types based on their spectral optical attenuation depth $z_k = 1/k_d$ was proposed by Jerlov in 1968. This classification was made in the upper portions of the ocean, and it was based on spectral irradiance transmittance measurements [29]. The four major water types that are usually referred in the literature are the following [30, 31]

- Pure deep ocean waters cobalt blue where the absorption is high and the scattering coefficient is low.

http://dx.doi.org/10.5772/67915

- Clear sea waters with higher scattering due to many dissolved particles.
- Near coasts ocean waters with absorption and scattering due to planktonic matters, detritus, and mineral components.
- Harbor murky waters, which are quite constraining for optical propagation due to dissolved and in-suspension matters.

3.2. Absorption

The absorption coefficient, $\alpha(\lambda)$, is the ratio of the absorbed energy from an incident power per unit distance due to various dissolved particles such as phytoplankton, detritus, etc. [19, 23]

$$\alpha(\lambda) = \alpha_w(\lambda) + \alpha_c^0(\lambda)(C_c/C_c^0)^{0.602} + \alpha_f^0 C_f e^{-k_f\lambda} + \alpha_h^0 C_h e^{-k_h\lambda} \tag{7}$$

where $\alpha_w(\lambda)$ is the absorption by the pure water in m^{-1}, λ is the wavelength in nm, $\alpha_c^0(\lambda)$ is the absorption coefficient of chlorophyllin m^{-1}, C_c is the total concentration of chlorophyll per cubic meter ($C_C^0 = 1 mg/m^3$), $\alpha_f^0 = 35.959$ m^2/mg is the absorption coefficient of fulvic acid, $k_f = 0.0189$ nm^{-1}, $\alpha_h^0 = 18.828$ m^2/mg is the absorption coefficient of humic acid, and $k_h = 0.01105$ nm^{-1}. The concentrations C_f and C_c, are expressed through C_c as follows

$$C_f = 1.74098\ C_c e^{0.12327\left(\frac{C_c}{C_c^0}\right)} \tag{8}$$

$$C_h = 0.19334\ C_c e^{0.12343\left(\frac{C_c}{C_c^0}\right)} \tag{9}$$

3.3. Scattering

Scattering coefficient, $b(\lambda)$, is the ratio of energy scattered from an incident power per unit distance. It is the sum of backward scattering, $b_b(\lambda)$, and forward scattering coefficient, $b_f(\lambda)$.

Scattering is caused by small and large particles. Small particles are the particles with refractive index equal to 1.15, whereas large particles have a refractive index of 1.03.

The scattering and backscattering coefficients are calculated as follows [23]

$$b(\lambda) = b_w(\lambda) + b_s^0(\lambda)C_s + b_l^0(\lambda)C_l \tag{10}$$

$$b_B(\lambda) = 0.5b_w(\lambda) + B_s b_s^0(\lambda)C_s + B_l b_l^0(\lambda)C_l \tag{11}$$

For small and large particulate matter, b_s^0 and b_l^0 are given as follows

$$b_s^0(\lambda) = 1.151302\ (m^2/g)\left(\frac{400}{\lambda}\right)^{1.7} \tag{12}$$

$$b_l^0(\lambda) = 0.341074\ (m^2/g)\left(\frac{400}{\lambda}\right)^{0.3} \tag{13}$$

and the concentrations are expressed through the chlorophyll concentration as follows

$$C_s = 0.01739(g/mg)\ C_c e^{0.11631\left(\frac{C_c}{C_c^0}\right)} \tag{14}$$

$$C_l = 0.76284(g/mg)\ C_c e^{0.03092\left(\frac{C_c}{C_c^0}\right)} \tag{15}$$

3.4. Oceanic turbulence

Optical wireless communications are greatly affected by optical turbulence, which refers to random fluctuations of the refraction index. In the case of underwater systems, these fluctuations are mainly caused by variations in temperature and salinity of the oceanic water.

An important parameter for the description of oceanic turbulence is the scintillation index, which expresses the variance of the wave intensity. As shown in Ref. [32], when a Gaussian beam propagates through weak turbulence and without taking into consideration the scattering phenomenon, the scintillation index is expressed as the sum of two components

$$\sigma_I^2(\vec{r}\ , L, \lambda) = \sigma_{I,l}^2(0, L, \lambda) + \sigma_{I,r}^2(\vec{r}\ , L, \lambda), \tag{16}$$

where

$$\sigma_{I,l}^2(0, L, \lambda) = 8\pi^2 k^2 L \int_0^1 \int_0^\infty \kappa \Phi_n(\kappa) \exp\left(-\frac{\Lambda L \kappa^2 \xi^2}{k}\right) \times \left[1 - \cos\left(\frac{L\kappa^2}{k}\xi\left(1 - (1 - \Theta)\xi\right)\right)\right] d\kappa\, d\xi \tag{17}$$

is the longitudinal component and

$$\sigma_{I,r}^2(\vec{r}\ , L, \lambda) = 8\pi^2 k^2 L \int_0^1 \int_0^\infty \kappa \Phi_n(\kappa) \exp\left(-\frac{\Lambda L \kappa^2 \xi^2}{k}\right) [I_0(2\Lambda r \xi \kappa) - 1] d\kappa\, d\xi \tag{18}$$

is the radial component. In the above equations, λ is the wavelength, k is the wave number, L is the direction of propagation, κ is the magnitude of spatial frequency, ξ is the normalized path length, $I_0(x)$ is the zero order Bessel function, and Λ, Θ are parameters of the Gaussian beam given by

$$\Lambda = \frac{\frac{2L}{kW_0^2}}{\left(\frac{2L}{kW_0^2}\right)^2 + \left(1 - \frac{L}{F_0}\right)^2} \tag{19}$$

and

$$\Theta = \frac{1 - \frac{L}{F_0}}{\left(\frac{2L}{kW_0^2}\right)^2 + \left(1 - \frac{L}{F_0}\right)^2} \tag{20}$$

where W_0 is the radius of the Gaussian beam at the 1/e amplitude at the plane of propagation and F_0 is the radius of curvature of the beam wavefront. The function $\Phi_n(\kappa)$ is the power spectrum of turbulence and for homogeneous and isotropic oceanic waters takes the form [32]

$$\Phi_n(\kappa) = 0.38{\cdot}10^{-8}\varepsilon^{-\frac{1}{3}}\kappa^{-\frac{11}{3}}\left[1 + 2.35(\kappa\eta)^{\frac{2}{3}}\right]\frac{\chi_T}{w^2}\left(w^2 e^{-A_T\delta} + e^{-A_S\delta} - 2we^{-A_{TS}\delta}\right) \tag{21}$$

where $\eta = 10^{-3}$ m is the Kolmogorov microscale, ε is the rate of dissipation of turbulent kinetic energy per unit mass of fluid, χ_T is the rate of dissipation of mean-square temperature, $\delta = 8.284(\kappa\eta)^{4/3} + 12.978(\kappa\eta)^2$, and A_T, A_S, and A_{TS} are constants with values $A_T = 1.863\times10^{-2}$, $A_S = 1.9\times10^{-4}$, and $A_{TS} = 9.41\times10^{-3}$. The unitless variable w is the relative strength of the fluctuations caused by either temperature or salinity.

The separation between weak and strong turbulence is usually done via Rytov variance σ_R^2, which is the scintillation index when a plane wave is considered. More specifically, values of $\sigma_R^2 \ll 1$ correspond to weak fluctuations, whereas $\sigma_R^2 \gg 1$ indicates a strong turbulence regime. In contrast to atmospheric studies, strong oceanic turbulence appears at distances shorter than 100m [32].

The fading coefficient can be computed statistically using a proper distribution depending on the turbulence regime. As discussed in Ref. [33], the log-normal distribution is used in situations of weak fluctuations, and its probability density function (PDF) can be written as follows [34]

$$f_{\tilde{h}}(\tilde{h}) = \frac{1}{2\tilde{h}\sqrt{2\pi\sigma_X^2}}\exp\left(-\frac{(\ln\tilde{h} - 2\mu_X)^2}{8\sigma_X^2}\right) \tag{22}$$

The fading coefficient in the above Eq. (22) takes the form $\tilde{h} = e^{2X}$, where X is the fading log-amplitude with mean μ_X and variance σ_X^2. Normalizing the fading coefficient so as to ensure that fading does not affect the average power leads to $\mu_X = -\sigma_X^2$.

For strong turbulence, the distribution that is usually used is the K distribution with PDF [34]

$$f_{\tilde{h}}(\tilde{h}) = \frac{2\alpha}{\Gamma(\alpha)}(\alpha\tilde{h})^{\frac{(\alpha-1)}{2}}K_{\alpha-1}\left(2\sqrt{\alpha\tilde{h}}\right) \tag{23}$$

where $\Gamma(\alpha)$ is the Gamma function, $\alpha = 2/(\sigma_I^2 - 1)$ is a positive parameter, and $K_p(x)$ is the pth order modified Bessel function of the second kind.

Jamali et al. [34] also proposed the combination of exponential and log-normal distributions, creating a two-lobe distribution, to better describe situations of received signals with a large dynamic range. Such a case corresponds to the extensive intensity fluctuations of the received signal, due to the presence of air bubbles. The PDF of this two-lobe distribution is as follows

$$f_{\tilde{h}}(\tilde{h}) = \frac{k}{\gamma}\exp\left(-\frac{\tilde{h}}{\gamma}\right) + \frac{(1-k)\exp\left(-\frac{(\ln\tilde{h}-\mu)^2}{2\sigma^2}\right)}{\tilde{h}\sqrt{2\pi\sigma^2}} \tag{24}$$

where k determines the balance between the two distributions, γ is the mean of the exponential distribution, and μ, σ^2 are the mean and variance of the log-normal distribution, respectively.

4. Link budget and modulation techniques

4.1. Link budget

Empirical path loss models are effective enough to estimate the received optical power for underwater communications under LOS conditions. Several models appeared in the literature but the most generic is the following which considers the transmitter power, telescope gain, and losses [6, 19]

$$P_r = P_t \eta_t \eta_r e^{\frac{-c(\lambda)R}{\cos\theta}} \frac{A_R \cos\theta}{2\pi\, R^2(1-\cos\theta_0)} \tag{25}$$

where P_t is the transmitted power, η_t and η_r are the optical efficiencies of the T_x and R_x correspondingly, $c(\lambda)$ is the extinction coefficient, R is the perpendicular distance between the T_x plane and the R_x plane, θ_0 is the T_x beam divergence angle, θ is the angle between the perpendicular to the R_x plane and the T_x-R_x trajectory, and A_r is the receiver aperture area [19].

4.2. Optical modulation techniques

The most simple and widespread modulation technique is the indirect modulation with direct detection (IM/DD) with on-off keying (OOK). OOK is divided into two categories, the return to zero (RZ) OOK and the non-return to zero (NRZ). The expressions for the anticipated bit error rate (BER) are usually produced using the Poisson model for photon arrival in photon counters [6, 19].

The main disadvantage of OOK modulation is that it requires dynamic thresholding for the direct detection; something not necessary for pulse position modulation (PPM). Here, the transmitter transmits a light pulse and creates a specific time slot, while the receiver detects the pulse, calculates its position, and reforms the signal to the original pulse [35]. PPM consumes lower energy but it has lower bandwidth efficiency. For the L-PPM scheme, the symbol corresponds to $M = \log_2 L$ bits. Other modified PPM modulations have been proposed such as differential pulse position modulation (DPPM), dual pulse interval modulation (DPIM), and dual header pulse interval modulation (DH-PIM).

By exploiting the different polarization states of light, we can improve the performance of UOWC systems in case of intense backscattering, turbulence, and ambient light. Such a way was proposed in Ref. [36], where a binary polarization shift keying (BPolSK) system was employed. Another modulation technique with better numerical results is the polarized PPM (P-PPM) [37]. P-PPM is a combination of PPM and PolSK and manages to maintain the benefits of both schemes improving data rate, BER, and link distance.

5. Research on UOWC systems

Research on UOWC is a quite intriguing topic, and several experiments were carried out worldwide. The following are some indicative of the many studies presented in the open technical literature. For more information, the readers are referred to the complete surveys of Refs. [38, 39], which examine the UOWC area in more details.

An optical modem capable of optical data mulling was designed and implemented in 2005 at Massachusetts Institute of Technology by Vasilescu et al. [40]. They used a LED with 532nm, 700mW radiant power and a PIN photodiode receiver with surface area of 8 mm^2. They employed DPIM at an average data rate of 320kb/s with 2bits per pulse. A 4% packet loss communication over 7m distance and a close to error-free over 6.4m through clear waters was noticed.

Two early versions of AquaOptical, that is, a lightweight device for high rate long range underwater point-to-point communication are shown in Ref. [41]. The researchers used a Luxeon V LED at 470nm (480 mW) or 532nm (700mW) and a PDB-C156 photodiode in the first version. This is actually a miniaturized version capable of error-free communication in ocean water at data rates of 4Mbps over a distance of 2.2m (470nm) and 2.4m (532nm). In the second one, they employed an array of six such LEDs and an avalanche photodiode, and they managed to achieve a BER of 0.05 at 8m. The improved AquaOptical II with 18 LEDs and APD at the receiver was demonstrated in 2010. It operated at 470nm with total transmit power of 10W. Data rates of 4Mb/s at 50m in a swimming pool were achieved using NRZ amplitude modulation with 2bits per symbol [42, 43]. In Ref. [44], a robot estimating its position according to a sensor node in a data mulling application—another usage of Aquaoptical II—was described.

A link budget analysis of NLOS geometries and variations of point-to-point links performed in 2005 and 2006 at Woods Hole Oceanographic Institution (Woods Hole MA USA), by Farr et al. [45]. Their work is only valid in clear waters because they do not take spatial dispersion and multiple scattering into account even if they consider the attenuation coefficient. Their analysis showed that a LOS link could achieve 10Mbps at 100m with 6mW of transmitted power. However, for the NLOS link, a 50-fold increase in transmit power is necessary to achieve the same data rates at similar range due to broader transmit beams. In 2008, they achieved transmission, at a distance of 200m for a data rate of 5Mbps, and in 2010, at a distance of 108m for a data rate of 10Mbps [46, 47].

In 2006, Cochenour et al. used BPSK with a 70-MHz radiofrequency (RF) carrier at Naval Air Warfare Center to accomplice error-free communication at a data rate of 1 Mb/s in deep turbid water. Emission wavelength was 532nm and output power 500mW [48]. The work was extended in 2007, to 5Mbps with QPSK, 8-PSK, 16-QAM and 32-QAM modulations and output power 750mW [49]. Results showed that even in turbid waters (c=3.0/m and cz=11), no loss of modulation depth, that is, no temporal dispersion of the 70MHz carrier was observed, even with a large receiver FOV of approximately 100degrees. As such, the only channel impairment was due to attenuation. Therefore, with enough optical power (~3W), a 32-QAM link achieved almost 5Mbps.

Hanson and Radic [30] presented a Monte Carlo model to predict the spreading of an optical pulse for different water types for various FOVs. They managed to improve the signal to noise ratio (SNR) by increasing the FOV for low frequencies. They also observed temporal dispersion at high frequencies. After some manipulations, they eventually achieved 1GHz transmission over a 2m water pipe that simulated coastal ocean waters with 36dB of attenuation. Moreover, Jaruwatanadilok [50] tried to predict temporal dispersion by following the vector radiative transfer theory to model the UOWC channel which includes the multiple scattering effects and polarization. He examined the effect of the transmission distance on channel dispersion and showed a diminishing value of ISI for high data rates over a long distance.

In 2008, Cochenour et al. [31], conducted an analysis of multiple scattering special effects. They measured the beam-spread function experimentally in a large test tank, by scanning a photodetector laterally to the beam axis after the beam has propagated some distance. Results showed that in clear waters, the intensity falls off rapidly with lateral distance from the beam. However, their estimation ignored any impact of temporal dispersion. In another study on spatial dispersion, they investigated the impact of absorption on forward scattered light [51]. Later the authors also launched a set up which measured the transmission of light modulated at frequencies up to 1GHz in simulated ocean water up to 55 attenuation lengths [52–54]. Finally, the authors examined a retro-reflecting link and managed to reduce the contribution of backscattered light using polarization discrimination techniques [55].

An experiment conducted in North Carolina State University where a 405nm laser with OOK was tested in a water tank of 3.66m long. At first, a reliable transmission at 500kbps was achieved [56]. Some years later, a duplex communication at 1Mbps was attained [57]. Moreover, Simpson et al. created two separated optical beams of 405nm with two transmitters and two receivers combined with a beam splitter [58]. In a later work, they presented two set ups, one in 3m through turbid water and the other over a length of 7.7m through less turbid water and achieved 5Mbps throughput [59].

Dalgleish et al. [60] developed a Monte Carlo simulation method to estimate impulse responses for a particular system hardware design over a large range of environmental conditions in turbid ocean environments. Furthermore, a powerful tool to evaluate the performance of an underwater system for autonomous underwater vehicle communications was presented in Ref. [61]. Experimental tests were also carried out. Hardware modules and circuits design for underwater optical point to point links are also discussed in Ref. [62].

NLOS arrangements where optical beams are reflected in the atmosphere-water interface were discussed in Ref. [6]. These geometries have the benefit to avoid occasional obstructions but they do not have the same performance as LOS links. Gabriel et al. [27] assumed a Monte Carlo model of an underwater channel and simulated the trajectories of the emitted photons. They considered several parameters, for example, the transmitter beam width and beam divergence, beam wavelength, link distance, FOV, water types and turbidity, and aperture size. They reached the conclusion that the channel delay spread is nearly zero when the scattering albedo is moderate in small distances.

In the last years, the research is focused on spatial diversity, as a way to improve overall the performance contaminated by turbulence. This is usually accomplished by adding more sources or detectors. Moreover, aperture averaging methods were proposed. In Ref. [63], Yi et al. implemented an aperture averaging technique on the scintillation index for either plane

http://dx.doi.org/10.5772/67915

or spherical waves, in order to annihilate the negative effects of fading resulting from oceanic turbulence. A multiple-input single-output (MISO) link was described in Ref. [64], where the source is replaced by a uniform circular array of sources, and the photons are collected by a single detector. The numerical results indicated a degradation in BER, under turbulence conditions. A single-input multiple-output (SIMO), consisting of a LED as a source and a concentric circular array of lenses as a detector, was discussed in Ref. [65]. The system manages to extend the effective communication range by about 20 m, even in the case of strong turbulence. A more general set up of the multiple-input multiple-output (MIMO) systems was presented in Ref. [66]. Finally, in Ref. [67], a spatial diversity technique in combination with optical amplification to further improve performance was proposed.

6. Conclusions

The need to provide broadband wireless communications for underwater applications will be increasing in the forthcoming years. UOWC constitutes an alternate and effective transmission technique which can attain this scope, instead of the traditional acoustic one. Due to this fact, several studies on UOWC systems were devised worldwide in the recent years. The present study provided a concise review of the key advances in channel modeling and experimental works reported so far in the technical literature.

Author details

Lydia K. Gkoura[1], George D. Roumelas[1], Hector E. Nistazakis[1]*, Harilaos G. Sandalidis[2], Alexander Vavoulas[2], Andreas D. Tsigopoulos[3] and George S. Tombras[1]

*Address all correspondence to: enistaz@phys.uoa.gr

1 Department of Electronics, Computers, Telecommunications and Control, Faculty of Physics, National and Kapodistrian University of Athens, Athens, Greece

2 Department of Computer Science and Biomedical Informatics, University of Thessaly, Lamia, Greece

3 Department of Battle Systems, Naval Operations, Sea Studies, Navigation, Electronics and Telecommunications, Hellenic Naval Academy, Piraeus, Greece

References

[1] Stojanovic M. Preisig J., (2009), "Underwater Acoustic Communication Channels: Propagation Models and Statistical Characterization", IEEE Communications Magazine Volume: 47 Issue: 1

[2] Vall L.D., (2011), Towards Underwater Video Transmission, MSc Thesis, Massachusetts Institute of Technology. https://seagrant.mit.edu/publications/

[3] Ainslie M. A., McColm J. G., (1998), "A simplified formula for viscous and chemical absorption in sea water", Journal of the Acoustical Society of America, Vol. 103, No. 3, pp. 1671–1672

[4] Stojanovic M., (2008), "Underwater acoustic communications: Design considerations on the physical layer", IEEE/IFIP Fifth Annual Conference on Wireless On demand Network Systems and Services (WONS 2008), Garmisch-Partenkirchen Germany, January 2008, pp. 2–3

[5] Urick Robert J., (1983), "Principles of underwater sound", Book, Peninsula Publishing, Science , Los Altos CA

[6] Arnon S. and Kedar D., (2009), "Non-line-of-sight underwater optical wireless communication network", Journal of the Optical Society of America A, Vol. 26, No. 3, pp. 530–539

[7] Li B., Stojanovic M., Freitag L., and Willett P., (2008), "Multicarrier communication over underwater acoustic channels with nonuniform Doppler shifts". IEEE Journal of Oceanic Engineering, Vol. 2, No. 33, pp. 198–209

[8] Akyildiz I.F., Pompili D., and Melodia T., (2005), "Underwater acoustic sensor networks: Research challenges", Ad Hoc Networks, Vol. 3, No. 3, pp. 257–279

[9] Kilfoyle D. B., and Baggeroer A. B., (2000), "The state of the art in underwater acoustic telemetry", IEEE Journal of Oceanic Engineering, Vol. 25, No. 1, pp. 4–27

[10] Glegg S.A.L., Pirie R., LaVigne A.,(2000) "A study of ambient noise in shallow water", Florida Atlantic University Technical Report

[11] Stojanovic M.,(2003) "Acoustic (underwater) communications", in J.G. Proakis (Ed.), Encyclopedia of Telecommunications, Wiley, New York.

[12] Stojanovic M., (1996), "Recent advances in high-speed underwater acoustic communications", IEEE Journal of Oceanic Engineering, Vol. 21, No. 2, pp. 125–136

[13] Pelekanakis C., Stojanovic M., Freitag L., (2003), "High rate acoustic link for underwater video transmission", OCEANS 2003, 22–26 September 2003, Vol. 2, pp. 1091–1097

[14] Balanis C. A., (1989), Advanced Engineering Electromagnetics, John Wiley & Sons, New York, NY.

[15] Pei L.V., Junhua H., Renkui Z., Haiying L., (2008), "The application of underwater optics and its development", Information Optics and Photonics Technologies II, Vol. 6837, pp. 7–8

[16] Gkoura L.K., Nistazakis H.E, Vavoulas A., Tsigopoulos A.D., and Tombras G.S., (2014) "Underwater optical wireless communications possibilities disadvantages and possible solutions" 6th International Conference from Scientific Computing to Computational Engineering IC-SCCE, 9–12 July Athens

[17] Lanbo L., Shengli Zhou, and Jun-Hong Cui, (2008) "Prospects and problems of wireless communication for underwater sensor networks", Wireless Communications and Mobile Computing, Vol. 8, No. 8, pp. 977–994

[18] Andrews L.C., Phillips R.L., and Hopen C.Y.,(2001), Laser Beam Scintillation with Applications, SPIE Optical Engineering Press The International Society of Optical Engineering Bellingham Washington USA

[19] Vavoulas A., Sandalidis H.G, and Varoutas D., (2012) "Weather effects on FSO network connectivity", Journal of Optical Communications Networking, Vol. 4, No. 10, pp. 734–740.

[20] Nistazakis H.E., Karagianni E.A., Tsigopoulos A.D., Fafalios M.E., and Tombras G.S., (2009) "Average capacity of optical wireless communication systems over atmospheric turbulence channels", Journal of Lightwave Technology, Vol. 27, No. 8, pp. 974–979

[21] Young E.Y.S, Bullock A.M., (2003), "Underwater airborne laser communication system characterization of the cannel" Free-Space Laser Communication Technologies XV, G. Stephen Mecherle, Editor, Proceedings of SPIE, Vol. 4975

[22] Sathyendrenath S.,(1984) "Inherent optical properties of natural seawater", Defence Science Journal, Vol. 34, No. 1, pp. 34, 1–18

[23] Haltrin V., (1999) "Chlorophyll-based model of seawater optical properties", Applied Optics, Vol. 38, No. 33, pp. 6826–6832

[24] Cochenour B., and Mullen L., (2012) "Free-space optical communications underwater", in Advanced Optical Wireless Communication Systems, (ed). S. Arnon, J.R. Barry, G.K. Karagiannidis, R. Schober, and M. Uysal, Advanced Optical Wireless Communication Systems Pages 273–302 Cambridge, UK: Cambridge Univ. Press

[25] Cochenour B., (2013) "Experimental measurements of temporal dispersion for underwater laser communications and imaging", PhD Thesis, North Carolina State University USA

[26] Gupta S.K.,(1984) "Spectral transmission studies of ocean water under different sea conditions", Defence Science Journal, Vol. 34, No. 1, pp. 19–28

[27] Gabriel C., Khalighi M.A., Bourennane S., Pierre Leon, Vincent Rigaud, (2013) "Monte-carlo-based channel characterization for underwater optical communication systems" Journal of Optical Communications and Networking January 2013, Vol. 5, Issue 1, pp. 1–12 The Optical Society

[28] Mobley C.D., (1994), "Light and Water: Radiative Transfer in Natural Waters", Book, Academic Press Inc. A division of Harcourt Brace and Company 525B Street Suite 1900, San Diego, California 92101-4495 United Kingdom Edition published by: Academic Press Limited, 24-28 Oval Road, London NW1 7DX

[29] Arnon S., (2003), "Optical wireless communications", Book Chapter, Encyclopedia of Optical Engineering (EOE), Edited by R. G. Driggers, 3 Marcel Dekker, pp. 1866–1886

[30] Hanson R., and Radic S,(2008) "High bandwidth underwater optical communication", Applied Optics, Vol. 47, No. 2, pp. 277–283

[31] Cochenour B., Mullen L., Laux A., (2008), "Characterization of the beam-spread function for underwater wireless optical communications Links", IEEE Journal of Oceanic Engineering, Vol. 33, No.4, pp. 513–521

[32] Korotkova O., Farwell N., and Shchepakina E., (2012) "Light scintillation in oceanic turbulence", Waves in Random and Complex Media, Vol. 22, No. 2, pp. 260–266

[33] Andrews L.C. and Phillips R.L., (2005) Laser Beam Propagation through Random Media, Date Published: 16 September 2005, Pages: 808, ISBN: 9780819459480, Volume: PM152, Published by SPIE The International Society of Optical Engineering, Bellingham Washington 98227-0010 USA

[34] Jamali M.V., Khorramshahi P., Tashakori A., Chizari A., Shahsavari S., Abdollah-Ramezani S., Fazelian M., Bahrani S., and Salehi J.A., (2016) "Statistical distribution of intensity fluctuations for underwater wireless optical channels in the presence of air bubbles", Iran Workshop on Communication and Information Theory IWCIT 2016, 3–4 May 2016

[35] Meihong S., Xinsheng Y. and Fengli Z., (2009), "The evaluation of modulation techniques for underwater wireless optical communications", International Conference on Communication Software and Networks, Chengdu Sichuan China, 27–28 February 2009, pp. 138–142

[36] Cox W.C., Hughes B.L, and Muth J.F., (2009) "A polarization shift-keying system for underwater optical communications", OCEAN 2009, Biloxi MS USA, 26–29 October 2009

[37] Dong Y., Zhang T., and Zhang X., (2013) "Polarized pulse position modulation for wireless optical communications", 2013 47th Annual Conference on information Sciences and Systems (CISS), Baltimore Maryland, 20–22 March 2013

[38] Kaushal H. and Kaddoum G., (2016), "Underwater optical wireless communication", IEEE Access, Vol. 4, pp. 1518–1547

[39] Zeng Z., Zhang H., Dong Y, and Cheng J., (2017) "A Survey of underwater wireless optical communication," IEEE Communication. Survey Tutorials, to appear

[40] Vasilescu I., Kotay K., Rus D., Dunbabin M., Corke P., (2005), "Data collection, storage, and retrieval with an underwater sensor network", ACM Proc. 3rd International Conference on Embedded Networked Sensor Systems (SenSys), San Diego California USA, 2–4 November 2005, pp. 154–165

[41] Doniec M., Vasilescu I., Mandar C., Detweiler C., Hoffmann-Kuhnt M., Rus D., "AquaOptical: A Lightweight Device for High-rate Long-range Underwater Point-to-Point Communication" Marine Technology Society Journal, Vol. 44, No. 4, pp. 55–65

[42] Doniec M., Detweiler C., Vasilescu I, and Rus D., (2010), "Using optical communication for remote underwater robot operation", 2010 IEEE/RSJ International Conference on Intelligent Robots and Systems (IROS), Taipei Taiwan, 18–22 October 2010, pp. 4017–4022

[43] Doniec M. and Rus D., (2010), "Bidirectional optical communication with AquaOptical II," in Communication Systems (ICCS), 2010 IEEE International Conference on, Nov. 2010, pp. 390–394

[44] Doniec M.W., Topor I., Chitre M., and Rus D.,(2012), "Autonomous, localization-free underwater data muling using acoustic and optical communication", The 13th International Symposium on Experimental Robotics ISER 2012, Québec City, Canada, June 18–21 2012

[45] Farr N., Chave A., Freitag L., Preisig L., White S., Yoerger D., and Sonnichsen F., (2006), "Optical modem technology for seafloor observatories", OCEANS 2006, Singapore, 16–19 May 2006

[46] Pontbriand C., Farr N., Ware J., Preisig J., Popenoe H., (2008) "Diffuse high-bandwidth optical communications", OCEANS 2008, Quebec City QC, Canada, 15–18 September 2008

[47] Farr N., Ware J., Pontbriand C., Hammar T., Tivey M.,(2010) "Optical communication system expands cork seafloor observatory's bandwidth", OCEANS 2010, Sydney Australia, 24–27 May 2010

[48] Cochenour B., Mullen L., Laux A., and Curran T. (2006), "Effects of multiple scattering on the implementation of an underwater wireless optical communications link," Singapore, 16–19 May 2006

[49] Cochenour B., Mullen L., and Laux A., (2007) "Phase coherent digital communications for wireless optical links in turbid underwater environments," OCEANS 2007, Aberdeen Scotland UK, 18–21 June 2007

[50] Jaruwatanadilok S., (2008),"Underwater wireless optical communication channel modeling and performance evaluation using vector radiative transfer theory", IEEE Journal on Selected Areas in Communications, Vol. 26, No. 9, pp. 1620–1627

[51] Cochenour B., Mullen L., and Muth J., (2010) "Effect of scattering albedo on attenuation and polarization of light underwater", Optics Letters, Vol. 35, No. 12, pp. 2088–2090

[52] Mullen L., Alley D., Cochenour B., (2011) ''Investigation of the effect of scattering agent and scattering albedo on modulated light propagation in water'' Applied Optics Vol. 50, Issue 10, pp. 1396–1404. https://doi.org/10.1364/AO.50.001396

[53] Mullen L.J., Laux A.E., Cochenour B.M., (2009) "Propagation of modulated light in water: Implications for imaging and communications systems", Applied Optics, Vol. 48, No. 14, pp. 2607–2612

[54] Mullen L., Cochenour B., Laux A., and Alley D., "Optical modulation techniques for underwater detection, ranging, and imaging", In Proc. of SPIE Ocean Sensing and Monitoring III, pp. 1–9. SPIE, 2011

[55] Mullen L., Cochenour B., Rabinovich W., Mahon R., Muth J., (2009) "Backscatter suppression for underwater modulating retroreflector links using polarization discrimination", Applied Optics, Vol. 48, No. 2, pp. 328–337

[56] Cox W., (2008), "A 1Mbps underwater communication system using a 405nm laser diode and photomultiplier tube," MSc Thesis, North Carolina State University USA

[57] Cox W., Simpson J., and Muth J., (2011), "Underwater optical communication using software defined radio over led and laser based links," Military Communications Conference, 2011 MILCOM 2011, Baltimore MD USA, pp. 2057–2062

[58] Simpson J., Hughes B., and Muth J., (2009), "A spatial diversity system to measure optical fading in an underwater communications channel," OCEANS 2009, Biloxi MS USA, 26–29 October 2009

[59] Simpson J, Cox W., Krier J, Cochenour B., Hughes B., and Muth J., (2010), "5 Mbps optical wireless communication with error correction coding for underwater sensor nodes" Sydney Australia, 24–27 May 2010

[60] Dalgleish F.R., Caimi F.M., Vuorenkoski A. K., Britton W.B., Ramos B, (2010),"Efficient laser pulse dispersion codes for turbid undersea imaging and communications applications", Proc. SPIE, Vol. 7678

[61] Anguita D., Brizzolara D., Parodi G., and Hu Q., (2011), "Optical wireless underwater communication for AUV: Preliminary simulation and experimental results", OCEANS 2011, Waikoloa HI USA, 19–22 September 2011

[62] Anguita D., Brizzolara D., and Parodi G., (2010), "Optical wireless communication for underwater wireless sensor networks: Hardware modules and circuits design and implementation", OCEANS 2010, Sydney Australia, 24–27 May 2010

[63] Yi X., Li Z., and Liu Z., (2015) "Underwater optical communication performance for laser beam propagation through weak oceanic turbulence", Applied Optics, Vol. 54, No. 6, pp. 1273, 2015

[64] Dong Y. and Liu J., (2016) "On BER performance of underwater wireless optical MISO links under weak turbulence", OCEANS 2016, Monterey CA USA, 19–23 September 2016

[65] Liu W., Xu Z., and Yang L., (2015) "SIMO detection schemes for underwater optical wireless communication under turbulence", Photonics Research, Vol. 3, No. 3, pp. 48–53

[66] Jamali M.V. and Salehi J.A., (2015) "On the BER of multiple-input multiple-output underwater wireless optical communication systems," 4th International Workshop on Optical Wireless Communications IWOW 2015, Istanbul Turkey, 07–08 September 2015, pp. 26–30

[67] Boucouvalas A., Peppas K., Yiannopoulos K., and Ghassemlooy Z., (2016) "Underwater optical wireless communications with optical amplification and spatial diversity," IEEE Photonics Technology Letters, Vol. 28, No. 22, pp. 2613–2616